The Wellborn Science

Monographs on the History and Philosophy of Biology

RICHARD BURIAN, RICHARD BURKHARDT, JR.,
RICHARD LEWONTIN, JOHN MAYNARD SMITH
EDITORS

The Wellborn Science

Eugenics in Germany, France, Brazil, and Russia

Edited by

MARK B. ADAMS

New York   Oxford
OXFORD UNIVERSITY PRESS
1990

Oxford University Press

Oxford   New York   Toronto
Delhi   Bombay   Calcutta   Madras   Karachi
Petaling Jaya   Singapore   Hong Kong   Tokyo
Nairobi   Dar es Salaam   Cape Town
Melbourne   Auckland

and associated companies in
Berlin   Ibadan

Copyright © 1990 by Oxford University Press, Inc.

Published by Oxford University Press, Inc.,
200 Madison Avenue, New York, New York 10016

Library of Congress Cataloging-in-Publication Data
The Wellborn science : eugenics in Germany, France, Brazil, and Russia
/ edited by Mark B. Adams.
p. cm.—(Monographs on the history and philosophy of biology)
Bibliography: p. Includes index.
ISBN 0-19-505361-3
1. Eugenics—History—20th century—Cross-cultural studies.
2. Eugenics—Germany—History—20th century. 3. Eugenics—France—
History—20th century. 4. Eugenics—Brazil—History—20th century.
5. Eugenics—Soviet Union—History—20th century.
I. Adams, Mark B. II. Series.
HQ751.W46 1989
362.1—dc19   88-38434 CIP

9 8 7 6 5 4 3 2 1

Printed in the United States of America
on acid-free paper

Preface

In the early part of this century, eugenics (from the Greek *eugenēs*, "wellborn") arose as a science of human hereditary improvement in more than thirty countries. Until very recently, however, almost all published research on its history has dealt with the United States or Britain. The present volume challenges common stereotypes by presenting histories of eugenics in four countries outside of the Anglo-Saxon world—Germany, France, Brazil, and Russia—where events unfolded very differently. In addition to providing a new perspective on the social dimensions of biology and medicine in continental Europe, Latin America, and the Soviet Union, these studies revise our picture of eugenics as a whole and suggest an agenda for international comparative research.

This book was a long time in the making. I first became interested in the history of eugenics in 1963, as a Harvard undergraduate, while taking Everett Mendelsohn's courses on "The History of Biology" and "The Social Context of Science." Fascinated by science, and stirred by the Sixties, I found his lectures a revelation. Looking back, they were years ahead of their time in focussing on the subtle and complex interactions between science and society. I was particularly taken with the material on evolution, genetics, and eugenics. The courses stimulated me to major in the history of science, to pursue graduate work, and, ultimately, to take up the field professionally. Soon I became involved in other research, on the history of science in Russia, the history of evolutionary theory and genetics, the rise of molecular biology, and the problem of Lysenkoism. But eugenics kept coming up. During a summer NSF workshop in 1966, I roomed with an undergraduate, Kenneth Ludmerer, and enjoyed discussing his research on Madison Grant's book, *The Passing of the Great Race* (New York, 1916). I still remember the day he returned from a research trip to Philadelphia, full of enthusiasm at what he had found in the Davenport papers of the Library of the American Philosophical Society. Soon his research led to a *summa* undergraduate thesis (written under Yehuda Elkana) and a pathbreaking book, *Genetics and American Society* (Baltimore, 1972).

Then, in 1969, I was offered a job at the University of Pennsylvania and, while visiting, met Lyndsay Farrall. He was returning to Australia, and I was to replace him as Penn's historian of biology. I still remember a pleasant dinner at his Haverford home where we talked about his research. At the time, he had just completed his dissertation at Indiana University on the British eugenics movement. He kindly provided me with a copy, and upon reading it I was struck not only by the differences between the American and British movements, but also by the analytical approaches employed: while Ludmerer had looked at American eugenics as *science* (and *pseudo*science), focussing on the founding role of American geneticists and their subsequent disenchantment, Farrall had looked at British eugenics as a *social movement*, constructing a "collective biography" (prosopography) of members and patrons by analyzing their professional, class, and family backgrounds and interconnections.

From studying Russian science, and especially Russian biology, I had become convinced that national differences can be significant. Soviet Russia had had a eugenics movement in the 1920s, I knew, and so too did dozens of other countries. If eugenics differed in Britain and the United States—two countries with a common language and closely related political, scientific, and cultural traditions—what might eugenics look like when its full international diversity was studied? Furthermore, as an area on the interface of science and society, eugenics looked like a good place to examine the interactions of the "internal" and "external" history of science (a distinction then in vogue). Eugenics seemed an especially promising area for research in yet another respect. In 1967, as a result of taking concurrent graduate seminars given by Ernst Mayr (on evolutionary theory) and Everett Mendelsohn (on the social context of science), I had become convinced that an evolutionary approach to the history of science would prove illuminating. I set forth its outlines in a hundred-page draft, but did not move to publication: resistance among historians to such an approach was (and continues to be) deep-seated. (Only a decade later did I summon enough nerve to publish something on it.) But in the late 1960s and early 1970s, I was hunting for some way to "test" the evolutionary model against the historical record. In this connection, the breadth and diversity of international eugencis was attractive. Unfortunately, I did not know thirty languages.

As the number of graduate students in the history of science at Penn began to increase, I decided on a modest project: a seminar, given in fall 1973, on the comparative history of eugenics. We would begin by reading the available literature on America and Britain, then move on to broader methodological and comparative analysis. Finally, working in a foreign language, each student would do original research on the history of eugenics in some country *other than* Britain and the United States, producing a seminar-length paper (roughly 60–75 pages). The result, I hoped, would be a volume covering the history of eugenics in fifteen or twenty countries and setting forth a methodological framework that was generalizable beyond eugenics. What I was aiming for, at least according to what the course syllabus said, was to use eugenics as a strategic research site in order to understand "the 'ecology' of ideas and their evolution, how they are picked up by different groups and become adapted to different intellectual and social environments." Remarkably, fifteen students took the course, eight of them undergraduates.

In retrospect, the plan was a bit over-optimistic: in my innocence, I struggled to extract from each student a definitive case study, making the seminar trying for teacher and student alike. Alas, not all of their papers were publishable. Nonetheless, I am astonished in retrospect at what was produced: a remarkable set of studies on the eugenics movements in Argentina (Marc Machiz), Austria-Hungary (Eric Lipman), Canada (Joseph Maline and Daniel P. Todes), China and Japan (John Lorwin), France (Nancy Mautner and William Schneider), Germany (Bonnie Blustein, Stephen Perloff, and John Pitts), Italy (George MacPherson), Norway (Eric Rosenbaum), Spain (Dolores Maria Avalos), and Sweden (P. Thomas Carroll). Two seminar members (Mautner and Rosenbaum) presented their research at the Joint Atlantic Seminar on the History of Biology, held at the University of Pennsylvania in spring 1974.

I had hoped to press forward with a volume at that time, but soon the students dispersed and other pressing matters diverted attention. During the following decade, the seminar papers were made available to those with a serious research interest in the subject. They were perused by scholars visiting our department, notably Dan Kevles and Donald MacKenzie, and also by graduate students who were encouraged to take up where their predecessors had left off. A number did, notably Martha Bettes, who completed a pioneering study of women in the American eugenics movement; David van Keuren, who studied eugenics and British anthropology; Donna Houck, who wrote a paper on Paul Kammerer's eugenic ideas; Christine Hoepfner, who studied the Pan-American eugenics congress; Eva Artschwager, who investigated Viennese eugenics; and Barbara Kimmelman, who studied the relationship between eugenics and Mendelism in American agricultural settings.

In the meantime, I assumed that the comparative international history of eugenics would soon attract increasing scholarly attention from historians who had the requisite language skills. This was only partly right. True, eugenics became an increasing focus of attention—but not its comparative history. Furthermore, perhaps because of the declining foreign language skills and requirements in universities, perhaps because of the increasing preoccupation with America among science historians, most of the rapidly expanding new literature dealt with the two most familiar cases: the United States and Britain.

For this reason, I decided to try again in connection with the Darwin centennial. Thanks to support from the Humanities Coordinating Committee of the Faculty of Arts and Sciences of the University of Pennsylvania, the Department of the History and Sociology of Science hosted a three-day conference (16–19 May 1983) entitled "The History of Eugenics: Work in Progress." Its purpose was to stimulate work on international eugenics and to acquaint scholars working on Britain and America with other national cases. Participants included Garland Allen, Richard Burian, Tom Carroll, Linda Clark, Lyndsay Farrall, Bentley Glass,

Loren Graham, Greta Jones (who was at Penn as a Mellon Fellow), Daniel Kevles, Henrika Kuklick, Pauline Mazumdar, William Montgomery, B. J. Norton, Diane Paul, Jack Pressman, Eli Rosenbaum, Charles Rosenberg, William Schneider, Nancy Stepan, and Sheila Weiss. The result was a series of papers and discussions and, ultimately, this book, thanks in no small measure to the enthusiasm and steadfast support of Dick Burian.

In addition, stimulating discussions with colleagues from Scandinavia, France, Germany, and England attending the XVIIth International Congress of the History of Science in 1985 at Berkeley convinced me of the timeliness and importance of the project, and this conviction was reinforced by subsequent discussions in Paris, Uppsala, Berlin, Ischia, and Toronto with Gunnar Broberg, Nils Roll-Hansen, Bent Sigurd Hansen, Marjetta Hietala, Bob Olby, Jan Sapp, Paul Weindling, Peter Weingart, and others, and by the lively interest in the topic shown at the XVIth International Congress of Genetics (August 1988, Toronto). I am indebted to them, and to Jean Louis-Fischer, Tore Frangsmyr, Mirko Grmek, Bernardino Fantini, Michel Morange, and Bob Haynes for making these meetings possible.

As is evident above, there are many people who should be thanked, but foremost among them are the members of the 1973 seminar, for their pioneering spunk and hard work; the participants in the 1983 workshop, who shared graciously their time and knowledge; and my co-contributors, for their cooperative spirit and abiding patience.

Philadelphia   M. B. A.
July 1989

Contents

The Wellborn Science

CHAPTER 1

Eugenics in the History of Science

Mark B. Adams

The word *eugenics* was coined by the Englishman Francis Galton in 1883 (from the Greek *eugenēs,* "wellborn") to denote the "science" of the biological improvement of the human kind. Galton was convinced that a wide range of human physical, mental, and moral traits were inherited. If this were so, he reasoned, human progress depended on improving the human stock, and social measures were likely to be truly progressive only by virtue of their effects on the selective transmission of the population's hereditary endowments to future generations.

In recent years eugenics has emerged as a major topic of research in the history of science. There are both disciplinary and social reasons for this persistent and growing interest. Over the last two decades, historians of science have sought to integrate the so-called internalist account of the evolution of scientific ideas with the so-called externalist account of its social context. As a mix of science and policy, and as a would-be discipline as well as a social movement, eugenics lay at the interface of biological science and society, and for that reason has proved an especially apt and intriguing research site. The history of eugenics has also been seen as relevant to vexing social issues. In recent years controversy has surrounded IQ tests, race and intelligence, sociobiology, genetic engineering, in vitro fertilization, cloning, and the relationship between crime and the "XYY" karyotype. The scientific validity and the ethical, legal, social, and political implications of this work have received much public attention. For these and other reasons, it is interesting and important for us to understand the history of eugenics, and the new interest in the subject has been reflected in a new historical literature.

Unlike earlier treatments of eugenics, this new literature has been produced principally by historians and historians of science, whose work has reflected the contemporary developing interest in science and society. One of the earliest studies was written by the American historian Mark Haller, and reflected and devel-

---

The references cited in this chapter can be found in the bibliography at the end of Chapter 6.

oped the perspective of Richard Hofstadter on American social Darwinism (Haller 1963). Donald Pickens wrote a book portraying eugenics in the context of American Progressivism (Pickens 1968). One of the first figures trained in the history of science to become engaged in the subject was Kenneth Ludmerer; he wrote an undergraduate thesis on American eugenics and genetics, using the archives of the American Philosophical Society, and later transformed it into an important book (Ludmerer 1972). Garland Allen has been doing research on the subject for several decades: first becoming interested in the subject through his work on T. H. Morgan and American genetics, Allen has emphasized the role of racism and economics in the development of the movement. Quite recently, Daniel Kevles's articles in the *New Yorker* and his subsequent book, *In the Name of Eugenics* (Kevles 1985), have attracted public interest. The book broke new ground by presenting the history of eugenics and that of human and medical genetics as part of the same story, and by dealing with both the United States and Britain. The history of British eugenics has received considerable attention in its own right. Within the history of science, Ruth Schwartz Cowan's dissertation and subsequent work on Francis Galton was a major contribution to both the history of biology and the history of eugenics. Lyndsay Farrall's dissertation at Indiana University pioneered the study of the British eugenics movement (1970) by analyzing its evolving patronage, politics, and professional (and class) makeup. In subsequent years, this style of social analysis has developed great sophistication in works by Searle (1976a, 1976b, 1979), MacKenzie (1976, 1981), Jones (1980, 1982, 1983), and Mazumdar (1980).

Until quite recently, these works on the U.S. and British movements made up almost all of the published literature on the history of eugenics. Perhaps the most glaring omission was the absence of proper treatment of German developments. Now, in recent months, first-rate books on German "race hygiene" have begun to appear. An early sketch by German geneticist Benno Müller-Hill (1984), now available in English (1988), used archives and interviews to chronicle for the German public the nature of the involvement of biologists in the Nazi exterminations. Historians have subsequently produced a series of major monographs based on years of research, including a biography of Wilhelm Schallmayer elucidating his concept of eugenics as a form of biomedical technocracy (Weiss 1987); a 725-page tome setting forth the history of German eugenics in the context of German culture from the time of Weismann and Nietzsche (Weingart, Kroll, and Bayertz 1988); a lively indictment of German medical scientists as creators and executors of Nazi racial policies (Proctor 1988); and a sophisticated new history that places race hygiene in the context of the evolution of state medicine, institutions, ideology, and health policy from the time of German unification through the 1930s (Weindling 1989). Historians will need time to appreciate fully the implications of these new studies. Already they are beginning to change our picture of eugenics in some important respects (Anglo-American eugenics was largely private and supported by philanthropy, for example, whereas German race hygiene evolved within a statist medical tradition); but in other respects the German case may tend to reinforce certain preexisting stereotypes (e.g., that eugenics was essentially racist or right-wing). In any case, these new books have now made Germany the third major country whose eugenics movement's history has been thoroughly treated.

But the history of eugenics in the United States, Britain, and Germany does not begin to exhaust the subject. In the decades between 1890 and 1930, eugenics movements developed in more than thirty countries, each adapting the international Galtonian gospel to suit local scientific, cultural, institutional, and political conditions. In some places eugenics was dominated by experimental biologists, in others by animal breeders, physicians, pediatricians, psychiatrists, anthropologists, demographers, or public health officials. In some places it was predominantly Lamarckian, in others Mendelian. Patterns of patronage also varied from place to place. Indeed, a report on the International Commission of Eugenics published in 1924 in *Eugenical News* listed fifteen full members: Argentina, Belgium, Cuba, Czechoslovakia, Denmark, France, Germany, Great Britain, Italy, Netherlands, Norway, Russia, Sweden, Switzerland, and the United States. In addition, seven other countries were eligible for cooperation: Brazil, Canada, Colombia, Mexico, Venezuela, Australia, and New Zealand.

In recent years scholars have begun to explore a number of these movements; still others have yet to be studied. Such cases will, of course, illuminate the history of science and society in the various countries. More important, as other cases develop, we can gain some comparative perspective—and that, in turn, should help us to understand each individual case more clearly. It is a perspective the field sorely needs: aside from a suggestive article by Graham (1978) and Kevles's discussion of both the British and American movements (1985), comparative studies have been conspicuously absent. For some, the virtue and necessity of such a comparative perspective may not be self-evident; if one wishes to understand the development of eugenics in America, they may wonder, is it not sufficient to study American eugenics? The answer, I think, is "no."

If we look to other research areas, we may see the different functions that comparison can serve. Comparative approaches have played an important role over the last two centuries in a number of fields: linguistics, philology, morphology, physiology, biochemistry, anthropology, sociology, and religion are but a few. Each of these disciplines has found that the understanding of an individual object of study—be it a single language, organism, species, society, or religion—is greatly enriched and clarified by its comparison with others. Furthermore, unless one adopts a position of extreme nominalism and holds that only individuals are real (in which case even the notion of "eugenics" is problematic), the importance of comparison is surprisingly independent of one's intellectual or methodological orientation.

Linguistics may serve as a case in point. The great evolutionary "language trees" that decorated the inside covers of dictionaries until recently were based on comparative linguistic analysis. But even Chomsky's "Cartesian" attempt to establish the general "transformational grammar of the mind" that putatively underlies all human languages and linguistic behaviors must ultimately rely on the comparative study and knowledge of all available languages. And even if this proposed "mind grammar" should be found to correlate with certain structures or organizations in "the" human brain, we would still need to know whether this applies to all human brains. Comparative anatomy and morphology may serve as a second case in point. Cuvier used comparative methods to great effect as part of a program to determine the stable and essentially unchanging archetypes. Haeckel used similar methods in an attempt to understand patterns of organic evolution. D'Arcy Thompson employed such methods in an attempt to arrive at general laws of form.

These examples illustrate the fact that comparative analysis is vital to research whether the objects of study are seen as the products of an evolutionary process or the manifestations of invariant laws of type and form. Even the most exhaustive study of an individual cannot establish its uniqueness in any particular respect, for asserting uniqueness is ultimately an assertion about all the other roughly comparable cases or entities. Thus, even if our ultimate goal is to comprehend the "essence" of eugenics as a phenomenon, or to find the invariant laws or processes underlying the character of knowledge, or even to ascertain what is unique or atypical in a given movement or development, we cannot hope to do so without comparative studies. And this is as true for eugenics and the history of the sciences generally as it is for embryology, molecular biology, or linguistics.

Very recently, historians of science and technology have begun to explore the power of an explicitly evolutionary model (e.g., Adams 1979, Richards 1987, Basalla 1988, Hull 1988). From an evolutionary viewpoint, of course, a comparative approach to the history of eugenics is especially informative. Can the various national eugenics movements be understood as a series of interrelated but distinct geographical "varieties" arising through migration, diffusion, selection, mutation, recombination, and adaptation? Why the sudden "adaptive radiation" of eugenics into so many different settings? Did it then, even more suddenly, go extinct, or did it take on protective coloration, or perhaps evolve into something else? The answers can come only through comparative research. For those who share this populational, geographical, "evolutionary" orientation, then, the historical phenomenon of international eugenics is especially inviting.

But for those who seek to find the "essential" ideological or intellectual core of eugenics, comparative analysis is no less important. Those who study single cases run the danger of falsely generalizing from particulars; only comparison of the broadest possible number of cases can reveal what is invariant. For example, according to one persistent stereotype, eugenic sterilization is seen as a right-wing, reactionary policy based on a "biological determinism" reinforced by Mendelism. Against this view, we have the example of Russia's leading advocate of eugenic sterilization in the 1920s: a young Communist, M. V. Volotskoi, who was a strong advocate of the interests of the working class—and also an opponent of Mendelism and a devout advocate of a Lamarckian inheritance of acquired characteristics. This example may serve as a single illustration of how complicated and unexpected the real history of eugenics can be, and of the kind of surprises its study can bring. Bowler's recent study of MacBride may serve as another, slightly different example—that of a politically right-wing Lamarckian who also saw virtue in sterilization (Bowler 1984). Individually, such examples may seem anomalous. When many examples are studied, however, and many national case studies examined, new patterns may begin to emerge that do not fit our stereotypes.

The comparative dimension is also important if we are to get the most out of national case studies. When eugenics is studied in only one country, no matter how thoroughly, outcomes can appear to be overdetermined by a plethora of contextual religious, cultural, social, economic, institutional, and scientific variables that are difficult to rank in terms of their relative importance. When comparably sensitive case studies are set side by side, however, we may note broader patterns that would not be apparent from any single national case. This broader picture may, in turn, lead us to reevaluate and better understand eugenics in any particular country. Thus, even though they have been thoroughly studied, the eugenics movements in Britain, America, and Germany may be illuminated by the history of eugenics in other countries.

This book seeks to supplement our historical understanding of eugenics through four national case studies. The idea for this volume originated with a seminar on the comparative history of eugenics taught at the University of Pennsylvania in the fall of 1973. The prospect of publishing a comparative volume of papers emerged in 1982 in connection with the University of Pennsylvania's Darwin Centennial Celebration. A three-day conference was held entitled "The History of Eugenics: Work in Progress." The international gathering brought many scholars working on Britain and America into contact with other scholars (largely unbeknownst to them) who had done or were doing research on the history of eugenics in other countries. Thus, the first day was devoted to Britain and America, and the second to Germany, Scandinavia, France, Russia, and Brazil. The third day was devoted to rather lively discussion.

The chapters in this book were developed from presentations at that conference. Effort was given to selecting cases that would highlight the diversity of different experiences, ones that were still insufficiently explored in the literature. (Unbeknownst to us at the time, Germany was just beginning to be the subject of great interest and much work.) In order to produce parallel case studies, all authors subsequently agreed to cover the period 1900–1940; to develop their material into a national case study; to treat the key figures, eugenic institutions, and politics in a roughly parallel way; and to provide an authoritative bibliography on eugenics in their particular country. In addition, the authors discussed the materials among themselves over the course of many months, and we have learned from these exchanges. We share the conviction that each of our studies gains in important ways from the presence of the others.

Sheila Weiss's contribution analyzes the ideological, institutional, and intellectual origins of German eugenics, the degree to which the race hygiene movement did or did not participate in the barbarous programs of the Nazis, and the ways in which those programs related to German eugenic doctrines. William Schneider documents the variety of approaches to eugenics that developed in the French academic and scientific context, and their relation to Lamarckism, puericulture, and French concerns with the prospect of underpopulation. Nancy Stepan demonstrates the importance of the French training of many Brazilian physicians in developing a Lamarckian movement and the effects of the special racial and political situation in Latin America's largest country. My own contribution stems from many years' work on the history of Russian biology and Soviet science. It chronicles the creation of eugenics in a postrevolutionary state, the attempts to create a truly Bolshevik eugenics, and the interrelations of eugenics, genetics, "medical genetics," and Lysenkoism during the Stalin era.

The final chapter draws upon these case studies and others to show how a comparative international perspective changes our understanding of the "wellborn science." After surveying the current state of international research on the history of eugenics, the book concludes by setting forth an agenda for comparative analysis and future investigation. For those who wish to sample the national similarities and differences on a topic of special interest—Lamarckism, sterilization, racial politics, women's issues, or leftists who supported eugenics, for example—indexes have been provided.

CHAPTER 2

The Race Hygiene Movement in Germany
1904–1945

Sheila Faith Weiss

Although the histories of all national eugenics movements raise both difficult and controversial questions, the case of German eugenics, or "race hygiene" *(Rassenhygiene),* is by far the most troubling.[1] For many people the term *German eugenics* immediately brings to mind visions of the Nazi death camps and the "final solution." This presumed connection between German eugenics and the racial policies of the Third Reich makes a sophisticated analysis of German eugenics especially urgent, and also especially problematic. Indeed, until recently many German and non-German historians have simply subordinated race hygiene to the larger themes of either the history of European racism or the development of *völkisch* thought and, by so doing, have done little to challenge the "gut reaction"

---

1. It should be noted that the German term *Rassenhygiene* (race hygiene) had a broader scope than the English word *eugenics*. It included not only all atttempts aimed at "improving" the hereditary quality of a population, but also measures directed toward an absolute increase in population. Despite these differences, I will employ the two terms interchangeably throughout the essay. Even when German eugenicists limited themselves to measures that fall under the more limited term *Eugenik* (the Germanized form of the English word), they almost always employed the term *Rassenhygiene*.

This chapter was made possible in part by a Mellon Foundation research grant and release time awarded by the Faculty of Liberal Studies, Clarkson University, and by a National Science Foundation Summer Grant. Special thanks goes to Mark Adams of the University of Pennsylvania for initiating the writing of this article, and for bringing it to its final form. I am also indebted to my husband, Michael Neufeld, for his numerous suggestions and editorial assistance, and for his painstaking efforts in compiling and typing the bibliography. A modified version of this article has appeared in the third volume of *Osiris.* I wish to thank Arnold Thackray for graciously permitting the essay to be reprinted in this book.

that the entire history of German eugenics was a mere prelude to the Holocaust.[2] Insofar as they neglect to treat the German movement and the writings of its leaders in any detail, these authors have unwittingly obscured its context, logic, and aims (Gasman 1971; Mosse 1964, 1978; Altner 1968; Poliakov 1977; Mühlen 1977).

During the past decade, West German historians have begun to treat eugenics as a subject worthy of study in its own right. In particular, the work undertaken by Gunter Mann and his students in the Institute for the History of Medicine in Mainz has done an important service by rejecting race theory as the prime motive force behind the development of German race hygiene before 1933. They have also provided useful monographs on several leading German eugenicists. Following the approaches of traditional intellectual history, they and other historians have stressed its roots in social Darwinism—but at the price of ignoring the equally important social and professional contexts out of which it arose (Conrad-Martius 1955; Zmarzlik 1963; Bolle 1962; Mann 1973, 1977, 1978; Doeleke 1975; Rissom 1983). At least two of these authors also suggest that German eugenics can best be understood as a humanitarian attempt to improve the human race (or at least the Caucasian portion of it) that was unfortunately perverted by the Nazis (Doeleke 1975; Lilienthal 1979). This understanding, like that emphasizing social Darwinism, is at best very limited. While it would be hard to deny the importance of either social Darwinist ideology or the humanitarian impulse for the origins and appeal of race hygiene, these factors alone do not explain the phenomenon. Fortunately, very recent studies, such as those of Gerhard Baader, Paul Weindling, and Gisela Bock, to name only a few, are beginning to correct the deficiencies of the historiography of German eugenics.

Race hygiene in Germany was far more heterogeneous in its politics and ideology than is generally assumed. Although its advocates were overwhelmingly recruited from the ranks of the *Bildungsbürgertum* (educated middle classes), they embraced no single political outlook. Until Hitler's seizure of power in 1933 precluded the possibility of any visible political diversity within the movement, German eugenics captured the interest of individuals whose allegiance spanned the breadth of the Wilhelmine and Weimar political spectrum. While there were few committed Communists associated with the movement, the important position played by the socialist Alfred Grotjahn and the large number of members affiliated with the Weimar left-center eugenics society, the Deutscher Bund für Volksaufartung und Erbkunde (German Alliance for National Regeneration and the Study of Heredity) render it impossible to view German race hygiene as solely or even primarily a right-wing phenomenon. About the only unanimity discernible between such men as Grotjahn and the political conservative Fritz Lenz was on the question of laissez-faire capitalism: like the vast majority of other German race hygienists, both men viewed it as dysgenic. Their consistent critique of cap-

2. Throughout this essay the term *nonracist* will be applied to those race hygienists who rejected ideologies of Aryan supremacy. This is admittedly a very narrow definition of the word. Like most European intellectuals of their day, all eugenicists were racist in the sense that they believed in the "natural" inferiority of blacks and most other so-called nonwhite races.

italism should make us suspicious of interpretations that see race hygiene as just another intellectual prop of corporate capital (Graham 1977).

Just as German eugenicists varied greatly in their political orientation, they differed in the degree to which they accepted and promoted racist ideologies. Like the great majority of educated whites in Europe and North America of their day, all race hygienists accepted as a matter of course the racial and cultural superiority of Caucasians. From today's vantage point, then, all German eugenicists would be considered racist; however, since this type of racism was shared by most eugenicists everywhere, emphasizing it in the case of Germany, where the population was relatively homogeneous, tells us very little. The situation is more complicated with regard to ideologies of Aryan or Nordic supremacy. It is undeniable that many race hygienists, including several in the vanguard of the movement such as Alfred Ploetz (1860–1940), Max von Gruber (1853–1927), Ernst Rüdin (1874–1952), and Fritz Lenz (1887–1976), were Aryan enthusiasts. Indeed, among the prominent Aryan-minded eugenicists, there were those who were sometimes secretly, sometimes openly, in favor of using race hygiene to promote the so-called Nordic race. However, extreme caution must be taken not to equate these pro-Aryan sentiments made by a handful of German eugenicists with the aims of the movement as a whole. Many of Germany's leading eugenicists, such as Wilhelm Schallmayer (1857–1919), Hermann Muckermann (1877–1962), Artur Ostermann (1876–?), and Alfred Grotjahn (1869–1931), were uncompromising in their critique of all Aryan ideologies. Together with large segments of the Deutsche Gesellschaft für Rassenhygiene (German Society for race Hygiene), and virtually all of the members of the Bund, they rejected out of hand the desirability of a "Nordic race hygiene." In addition, anyone who examines the content of the two major Wilhelmine and Weimar eugenics journals, and looks at the platform of the Deutsche Gesellschaft, will be struck by the relative lack of space devoted to *völkisch* ideologies or "Nordic eugenics" as compared with other issues. On the whole this is true even of the writings of those eugenicists who embraced the Aryan mystique.

What both Aryan apologists and those eugenicists rejecting Aryanism *did* stress were strategies designed to increase the number of Germany's "fitter" elements and eliminate the army of the "unfit"—fitness being defined in terms of social and cultural productivity. The eugenicists' equation of fitness with productivity and achievement, and degeneracy with asocial behavior and the inability to contribute to society, reflected their own middle-class prejudices. When all is said and done, German eugenics before (and to some extent even after) 1933 was not primarily concerned with replenishing and improving the Nordic stock of Europe; occasional public displays of pro-Aryan sentiment notwithstanding, *Rassenhygiene* was more preoccupied with class than with race. Prior to Hitler's seizure of power, the concerns of German race hygiene were not fundamentally different from those of many other Western eugenics movements.

Despite this unity of class bias, the diversity of political outlooks and racial attitudes among German eugenicists nonetheless appears to preclude any single goal common to all of them. At first glance it is easy to see differences and conflicts of interest—so much so, in fact, that it is tempting to try to divide the movement into right-wing racist and left-wing nonracist camps as a first attempt at some kind of organizational clarity. While the institutional development of Weimar race hygiene, to be discussed later in this essay, does offer some justification for this classification, the story is more complex. Viewing the movement as an uneasy union of two separate and competing "camps" obscures the underlying rationale and logic of German eugenics. Whatever additional reasons may have motivated them, all German race hygienists embraced eugenics as a means to create a healthier, more productive, and hence more powerful nation. It was, however, quite unlike the usual political and economic strategies designed by those in power for the same purpose. Eugenics embodied a technocratic, managerial logic—the idea that *power* was essentially a problem in the rational management of *population*. For its practitioners race hygiene was a sometimes conscious, often unconscious strategy to buttress the supposedly declining cultural and political hegemony of Germany and the West through the rational management and control of the reproductive capacities of various groups and classes. Such a rational administration of human resources, the eugenicists believed, would ensure the necessary level of hereditary fitness thought to be a prerequisite for the long-term survival of Germany and western Europe and the allegedly superior cultural traditions they embodied. This logic constituted the common bond that united all German eugenicists.

The Origins of German Eugenics, 1890–1903

Social, professional, and intellectual contexts

German eugenics cannot be understood without examining the conjunction of circumstances that collectively account for its origin as a movement. Three contexts stand out as being particularly significant in shaping the early development of race hygiene: (1) the *social* problems resulting from Germany's rapid and thoroughgoing industrialization; (2) the *professional* traditions of the German medical community; and (3) the *intellectual* currency of the "selectionist" variant of social Darwinism then fashionable among certain German biologists and self-styled social theorists. These three contexts will be dealt with in turn.

During the last quarter of the nineteenth century, the newly unified German Empire was transformed from an agricultural to an industrial society. The industrialization and urbanization process, expeditious and thorough as it was, produced profound changes in the social and economic structure of the young Reich, engendering a myriad of series social tensions and problems (Hoffmann 1963; Wehler 1977, pp. 24, 41–59; Köllman 1969, p. 62). Had imperial Germany not possessed a rigidly authoritarian political structure shaped primarily by the self-interest of preindustrial elites and their allies in heavy industry, the social dislocations precipitated by industrialization would not have appeared so threatening to the stability of the state and the social order. But the *Kaiserreich* was certainly no democracy, and given the stranglehold that the landed aristocracy, the military, the barons of industry, and high-ranking members of the bureaucracy had over politics, these tensions and problems could not be effectively remedied (Wehler 1977, pp. 60–140; Evans 1978, pp. 16–22; Mock 1981).

Foremost among the problems afflicting the Reich as a result of this combination of political immobility and rapid social change was the rise of a radical labor movement. The growing number of strikes, lockouts, and other forms of labor unrest, coupled with the growing success of the officially Marxist Social Democratic party at the polls, provoked fear and anxiety among many middle- and upper-class Germans regarding the seemingly hostile, uncontrollable, and ever-increasing industrial proletariat (Ringer 1969, p. 129; Stern 1975, p. 15; Roth 1979, pp. 85–101). In addition, there were other social problems that were viewed by Germany's *Bildungsbürgertum* as posing a threat to the proper functioning of the state. These included (1) an increase in various types of criminal activity; (2) a rise in prostitution, suicides, alcohol consumption, and alcoholism; and (3) a heightened awareness of the existence of large numbers of insane and feeble-minded individuals. This latter group, the so-called mental defectives, was singled out by both medical and lay observers as an especially grave social and financial liability for the new Reich (McHale and Johnson 1976–77, pp. 212–14; Grotjahn and Kaup 1912, 1: 14, 687–88, 2: 376, 643; Fuld 1885, pp. 453–84; Evans 1976b, pp. 106–8; Roberts 1980, pp. 226, 232, 237; Meyer 1885, p. 83; Oettingen 1882, p. 671).

These problems were hotly debated by many of Germany's academic social scientists and reform-minded religious leaders under the rubric of the *"soziale Frage"*—a term referring to the social and political consequences of unbridled economic liberalism and the industrialization process (Müssigang 1968, p. 4; Wagner 1872; Schraepler 1964, 2: 62–66, 79–84; Ringer 1969, pp. 145–47). Although those discussing the "social question" embraced different economic and political ideals, all agreed that some kind of *Sozialpolitik* (social policy) was necessary to integrate Germany's proletariat (and asocial subproletariat) into the Reich, thereby preventing the collapse of the state. Like most educated middle-class Germans, the early eugenicists were keenly aware of this debate and were fully cognizant of the serious social problems that plagued the Reich as a result of the Industrial Revolution (Vondung 1976; Stern 1975). The increased visibility of a number of asocial, nonproductive types—an important component of the much-debated social question—was the problem they set out to tackle using a new form of *Sozialpolitik:* race hygiene.

That these race hygienists would be inclined to offer a biomedical solution for social and political problems can be attributed to the second major influence that shaped their eugenics: the distinctive social, political, and intellectual traditions of the German medical community. All of the movement's important leaders were physicians by training and had studied medicine before turning their attention to eugenics. Moreover, fully a third of those affiliated with Deutsche Gesellschaft during its early years were medically trained. As physicians, the founders of German eugenics not only shared the prejudices and posture of the *Bildungsbürgertum* as a whole, but were also heirs to a well-defined set of assumptions about the hereditary nature of disease and the role of medical professionals in safeguarding the health of the nation.

The medical professionals' perception of themselves as custodians of national health, and hence national wealth and efficiency, has a long history. In Germany it dates at least as far back as the mid-nineteenth century, when German physicians demonstrated their responsibility to the state during the so-called health reform movement (Ackerknecht 1932; Rosen 1975, pp. 99–102). Later, during the third quarter of the nineteenth century, the rise of scientific medicine and hygiene bestowed upon academic physicians, and the medical profession in general, an unprecedented level of social esteem and, indirectly, political importance (Rosen 1958, p. 44; Seidler 1977, pp. 91–92; Eulner 1969, p. 18). At this time many young medical professionals anxious to make a contribution to national health turned their attention to bacteriology; others, like some of Germany's future eugenicists, adopted a different approach. Their exposure to fields of medicine emphasizing the role of heredity in the etiology of disease (e.g. neurology and psychiatry) led them to question the efficacy of concentrating solely on pathogens. Instead, they were convinced that serious disorders such as mental illness, feeblemindedness, criminality, epilepsy, hysteria, and the tendency to tuberculosis were often inherited and could quite frequently be traced back to a "hereditary diseased constitution" (Grassmann 1896; Aronson 1894; Wahl 1885; "Über die Vererbung" 1879; Ackerknecht 1968). Many medically trained race hygienists argued that the surest way to improve the general level of national health was to upgrade the bodily constitution of all individuals in society—a task to be accomplished by means of an energetic eugenics program.

In addition to the social question and the German medical tradition, a third influence greatly shaped the early development of the movement: the "selectionist" variety of "social Darwinism" popularized by Germany's most outspoken biologist, Ernst Haeckel (1834–1919), and later legitimated by the scientific writings of the Freiburg embryologist August Weismann (1834–1914).

Haeckel went far beyond Darwin in his attempt to flesh out the larger philosophical and social meaning of the evolutionary theory. Although, like Darwin, he believed in the inheritance of acquired characteristics, Haeckel always stressed the importance of Darwin's selection principle as the most important engine of forward-directed organic change; indeed, for Haeckel Darwinism was synonymous with selection (1876, p. 120). Weismann, who came to reject the possibility of an inheritance of acquired characteristics through his work on heredity, afforded Darwin's principle of natural selection an even greater role in organic and social evolution than did the author of the *Origin* himself. His famous mechanism of heredity, "the continuity of the germ-plasm," first articulated in 1883, challenged the basic tenets of the more optimistic, first-generation social Darwinists who assumed that new characteristics acquired by an organism as a result of environmental change would be transmitted to future generations (Weismann 1889; Churchill 1968). As one German social Darwinist and eugenicist expressed it:

It was Weismann's teaching regarding the separation of the germ-plasm from the soma, the hereditary stuff from the body of the individual, that first allowed us to recognize the importance of Darwin's principle of selection. Only then did

we comprehend that it is impossible to improve our progeny's condition by
means of physical and mental training. Apart from the direct manipulation of
the nucleus, only selection can preserve and improve the race. (Siemens 1917a,
p. 10)

Indeed, for those who accepted Weismann's views with respect to both heredity
and the "all-supremacy" of selection, eugenics was the only practical strategy to
ensure racial progress and avert racial decline.

If the ideas of Haeckel and Weismann encouraged many contemporaries to
view natural selection as the sole agent of all organic and social progress, the writings of the two biologists also emphasized that progress was not inevitable. Under
certain conditions the "unfit" might prosper, thereby posing a challenge to any
further evolutionary development. This "selectionist" perspective and language
provided Germany's future eugenicists with novel tools of analysis with which
they were able to come to grips with the social question by transforming it into a
scientific problem: the asocial individuals created by industrialization became for
them the biologically and medically "unfit." The only way to gradually eliminate
their number was through a policy of "rational selection" or race hygiene.

The significance of these three contexts is nowhere more clearly visible than in
the intellectual backgrounds and early writings of Alfred Ploetz and Wilhelm
Schallmayer. Working largely independently of one another during the prehistory
of the movement (1890–1903), both men laid the foundations for the future
course of race hygiene in their country. Ploetz's organizational talents and char-
ismatic personality allowed him to create almost single-handedly the institutional
basis for the young movement. Schallmayer's eugenic treatises defined the significant theoretical and practical problems that would occupy German eugenicists
for decades.

Alfred Ploetz

Ploetz was born in 1860 into an upper-middle-class family in Swinemünde on the
Baltic Sea. Although details of his early life remain sketchy, he had already
become acquainted with the works of Darwin and Haeckel while still at the *Gym-
nasium*. Even before he began his university training in economics at the University of Breslau in 1884, he developed a strong interest in the *soziale Frage;*
prior to his matriculation he had already devoured the works of Plato, Malthus,
and Rousseau. During his student days at Breslau, he became increasingly sympathetic to some form of socialism. Indeed, he transferred to the University of
Zurich in 1885 in order to become better acquainted with the various brands of
socialist theory. There he not only attended lectures on socialism, but became
personally acquainted with August Bebel, the leader of the German Social Democratic party, and other Socialists who found Zurich a convenient haven from
Germany's oppressive Anti-Socialist Law (1878–1890) (Doeleke 1975, pp. 4, 18;
Muckermann 1931a, p. 261).

Ploetz's interest in economic theory, particularly socialism, was not merely
theoretical: he was determined to establish a kind of pan-Germanic utopian com-
mune. As Ploetz himself stated in his memoirs only a few years before his death,

the popular novels of Felix Dahn, professor of early German history, as well as works of other enthusiasts of Germany's Teutonic past, awakened his interest in the Germanic race. Indeed, Ploetz became so obsessed with the glories of the old Teutonic tribes that, together with several friends, he took an oath under an oak tree to do everything in their power to elevate the Germanic race to the level it had allegedly attained a thousand years earlier. (Doeleke 1975, p. 5; Muckermann 1931a, p. 261). Owing to this passionate concern, he chose to study economics rather than his first love, biology, believing it would prove to be more useful in helping him accomplish his goal. While at the University of Breslau, Ploetz and a small circle of friends—including the writers Carl and Gerhart Hauptmann— formed a society with the expressed intention of establishing a colony or socialist cooperative in a country containing a large percentage of Germanic stock. The American Pacific Northwest was chosen as the best possible site (Doeleke 1975, p. 6).

Ploetz traveled to the United States to learn more about the written constitutions of such communities, to familiarize himself with the social and economic conditions of the region, and to experience firsthand life in one of the already-established utopian socialist colonies. He spent six months in Iowa living and working in the cooperative known as Icarus. Appalled at the amount of fighting, laziness, egotism, and infidelity he observed in a community whose economic organization was supposed to eradicate such behavior, Ploetz came to an unusual conclusion:

> The unity of such colonies, especially those offering a large amount of individual freedom, cannot be maintained owing to the average *quality* of human material at present. . . . I came to the conclusion that the plans we wished to execute would be destroyed as a result of the *low quality of human beings. . . .* (Doeleke 1975, p. 13)

His reaction to this revelation was no less surprising:

> For this reason I must direct my efforts *not merely toward preserving the race but also toward improving it* . . . My views . . . immediately led me to the field of medicine—which appeared to be relevant to the biological transformation of human beings. (Doeleke 1975, p. 13)

After his six-month stay in the United States, Ploetz returned to Zurich to begin his medical studies and, later, under the direction of the psychiatrist and future eugenicist Auguste Forel, began an internship in a Swiss mental hospital. Although he harbored certain proeugenic sentiments even before beginning his medical training, Ploetz moved a step closer to articulating the need for race hygiene as a result of it. His experiences in the psychiatric hospital acquainted him with the so-called mental defectives and focused his attention on one of the causes of mental illness: alcoholism. Largely as a result of his discussions with Forel on the subject of alcohol and heredity, Ploetz took an oath of abstinence. From that point on, the counterselective effects of alcohol consumption became one of his major eugenic concerns (Doeleke 1975, pp. 17–18).

After spending some time taking specialized medical courses in Paris in 1890, Ploetz returned to the United States and opened a medical practice in Springfield,

Massachusetts. However, he found this work very disappointing. Ploetz was dismayed not only by the lack of time available for the study of eugenic problems, but by the limitations of therapeutic medicine. Having read the works of Haeckel, he could not have overlooked Haeckel's own attack on medical science in his popular work, *The History of Creation.* By this time Ploetz had recognized the need for a separate discipline dedicated to the hereditary improvement of the race—a discipline more effective in eliminating disease than the thankless "Sisyphian labor" carried out by modern therapeutic medicine (Doeleke 1975, p. 19).

Influenced by this constellation of social, professional, and intellectual contexts, Ploetz completed *Die Tüchtigkeit unsrer Rasse und der Schutz der Schwachen* (The fitness of our race and the protection of the weak) in 1895. Although his book did not initially generate much public interest, it raised the broad biological, social, and ethical problems that created the need for race hygiene in the first place. It also revealed the technocratic logic underlying eugenic thought.

The major thrust of Ploetz's argument recalls Darwin's personal dilemma in *The Descent of Man:* how can human beings reconcile the inevitable conflict between the humanitarian ideals and practices of the noblest part of our nature with the interest of the race, whose biological efficiency is allegedly impaired by those very ideals and practices? Translated into concrete economic and political terms, Ploetz viewed the problem as follows: should the state continue to expand the social net and regulate various aspects of economic life in order to lessen the hardship of the weak and economically underprivileged, but at the risk of undermining the overall biological fitness of its citizens? Would not health, accident, and old-age insurance invariably lead to an increase in the number of unfit, perhaps at the expense of the fittest members of society (1894, pp. 989–97; 1895b, chap. 3)?

Ploetz was not oblivious to the serious moral and social issues raised by this alleged conflict. As important as preventing *Entartung* (degeneration) was for him, Ploetz did not believe in ignoring the needs of the present generation; the danger of *Entartung* was not a signal for Germany to abandon her health and welfare legislation, despite its counterselective effects. Nor did it mean that one must embrace capitalism, as the seemingly most "proselective" economic system, and relinquish all hope of creating a humane, socialist society. The solution to these pressing conflicts was the substitution of the inhumane and inefficient process of natural selection by a humane and scientific policy of "rational selection." However, unlike the already existing personal hygiene, with its concern for the health of the individual, the new hygiene would direct its attention to improving the hereditary fitness of the human race. Ploetz named it *Rassenhygiene* (1895b, p. 5).

Considering Ploetz's own enthusiasm for all things Teutonic and the heated controversy that later ensued over the use of the word *Rassenhygiene* as a synonym for eugenics, it is worth examining what he meant by the term. His definition of *Rasse* is ambiguous and difficult to translate into English: "einer durch Generationen lebenden Gesammtheit von Menschen in Hinblick auf ihre körperlichen und geistige Eigenschaften" (1895b, p. 2). Roughly speaking, Ploetz seemed to view as a *Rasse* any interbreeding human population that, over the course of generations, continues to demonstrate similar physical and mental traits. This imprecise term could denote any small ethnic community, a nation, an anthropological race, or the entire human race (Lilienthal 1979, pp. 115–16). Ploetz's use of his newly coined term *Rassenhygiene* was equally broad and vague, denoting the hygiene of any and all of the previously mentioned groups. Somewhat later he defined race hygiene as the measures needed to ensure "the optimal preservation and development of a race" (1940a, p. 11). Hence, the word referred to the hereditary improvement of such disparate populations as the Jews, the Germans, the "Aryans," and all humanity. Lenz later suggested that Ploetz was not familiar with Galton's term when he wrote his book and simply chose the word *Rassenhygiene* to signify approximately what Anglo-Saxons had in mind when they employed the word *eugenics*. In other publications Ploetz explained that *Rassenhygiene* had a much larger scope than the English term *eugenics,* embracing not only those measures designed to improve the hereditary quality of a population, but also those aimed at achieving its so-called optimal size (1906b, p. 865). As alternatives, he later could have used either the Germanized form of the word eugenics, *Eugenik*, or the term *Rassenhygiene* (hygiene of the human race), as Schallmayer suggested.

What Ploetz actually thought about race is most clearly revealed in *Die Tüchtigkeit unsrer Rasse,* the one published source in which he devotes a significant amount of time to a discussion of these questions:

> The hygiene of the entire human race converges with that of the Aryan race, which apart from a few small races, like the Jewish race—itself quite probably overwhelmingly Aryan in composition—is the cultural race *par excellence.* To advance it is tantamount to the advancement of all humanity. (1895b, p. 5)

Although he states elsewhere that Germanic stock probably represents the best portion of the "Aryan race," he is primarily concerned here with whites generally. His views regarding the alleged cultural superiority of white people, however outrageously chauvinistic, were not fundamentally different from those of Schallmayer (who was vehemently opposed to the Aryan mystique), or indeed from those of most European intellectuals of his time. Nor was his pro-Aryan sentiment in any way anti-Semitic. Ploetz was, if anything, pro-Semitic at the time he wrote his book (although his views appear to have changed later on). Not only did he stress the significant role played by Jews in the intellectual history of humanity, placing them on the same level as the Aryans in terms of their cultural capacity, but he also opposed all attempts to ghettoize or otherwise separate the former from the latter. He was strongly in favor of intermarriage between Jews and Aryans on the grounds that it would be both socially and *biologically* advantageous to do so. Ploetz wrote his treatise at a time when economic anti-Semitism was making a strong comeback in Germany. His favorable discussion of the Jews, he stated, was included in his work partially in order to combat the new trend. He had little patience, at least at the time, with Jew haters. "All anti-Semitism is a pointless pursuit—a pursuit whose support will slowly recede with the tide of scientific knowledge and humane democracy, indeed all the faster the less the reactionary-national wind blows" (1895b, p. 142). Hence, the ideas expressed in this book could later be incorporated into Nazi racial policy only by misrepresenting the views of the author.

Although Ploetz discussed at length the merits of the Aryan race and defined the terms *Rasse* and *Rassenhygiene,* the major purpose of his book lay elsewhere. Above all, Ploetz sought to reconcile the inherent conflict between the Darwinian worldview and the humanitarian-socialist ideal through a conscious policy of "control over variation" (1985b, p. 226). What he had in mind was a utopian vision of pushing selection back to the prefertilization stage—a form of germ-plasm selection. According to this plan, the genetically best germ cells of all married couples would be chosen as the hereditary endowment for the next generation. As a result, inhumane social measures and economic systems previously deemed necessary to avert biological decline would become superfluous. "The more we can prevent the production of inferior variations," Ploetz asserted, "the less we need the struggle for existence to eliminate them" (1895b, pp. 226–39; Graham 1977, p. 1137). Although Ploetz's particular solution to the "degeneration problem" was unfeasible and never seriously entertained by any of Germany's race hygienists, it embodied the view, shared by Schallmayer and later by other eugenicists, that population was a resource amenable to "rational management." As such it was a biomedical solution for sociopolitical problems: eugenic experts, armed with their knowledge of evolutionary theory and the laws of heredity, would solve the social question with the aid of science.

Although Ploetz's work was the first to employ the term *Rassenhygiene,* it was not the first eugenic treatise to be published in Germany. The author of the earliest eugenic tract was the Bavarian physician Wilhelm Schallmayer.

Wilhelm Schallmayer

Schallmayer's intellectual biography and early career closely parallel those of Ploetz. Schallmayer was born in Mindelheim, a small Swabian town in Bavaria about twenty miles southwest of Augsburg. Like Ploetz, Schallmayer enjoyed the comforts of middle-class life. His father was the owner of a prosperous carriage and wagon business (Gruber 1922, p.55). Before turning to medicine as a means of securing a livelihood, Schallmayer studied economics, sociology, and socialist theory for two years at the University of Leipzig; he found the works of Karl Marx and the German economist and social theorist Alfred Schaeffle especially interesting (Gruber 1922, p. 55; Lenz 1919, p. 1295). Yet despite the interest he shared with Ploetz in the *soziale Frage* and socialist theory, Schallmayer's concern was purely theoretical—at least until he realized that a faulty social and economic system could have grave eugenic consequences. Unlike Ploetz, he never harbored utopian dreams of building a socialist Germanic community in the New World. Indeed, throughout his life he refused to embrace those ideologies of Aryan supremacy so important to Ploetz and several other eugenicists.

In 1881 Schallmayer enrolled in the faculty of medicine at the University of Munich. Upon completion of his degree, he secured an internship in the university hospital's psychiatric clinic, where he worked under Bernhard von Gudden (1824–1886) (Gruber 1922, p. 55). It is possible only to speculate how Schallmayer's internship might have influenced his later eugenic thought. He undoubtedly witnessed some of the most severe forms of those mental disturbances classed under the heading "insanity," thus becoming directly acquainted with many "mental defectives." Regarding the treatment and care of the insane and retarded, the young physician probably came away with the views of his teacher, which, as one obituary of von Gudden reported, amounted to a "near complete resignation regarding the effectiveness of medical intervention" (Kraeplin 1886, p. 607).

Precisely when Schallmayer began to contemplate eugenics is unknown. It seems likely, however, that his work in the psychiatric clinic led him to doubt the value of medicine for the health of the race, as opposed to that of the individual. A self-proclaimed social Darwinist and admirer of Haeckel, he could not have failed to see the connection between his clinical experience and the articulation of the counterselective effects of medicine as presented in *The History of Creation*. Schallmayer's own experiences working with "mental defectives," coupled with his "selectionist" outlook, accounted for his own indictment of therapeutic medicine in his first eugenic treatise, a short work entitled *Über die drohende körperliche Entartung der Kulturmenschheit* (Concerning the threatening physical degeneration of civilized humanity). Published in 1891 and reprinted under a slightly different title in 1895, it was Germany's first eugenic tract.

Although Schallmayer's slim volume attracted even less attention than Ploetz's treatise did four years later, it touched on the social, economic, and political justifications for eugenics, and it offered such practical proposals as the creation of medical genealogies and health passports and the introduction of marriage restrictions (Schallmayer 1895, pp. 23–32). Schallmayer's book also stressed the role of physicians and the importance of education and propaganda as the most effective means of achieving eugenic goals—both hallmarks of German race hygiene policy until 1933. Most importantly from the standpoint of the future development of German race hygiene, however, Schallmayer's treatise stressed the technocratic logic and the cost-benefit analysis that later so colored the movement.

Schallmayer's frustrations over the limitations of therapeutic medicine also shaped his personal career. Soon after he began work as a general practitioner, he decided to specialize in urology and gynecology. At least in this area, Schallmayer thought, the prevention and treatment of disease would benefit future generations as well as the individual. Ultimately, however, he found even this work disappointing. Like Ploetz, Schallmayer was anxious to devote all of his time to the cause of eugenics. So, in 1897, after he had acquired sufficient means to give up his lucrative medical practice in Düsseldorf, Schallmayer settled down as *Privatgelehrter* (Lenz 1919, p. 1295). His newly won freedom afforded him the time to compose a second eugenic treatise with the specific intention of submitting it to the Krupp *Preisausschreiben.*

In 1900 Friedrich Krupp, son of Essen's munitions baron Alfred Krupp, set aside thirty thousand marks to be used in a contest to answer to the question, "What can we learn from the theory of evolution about internal political development and state legislation?" (Ziegler 1903, pp. 1–2). It seems likely that Friedrich Krupp, an amateur biologist, was convinced by Haeckel of the scientific and political desirability of such an undertaking; Haeckel greatly resented the ban on teaching Darwinism in the schools, which stemmed in large measure from Social Democratic attempts to use Darwin's theory in support of their left-wing politics (Weiss 1983, pp. 137–38). Wishing to remain anonymous, Krupp delegated most of the responsibility for the administration and execution of the competition to Haeckel, who arranged for the entries to be examined by two different panels. The first, the panel of judges *(Schiedgericht)*, was composed of three respected scholars: Heinrich Ernst Ziegler, zoologist; Johannes Conrad, economist; and Dietrich Schäfer, historian. These three men, representing three diverse and "relevant" fields, independently judged and ranked all manuscripts. A prize committee consisting of Haeckel, Conrad, and Stuttgart paleontologist Eberhard Fraas was also established to settle any disparities and deadlocks among the three judges (Ziegler 1903, p. 4).

On 7 March 1903, the prize committee announced the winners of the competition. In addition to a first prize and two second prizes, five lesser monetary awards were handed out to the best entries. All eight award-winning manuscripts were to be published both individually and as part of a series entitled *Natur und Staat: Beiträge zur naturwissenschaftlichen Gesellschaftslehre* (Nature and state: Contributions toward a scientific study of society). The contest and the series well served the purpose of Fritz Krupp and the judges to demonstrate that Darwin's theory neither "possessed the state-damaging character attributed to it by its opponents" nor in any way "destroyed morals." Darwinism was shown to be inimical to social democracy and, although opposed to Christian morality, could be said to lay the groundwork for a new ethics founded on "a scientific-sociological basis" (Ziegler 1903, pp. 15–16).

First prize in the contest was awarded to Schallmayer's *Vererbung und Auslese im Lebenslauf der Völker* (Heredity and selection in the life-process of nations), a densely packed 381-page treatise representing, as Ziegler appraised it, a "hygienic-sociological" approach to the question. Schallmayer certainly saw the practical and political aims of his book as timely. Whereas the nineteenth century had been concerned with Darwin's evolutionary hypothesis on a purely theoretical level, "the twentieth century," argued Schallmayer, "is called upon to apply the theory of descent to everyday life" (1903, p. x). Yet before he himself attempted to elucidate the practical and political meaning of Darwinism, he thought it important to provide the reader with the necessary biological background. Schallmayer devoted over ninety pages to the theories of Darwin and Weismann, stressing the significance of the selection principle and Weismann's hereditary theories for the evolutionary process. In Schallmayer's view heredity and selection accounted for the enormous organic and social progress visible on the globe.

The book's central theme was the rational management of national efficiency. The real political lesson to be learned from Darwin's theory was that long-term state power depended upon the biological vitality of the nation; neglect of the hereditary fitness of its population, such as might result from unenlightened laws and customs, was "bad politics" and would inevitably result in the downfall of the state. Hence, the wise politician "would recognize that the future of his nation is dependent on the *good management* of its human resources" (Schallmayer 1903, pp. 380–81). In the interest of self-preservation, he argued, it was imperative that Germany take an active part in regulating the overall biological efficiency of its citizens by embarking on a political program that would encourage the biologically best elements in society to reproduce more than those with objectionable hereditary traits. Eugenics, or *Vererbungshygiene* (hereditary hygiene) (1903, p. 354), as he still called it at the time, was the perfect tool to ensure a strong and healthy state; it went hand in hand with his political ideal—a meritocracy (1903, p. 373; 1907, p. 735).

Schallmayer also presented his readers with a series of eugenics reforms, but he was very cautious in the area of negative eugenics. Although Schallmayer clearly believed that marriage restrictions for the insane, the feebleminded, the chronic alcoholic, and other defectives were in the best interest of the state and the race, he refrained from openly supporting state legislation as a means to this end. Until such time as more exact information regarding the laws of heredity was known, and enough detailed genealogies could be amassed, eugenicists would have to concentrate on voluntary measures. Instead, he emphasized positive eugenics—convincing the "fitter" groups in society to increase their fertility rate. The question, of course, remained: which groups were, biologically speaking, the "fittest"? Schallmayer assumed that biology would one day decide the question objectively. "In the meantime," argued Schallmayer, "it would not be incorrect to view highly socially productive individuals, especially the better educated, as being, on the average, more biologically valuable" (1903, p. 338). Civil servants, officers, and teachers were encouraged to marry as early as possible. Those who chose to remain single should suffer some sort of financial disadvantage. To encourage civil servants to have larger families, Schallmayer suggested they be given a bonus for each school-age child (1903, p. 338). The class bias implicit in Schallmayer's criteria could hardly be more blatant; his own social group, the *Bildungsbürgertum,* turned out to be the "fittest." As we shall see, Schallmayer was not the only eugenicist to share this orientation.

The Wilhelmine Race Hygiene Movement, 1904–1918

The journal and the society

The Krupp competition marked a turning point both in Schallmayer's personal career and in the attention paid to eugenics in Germany. Prior to this time Schallmayer and Ploetz were virtually lone prophets in their eugenics crusade. To be sure, there undoubtedly were other Germans concerned with similar issues; yet insofar as they were not personal friends of either of the two cofounders, they remained in complete intellectual isolation. Only in the years immediately following the publication of *Vererbung und Auslese* in 1903 were the first institutional steps undertaken to transform an idea into a movement: the creation of Germany's most respected eugenics journal and the foundation of a race hygiene society.

The *Archiv für Rassen- und Gesellschafts-Biologie,* the first journal dedicated to eugenics anywhere in the world, was founded by Ploetz in 1904. Although there is no direct evidence linking the creation of the journal to the results of the Krupp contest, it seems likely that the scientific recognition and public attention given eugenics in the immediate aftermath of the *Preisausschreiben* at least suggested to Ploetz and his two assistant editors, the sociologist and economist Anastasius Nordenholz and the zoologist Ludwig Plate, that a more organized and "strictly scientific" manner of discourse on the subject was possible. Without openly admitting it, the editors of the *Archiv* sought to establish a more clearly focused and academically prestigious form of the *Politisch-anthropologische Revue*—a journal that occasionally carried articles on eugenics-related issues but was not taken seriously by most professionals owing to its unmistakenly *völkisch* tone (Weiss 1983, p. 190). During the first four years, the *Archiv* was financed by the publishers themselves. By 1908 the journal had proved to be marketable enough to attract a publishing company to underwrite the cost of its production; whether Ploetz, who was independently wealthy owing to his marriage to Nordenholz's sister, continued to help finance it is unknown (Reichel 1931, p. 6).

The *Archiv* sought to attract a wide variety of articles bearing on the "optimal preservation and development of the race" (Ploetz, Nordenholz, and Plate 1903, p. iv). It included entries not only by Germany's prominent race hygienists, but also by individuals who in no sense considered themselves to be eugenicists. Most of the articles appearing in the journal during the Wilhelmine period fall into one of five categories: (1) technical articles by such leading biologists as Weismann, Plate, Ziegler, Richard Semon, Carl Correns, Hugo de Vries, Erich von Tschermak, and Wilhelm Johannsen dealing with genetics and evolution; (2) entries concerned with so-called degenerative phenomena (such as insanity, alcoholism, and homosexuality); (3) articles preoccupied with the alleged dysgenic effects of certain social institutions and practices (such as medicine and welfare) and the social and economic costs of "protecting the weak"; (4) studies pertaining to the need for population increase and the hazards of neo-Malthusianism; and (5) a potpourri of anthropological contributions including many racialist, but not always racist, articles as well as high-quality entries from the eminent anthropologist Franz Boas.

Besides publishing rather specialized and lengthy articles, the *Archiv* also tried to keep its readers abreast of developments in eugenics through its numerous book reviews and announcements. Its volumes were substantial, indeed—the four quarterly issues together often totaled more than six hundred pages. Although most educated middle-class Germans could have "ploughed through" the *Archiv,* the long, dry, and technical articles made neither enjoyable nor easy reading. Its national and international reputation as a highly respected scholarly publication notwithstanding, the journal did little to spread the eugenics gospel in Germany beyond the small group of professionals already committed to the new discipline.

The second institutional development was the formation of the Gesellschaft für Rassenhygiene (Society for Race Hygiene)—the world's first professional eugenics organization. Founded in Berlin on 22 June 1905, by Ploetz, Nordenholz (the psychiatrist and former brother-in-law of Ploetz), Rüdin, and the ethnologist Richard Thurnwald, the society had as its aim "the study of the relationship of selection and elimination among individuals as well as the inheritance and variability of their physical and mental traits" (Ploetz 1907, p. 3). Although there is some confusion as to the exact title of the organization during the first two years of its history, there is little doubt that Ploetz always intended the society, which had begun with only twenty-four members, to be international. Since the word *Rasse* was frequently used by Ploetz as synonymous with white race, any race hygiene society worthy of the name had to transcend national boundaries and embrace individuals from all white "civilized" nations. Yet it was not until 1907 that the Gesellschaft was able to attract anyone from other countries, at which time it became the Internationale Gesellschaft für Rassenhygiene. Two local groups of the Gesellschaft, in Berlin and Munich, were formed shortly thereafter (Internationale Gesellschaft 1907, pp. 1, 17).

The society wished not merely to spread the eugenics gospel, but also, and perhaps more importantly, to serve as a model for what rational selection could accomplish. By offering membership to only those white individuals who were "ethically, intellectually, and physically fit" and from whom "economic prosperity could be expected" (Internationale Gesellschaft 1907, p. 3), the society proposed to demonstrate from statistics collected on both the members and their progeny "how much better the vital statistics, the military fitness, and physical and intellectual efficiency are, compared to the population at large. . . . and how much more efficient the population of a state would necessarily be that followed race hygiene principles" (Ploetz 1907, p. 5). The society's understanding of "fitness" thus mirrored Schallmayer's own definition of the term: the most important criteria for eligibility were material success and social usefulness.

The Internationale Gesellschaft did not articulate any specific social policy or proposals. However, at a meeting of public hygienists in 1910, Ploetz set forth a list of practical goals, which included:

(1) Opposition to the two-child system, fostering "fit" families with large number of children, combatting luxury, reestablishment of the motherhood ideal, strengthening the commitment to the family;

(2) Establishment of a counterbalance to the protection of the weak by means of isolation, marriage restrictions, etc., designed to prevent the reproduction of the inferior; support of the reproduction of the fit through economic measures designed to make early marriages and large families possible (especially in the higher classes);

(3) Opposition to all germ-plasm poisons, especially syphilis, tuberculosis, and alcohol;

(4) Protection against inferior immigrants and the settlement of fit population groups in those areas presently occupied by inferior elements—to be accomplished, if need be, through the expropriation laws;

(5) Preservation and increase of the peasant class;

(6) Introduction of favorable hygenic conditions for the industrial and urban population;

(7) Preservation of the military capabilities of the civilized nations;

(8) Extension of the reigning ideal of brotherly love by an ideal of modern chivalry, which combines the protection of the weak with the elevation of the moral and physical strength and fitness [*Tüchtigkeit*] of the individual. (1911c, p. 165)

These proposals reflect the international orientation of Ploetz and the early movement while simultaneously demonstrating their concern with national efficiency. The explicit statement regarding the encouragement of marriages and large families among members of the upper class once again shows the tendency to equate fitness with class. Belonging to the so-called Nordic or Germanic race, interestingly enough, was not a criterion for *Tüchtigkeit;* indeed, the terms *Nordic* and *Germanic* are not even mentioned in the list of tasks and programs. Having mentioned in his speech the lack of any general consensus regarding what constitutes the best race, as well as the rarity of finding pure races anywhere in Europe, Ploetz was indeed reluctant to make special claims for the Nordic population. Since, as he argued, "all these races (Alpine, Jewish, etc.) are seldom found pure here, it is best . . . to rely on fitness as a guide. This is because fitness—both individual and social—is the true guiding star. What particular colors or shapes are attached to it [fitness] will reveal itself in the future" (1911c, p. 190). Although Ploetz may have had a particular interest in the Nordic race, his position does suggest that fitness, as defined in terms of social and cultural productivity, was the true measure of the worth of both individuals and races. It remained the cornerstone of both his and the pre-1933 movement's eugenic policy.

In 1910, the year in which Ploetz presented these proposals, the individual German *Ortsgruppen* (chapters) were brought under the banner of the Deutsche Gesellschaft für Rassenhygiene, initially a sort of national subdivision of the Internationale Gesellschaft für Rassenhygiene (Deutsche Gesellschaft 1910). Yet the international society was not to last. By 1916, in the wake of both World War I and the creation of numerous national eugenics societies in Europe and the United States, Ploetz was forced to give up his dream of a single "intellectual center" for the preservation of the race. At this time the German society officially supplanted the international society, although it had long since done so in practice (Fischer 1930, p. 3).

In the meantime, the total membership remained small, but also grew steadily, and the occupational and class backgrounds of the members of the two societies continued to mirror those of its founders and leaders (see Table). In both the international and the German society, *Bildungsbürger* dominated the membership. The Table also reveals that medical professionals constituted the single largest group in both organizations, accounting for approximately one-third of those affiliated with the two societies. It seems likely that the self-image of German physicians as custodians of the nation's health had much to do with the disproportionate number of prominent physicians, hygienists, and professors of medicine in the early movement. Of the academics from outside medicine enrolled in the two societies, most were professors of zoology and anthropology. In addition to Ernst Haeckel and August Weismann, who as honorary members probably did

Occupational Composition of the Society for Race Hygiene

Occupation	1907 (International Society)		1913 (German Society)	
	N	Percent	N	Percent
Physicians and medical students	27	32.5	136	33.4
Nonmedical academics	14	16.9	76	18.7
Writers and artists	10	12.1	22	5.4
Civil servants and teachers	3	3.6	29	7.1
Miscellaneous	7	8.4	78	19.2
Wives*	22	26.5	66	13.8
Totals	83	100.0	407	100.0

*Includes only those women listed with their husbands, who had no other listed occupation.

Sources: Internationale Gesellschaft für Rassen-Hygiene 1907; Deutsche Gesellschaft für Rassen-Hygiene 1913.

not participate much in its activities, the Deutsche Gesellschaft included such distinguished biologists as Ludwig Plate, Heinrich Ernst Ziegler, and Erwin Baur. The two societies also included members of virtually all German political parties; moreover, Jews, as well as Protestants and Catholics, were among the members (Ploetz 1909, p. 278). The only specific qualifications mentioned in the statutes were that members be both white and "fit."

The intellectual development of Wilhelmine eugenics

The writings of Wilhelmine Germany's race hygienists exhibit some common themes and concerns. The primary intellectual preoccupation of the early movement was collecting and analyzing data on degeneration. A study of the celebrated Family Zero—a kind of Swiss counterpart to the legendary American Jukes—was undertaken to demonstrate that central Europe had its own share of degenerate stock (Jörger 1905). The psychiatrist Ernst Rüdin wrote numerous articles dealing with the inheritance of insanity—emphasizing the Mendelian nature of the transmission of various kinds of mental disorders. Agnes Bluhm, Germany's only prominent female eugenicist, concentrated her efforts on proving the degenerative effects of alcohol on future generations and studying the alleged decreased ability of German women to breast-feed their infants (Ploetz 1932, p. 63). Other eugenicists reported on such topics as the increase in venereal disease in large cities and its impact on the race, the degenerative effects of homosexuality, and the need to reform Germany's penal code along eugenic lines (Lenz 1910; Rüdin 1904; Hentig 1914). By and large the tone of these studies was scientific, not popular; they seem to have been written less to stir people to action than to communicate abstract information.

Like eugenicists in both the United States and Britain, the Germans also analyzed the cost of maintaining the army of the unfit. The word most often used to describe these individuals was *Minderwertigen*—a term that literally means "the less valuable" and was frequently employed as a synonym for nonproductive people. Certainly, the *Umschau,* a popular science journal, used the word in this way when, in 1911, it sponsored a written contest entitled "What Do the Inferior Elements [*Minderwertigen*] Cost the State and Society?" (Kaup 1913, p. 723). Accepting the premise that "all efforts to improve the environment break down in the face of hereditary sickness and inferiority," the sponsors of the competition suggested to potential contestants that only a reduction in the number of "minus variants" would allow society to continue to preserve the life of all those living. Only five contestants applied for the prize, however, and the problem did not yet generate the great concern that it would during Weimar (Kaup 1913, p. 723).

In his commentary on the cost of the unfit, Ignaz Kaup, professor of hygiene and member of the Deutsche Gesellschaft, reported on the results of a seminar held to discuss the subject (1913, p. 748). Since he doubted that the German people were ready to accept American-style sterilization methods as a means of alleviating the problem, some way of physically separating the "unfit" from the rest of society was necessary. False humanitarian considerations were not appropriate since "all forward-striving nations had the duty to ease the burden of the cost of the inferior as much as possible" (1913, p. 747). Recognizing that the *Minderwertigen* were a financial burden to the state who, "despite the expenditure paid out on their behalf are almost never in the position during their working lives to repay the money spent on them," Kaup recommended the creation of work colonies where they could be prevented from having inferior children and could be made to earn their keep at the same time (1913, p. 748). At this time, however, most German eugenicists would have been satisfied with some form of permanent institutionalization.

As the First World War approached, a third emphasis of the Wilhelmine eugenics movement came to the fore: *Bevölkerungspolitik* (population policy). While Germany's eugenicists did, of course, aim at instituting a eugenically healthy qualitative population policy, there was a marked tendency throughout the last years of the empire to view the prevention of a decline in population growth as an important measure in its own right. As early as 1904, Alfred Grotjahn spoke of the "growth of population quantity" as the "*conditio sine qua non* of a rational prophylaxis against degeneration" (1904, p. 61). Later the issue had become more pressing: as Schallmayer put it in 1915, arresting population decline was nothing short of "a matter of survival for the German nation" (1914–1915, p. 729).

In order to understand why German eugenicists became obsessed with the population question, it is worth discussing briefly the prewar demographic, social, and political changes in Germany that colored their intellectual perspective. On the surface there seemed little cause for alarm. Wilhelmine Germany was the second most populous country in Europe; it had also witnessed a substantial population increase of twenty-four million people between 1871 and 1910. Yet this healthy population growth owed far more to the dramatic decline in the death rate, particularly the infant mortality rate, than to a growth in fertility (Hohorst, Kocka, and Ritter 1974, pp. 15, 29–30). Indeed Germany, like all Western industrialized nations, experienced a steady birthrate decline during the last third of the nineteenth century and first third of the twentieth century. Between 1902 and 1914, for example, the Reich suffered an 8.3 per thousand drop in the number of live births (Hohorst, Kocka, and Ritter 1974, pp. 29–30). This and the steady decline in the excess of births over deaths after 1902 gave statisticians and eugenicists cause to expect an eventual population standstill, or even population decline. Many sought to account for Germany's declining birth rate, yet however much their explanations differed, all investigators agreed on two points: (1) that the actual decline in population growth was less frightening than the prospect that Germany's situation might soon begin to mirror French demographic realities; and (2) that the drop in the birthrate was conscious and was directly related to the practice of birth control methods advanced by supporters of neo-Malthusianism (Brentano 1908–1909; Wolf 1912; Seeberg 1913).

Considered by Rüdin and the president of the Deutsche Gesellschaft, Gruber, to be even more dangerous than the "relative increase of the unfit" (Gruber and Rüdin 1911, p. 158), German neo-Malthusianism encouraged birth control as a means of eliminating poverty and its attendant social problems. Much work remains to be done on neo-Malthusianism in Germany, but it seems likely that the German movement received its impetus and theoretical direction from the English Malthusian League (Ledbetter 1976; Soloway 1982). As was the case in England, many German reform-minded liberals and socialists saw the movement as the only hope for improving the conditions of the working class. For active German feminists and the growing number of liberal, university-educated, middle-class women, however, birth control was also a means of emancipation from the drudgery of unwanted pregnancies and a prerequisite for more productive and intellectually meaningful lives (Evans 1976a, pp. 115–43).

What made the so-called antibaby and antimotherhood propaganda of the feminists and the neo-Malthusians so disturbing was the deterioration in the international political climate after 1900, particularly the direct challenges to western European hegemony and the growing belief in the possibility of war. The rise of Japan to a position as a world power and the fear of a revitalized China with its burgeoning population growth and its potential military superiority raised the specter of a Yellow Peril—a term coined by Emperor William II and popularized after the Japanese victory in the Russo-Japanese War of 1904–1905 (Barraclough 1981, p. 81). Even more disconcerting from the German eugenicists' point of view, however, was the Slavic threat: with its army of over a million men, Russia bordered the Reich and was allied with Germany's potential enemies, England and France. In a paper attacking neo-Malthusianism at the First International Eugenics Congress in London in 1912, Ploetz indicated that the Slavic threat was biological as well as political: while western Europeans and Americans exhibited a decline in fertility, Ploetz lamented, the "Poles, Hungarians, Russians, and South Slavs—nationalities with strong Asiatic traits—have an extremely high birth rate such that they are everywhere successfully pushing westward." "The preservation of the Nordic race," he argued, "is severely threatened as a result" (1913, p. 171). Schallmayer also expressed fear over the potential slavicization of Germany, although he viewed it more as a threat to the Reich than to the well-being of some mystical Nordic stock (1908b, p. 411). By 1910 most German race hygienists agreed that without some means of combating neo-Malthusianism, the biological efficiency of the empire would become severely impaired.

This new and unhealthy state of affairs required not only that Germany and western Europe produce more people, but also that they produce the right kind of people. What horrified eugenicists most was the alleged counterselective impact of birth control practice: the biologically "superior"educated upper and middle classes were limiting the size of their families while the defectives and "less fit" were not. Even before the war, German eugenicists, like their British and American counterparts, were bemoaning the inverse relationship between social usefulness and fertility. At the well-publicized International Hygiene Exhibition in 1911, a battery of charts and graphs plotting demographic trends in France, Denmark, Holland, England, and Germany pointed to a time in the not-too-distant future when western Europe would be without the standard-bearers of its culture: the upper and educated middle classes (Gruber and Rüdin 1911, pp. 158–63). "In order to flourish, indeed in order to survive," Gruber and Rüdin asserted, "a people needs a sufficient number of hands and a sufficient number of heads to rule those hands" (1911, p. 158).

After the outbreak of the war, the realization that the fighting was likely to continue far longer than originally expected only intensified German race hygienists' concern with the population question. For men such as Schallmayer, Ploetz, Gruber, and Lenz, World War I was a kind of necessary evil—necessary because they, like most Germans, believed that the empire was provoked into war; evil because they realized the unmeasurable biological damage it would inflict not only upon Germany, but also upon other Western nations. Most believed that Russia, never part of the West as far as German eugenicists were concerned, would come out of the war with its biological efficiency relatively unscathed (Schallmayer 1918a, p. 22; Lenz 1918, p. 444). Hence, it would continue to be Germany's greatest political and biological threat.

In the face of a war that would, according to Lenz, deplete the Reich of its "racial capital" (1918, p. 440), leaving it prey to its hostile enemies (primarily Russia and France), eugenicists sought to devise a series of reform plans and programs to offset the anticipated quantitative and qualitative population loss. Even before the actual fighting began, the Deutsche Gesellschaft issued a set of resolutions aimed at halting birthrate decline. They were published in the *Archiv* and then translated for the American *Journal of Heredity*. Among the most important were (1) an inner colonization (back-to-the-farm) movement with privileges for large families; (2) economic assistance to large families and consideration of the size of public and private employees' families in determining wages; (3) abolition of impediments to early marriage for army officers and government employees; (4) obligatory exchange of health certificates before marriage; (5) prizes to artists who glorified the ideals of motherhood, family, and the simple life; and (6) attempts to awaken a sense of duty toward the coming generation—including education of the youth in this direction ("Leitsätze der Deutschen Gesellschaft" 1914; G. Hoffmann 1914). Some eugenicists had additional plans for boosting Germany's national efficiency—plans that stressed qualitative population policy more than those officially adopted by the society. Gruber and Schallmayer, for example, adopted a plan to reform Germany's inheritance laws so that if a deceased father had left fewer than four children, a portion of the inheritance would be turned over to relatives (Gruber) or the state (Schallmayer) (Schallmayer 1918b, p. 336). Yet German eugenicists remained "all talk and no action"—neither the society's proposals nor those of any other eugenicist were adopted by the government. During the war new organizations preoccupied with the population problem proliferated, including the German Society for Population Policy (1915, about one hundred members) and the Alliance for the Preservation and Increase of German National Strength (1915, about one thousand members). Yet despite these organizations, and despite the interest in race hygiene displayed for the first time by some government officials (Weiss 1983, pp. 278–79), not one eugenics-related law was passed during the Wilhelmine period.

There were undoubtedly many reasons for this state of affairs. Among them were the reluctance of German eugenicists to push for sterilization laws or other forms of negative eugenics, and their emphasis on a rather abstract and diffuse set of positive eugenics proposals that would have been difficult to translate into concrete statutes. Initially, those in the vanguard of the movement had been content to educate the public as to the social and political need for eugenics. Having an exaggerated sense of their own importance as intellectual leaders of the nation, German race hygienists overestimated the power of their well-manicured public utterances. However, the social, political, and economic disaster brought on by the war both encouraged the growth of the movement and stimulated a bolder approach.

Eugenics in the Weimar Republic, 1918–1933

The three major concerns of Wilhelmine race hygiene—degenerative phenomena, analysis of the burden of the *Minderwertigen,* and population policy—continued to preoccupy the second generation of the movement. The Weimar years, however, witnessed an increased emphasis upon reducing the social cost of the unproductive. Whereas eugenicists had earlier spoken in very abstract terms about improving the "race"—however differently that term was understood by individual practitioners—race hygiene during the republic was far more concerned with preventing the decline of the German *Volk* and state. This does not mean that the movement lost its international orientation entirely. German eugenicists continued to correspond with their English and American colleagues and, after the early 1920s, participated in international eugenics and genetics conferences. One still finds talk about saving "civilized nations" from degeneration. Yet, on balance, German eugenicists were absorbed with the problems besetting their own country. Especially during the early Weimar years, eugenicists saw the fatherland as engaged in a life-or-death geopolitical and economic struggle for survival with its western European and Russian enemies (Ziegler 1922). Oppressed by the economic and psychological impact of the Versailles treaty and inflation, forced, as the geneticist Erwin Baur put it, to suffer foreign domination by people "culturally beneath them" (Glass 1981, p. 364), and consigned to live under an unstable and, for the most part, unloved republic, race hygienists realized that improving Germany's biological and national efficiency was no longer of mere intellectual interest.

Fritz Lenz

The one person who did more than any other to spell out the importance of eugenics during the Weimar years was Fritz Lenz. A decidedly complex individual, Lenz became Weimar Germany's most prominent and, in many ways most controversial, eugenicist. The death of Schallmayer in 1919 and Ploetz's growing reluctance to shoulder the burdens of discipline building left Lenz as the acknowledged leader of the Munich chapter of the society—much to the dismay of his less conservative, nonracist colleagues in Berlin. Lenz viewed himself as a student of Ploetz; he shared his mentor's enthusiasm for the Nordic race—an enthusiasm undoubtedly strengthened through his contact with the anthropologist Eugen Fischer (1874–1967), whom he met while enrolled as a medical student at the University of Freiburg. During his medical studies at Freiburg from 1906 until 1912, he also attended the lectures of August Weismann. From his own account, Weismann made a lasting impression on Lenz and was probably responsible for his lifelong interest in the inheritance of hereditary diseases and intelligence. Lenz's training made him particularly receptive to the ideas of Ploetz, whom he first met in 1909. From that time on he sought to devote his life to the "practical" application of the study of human heredity; even his medical dissertation, completed in 1912, stressed eugenic concerns (Rissom 1983, pp. 15–17).

Although Lenz was active in the Munich chapter of the society before the war, he first came to the attention of the international eugenics community in 1921 as coauthor of *Grundriss der menschlichen Erblichkeitslehre und Rassenhygiene* (Principles of human heredity and race hygiene). The treatise comprised two volumes. The first had a theoretical orientation and contained chapters by Erwin Baur on the principles of heredity, by Eugen Fischer on the world's racial groups, and by Lenz on human inheritance. The second volume, composed entirely by Lenz, dealt exclusively with race hygiene. Such respected American geneticists as Raymond Pearl and H. J. Muller considered the section written by Baur to be a clear and objective "state-of-the-art" summary; portions of Fischer's and Lenz's contributions, as they stood in the 1931 American edition of the text, were thought by Muller to be less so (Glass 1981, p. 357). Even discounting the current prevalence of typological thinking about race, there can be little doubt that Fischer's and Lenz's discussions of race were largely a collection of personal and social prejudices masquerading as science. Considering the important position Lenz held in the movement and the subsequent outcome of Nazi eugenics, it is worth examining his views on this subject further.

Like Galton and many other non-German eugenicists, Lenz believed in the reality of physical and mental racial traits. He understood these traits to be hereditary in ways that other traits common to all humans were hereditary. As such, their relative frequency in a population was not static, but rather was influenced, according to Lenz, by an all-powerful and ubiquitous selection process. Although he fully recognized physical differences between the world's races, they were in and of themselves uninteresting and sometimes unreliable when it came to assessing an individual's racial type (Baur, Fischer, and Lenz 1923, 1:406, 409). Lenz concentrated almost exclusively on what he called the *seelische* (spiritual) differences, by which he meant the sum total of all nonphysical qualities of the major races. He clearly accepted a hierarchy of races, despite his comments to the contrary. Moreover, all talk of a transcendental "racial principle" aside, Lenz held up Western culture as the yardstick by which he measured the "fitness" of races. Those races seen as having a high level of culture—by which he meant European and, in particular, German culture—were fitter and hence more worthy of preservation than others (1923, 1:423–27). Not surprisingly, the Negroid race stood at the bottom of the scale; the Nordics and the Jews (the latter themselves consisting of two main races, the Near Eastern and Oriental) were the most culturally productive (Lenz 1923, 1:417–27). According to Lenz, Nordic man was future-directed, steadfast, and prudent, and hence able to subordinate sensual pleasure to more long-term goals; he was not only the religious and philosophical man par excellence—always searching but never finding what he needs—but he also exceeded all others in objectivity. Of Nordic woman he has less to say except that she, like women of other races, was on the average less objective than men. That, however, was no great problem "since women have an entirely different mission to fulfill in the life of the race" (Lenz 1923, 1:419, 422).

Lenz's sexism and the almost laughable manner in which he projected German educated middle-class values onto "Nordic man" were, of course, not recognized by him as prejudices. Above all, Lenz thought of himself as an objective scientist who arrived at his conclusions after careful consideration of the facts. He found all demagoguery and emotionalism essentially "un-Aryan." Indeed, in his critique of the "emotional" anti-Semitism found in Theodor Fritsch's *Handbuch der Judenfrage,* Lenz accused the work of being too "Jewish" and not Germanic enough in its lack of "absolute objectivity" (Lenz 1923, p. 431). Given his temperament, he never could have written an inflammatory book such as Madison Grant's *The Passing of a Great Race,* though he was not reluctant to discuss its merits. This desire to remain *sachlich* (objective) undoubtedly colored his attitude toward Jews. Lenz's anti-Semitism was the subdued variety commonly found among conservative German academics. Insofar, however, as he believed in the reality of racial types and was hence forced to describe the "spiritual" elements of the Jewish race, his stereotypical caricature of Jews has occasionally led people to see him incorrectly as an intellectual forerunner of Hitler. Although hardly pro-Semitic, he considered the Jews to possess many of the admirable qualities that Nordics, as well as others, did not possess to the same degree—much to the dismay of Germany's numerous rabid anti-Semites. Indeed, he felt that Nordics and Jews were more similar than dissimilar. What he did not like about the Jews (e.g., their preoccupation with making money and their liberal politics) he of course also projected onto their list of racial qualities, which he then attempted to relate to his reader in a cool, objective manner (Müller-Hill 1984, pp. 37–38). It is revealing of the degree to which typological thinking about race was generally accepted to find Lenz's book praised even by Jewish authors. Lenz was proud that the respected Jewish sexologist M. Marcuse, a specialist in the area of venereal disease and prostitution, had apparently accepted his "very unprejudiced and purely objective depiction of the racial condition and psychic constitution of the Jews as compared to that of the Germans" (Müller-Hill 1984, p. 426). As contemptible as Lenz was (especially for his willingness to cooperate with the Nazis after it was clear to him what *their* policies were), he seems to have believed that the promotion of the Nordic race need not go hand in hand with anti-Semitism. Although he later saw Hitler as the only political leader who truly embraced a eugenic standpoint, and as a result was favorably disposed toward him as early as 1931, he found Hitler's maniacal anti-Semitism too extreme (Lenz 1931b, p. 302).

While Lenz's acceptance of ideologies of Nordic supremacy was clearly evident in virtually everything he wrote, it should be pointed out that of the more than six hundred pages he contributed to *Menschliche Erblichkeitslehre,* only about fifty dealt with the race question. The bulk of his work was concerned with such issues as the transmission of hereditary diseases, the inheritance of intelligence and talent, the methodology of genetic research, and the theoretical principles and practical teachings of race hygiene. In his discussion of the inheritance of disease and talent, Lenz sometimes cited the work and methodology of British and American geneticists and eugenicists, most frequently Galton. A convinced Mendelian, Lenz sought to demonstrate the Mendelian pattern of inheritance for various pathological traits; when focusing his attention on "metrical" traits such as intelligence, he naturally used the statistical tools developed by the British biometricians. Having at least some training in genetics, he was far more knowledgeable than most German race hygienists about the newest developments in the field. In general, however, the technicalities of genetics were important to Lenz only insofar as they could be used to support and legitimize his eugenic views.

Lenz's major eugenic aim was the preservation of his own class, the *Bildungsbürgertum,* from biological extinction. Perhaps more than anyone else, he viewed eugenics as a means of boosting Germany's level of cultural productivity. Although virtually all German eugenicists equated the fit with the educated and socially useful elements in society, nobody was more crass in his class prejudices than Lenz. "Productivity and success in social life," Lenz affirmed in his textbook, "serves as a measure of the worth of individuals and families" (Baur, Fischer, and Lenz 1923, 2:206). Indeed, for him, *Entartung* was virtually synonymous with a lack of culture. Lenz, even more so than Schallmayer, saw the real threat of degeneration not in the marginal increase in the number of those with serious hereditary diseases, but rather in the low birthrate of the educated middle class and the "extinction of highly talented and otherwise distinguished families" (Baur, Fischer, and Lenz 1923, 2:192). Contemptuous of the value of manual labor as compared with that of "mental labor," Lenz was particularly dismayed at the drop in the standard of living of academics during the early years of the republic, as well as by the supposedly preferential treatment shown to workers after 1918:

> The German revolution had an overwhelmingly unfavorable selective effect. As a result of the one-sided promotion of the interest of the manual workers, those who work with their brains are forced into a terrible struggle for survival. . . . If one views German society as a whole, there can be little doubt that the results

of the revolution will lead to the extinction of educated families—the primary standard-bearers of German culture. (Baur, Fischer, and Lenz 1923, 2:63)

Yet for Lenz the events of 1918–1919 were not only dysgenic, but also politically distasteful. A conservative academic, he belonged to the extreme right-wing German National Peoples' Party and had little tolerance for the republic and its allegedly untalented leaders (Baur, Fischer and Lenz 1923 2:70; HSAM MK 35575). He viewed the German revolution with horror, attributing it to the "extermination" of a large number of "socially-minded" individuals on the battlefield. He found the new democratic order, with its promise, however limited in reality, of increased social equality, both biologically and socially dangerous. It was a far cry from the political ideal he shared with Schallmayer: a meritocracy (Baur, Fischer, and Lenz 1923, 2:57, 247).

Berlin versus Munich

Lenz's influence and his Nordic sympathies were recognized and resented by the many nonracists in the movement. Had Lenz's position been an entirely idiosyncratic one, prominent eugenicists such as Schallmayer (while he was still alive) and Grotjahn might have been annoyed, but relatively unconcerned, about its impact on the long-term direction of the movement. They knew, however, that Lenz, while perhaps more extreme than most who adopted the "unscientific" doctrine of Aryan supremacy, was not alone. Gruber and Rüdin, for example, were known to be sympathetic to Lenz's point of view. Ploetz, it will be recalled, also embraced the Aryan-Nordic mystique. Almost from the beginning a largely unadmitted confict arose between those who believed that eugenics had nothing to do with ideologies of Aryan supremacy and those who, *in addition* to articulating the class-biased positions of their nonracist colleagues, also wished to leave the door open for an Aryan or Nordic race hygiene.

The conflict was both terminological and institutional. Schallmayer was so adamantly opposed to the racist connotation of *Rassenhygiene* that he never employed the word. During the early years of the movement, he urged the adoption of the words *Rassehygiene* and, somewhat later, *Rassedienst* (racial service)—both of which, because they avoided the plural *Rassen,* precluded the denotation of an anthropological race (Schallmayer 1910b, p. 352). While supporting Schallmayer's efforts to rid race hygiene of all racist connotations, Grotjahn encouraged the use of yet another term—*Fortpflanzungshygiene* (reproductive hygiene), a word that avoided all mention of *Rasse* (Baur, Fischer, and Lenz 1923, 2:162). Neither man favored the term *Eugenik,* despite its alleged neutrality and objectivity, since it was seen as excluding quantitative population policy. Of course, those whose sympathies for the "Nordic race" made the double connotation of the term *Rassenhygiene* desirable did what they could to defend its use. This was especially true of Lenz, who, in addition to admitting that its racial connotation was a point in its favor, offered self-serving linguistic and practical reasons for the continued use of Ploetz's term (Lenz 1915b, pp. 445–48; Baur, Fischer, and Lenz 1923, 2:161–62). His insistence that "race hygiene is of course advantangeous to all races" did little to reassure those supporting a "scientific" eugenics that the movement might not be overrun with racist fanatics (Baur, Fischer, and Lenz 1923, 2:162).

Pre-Weimar organizational developments reinforced these fears. Ploetz and his like-minded colleagues not only tolerated Aryan and Nordic enthusiasts, they also catered to them. As early as 1911 Ploetz, Lenz, and a physician named Arthur Wollny founded a secret "Nordic Ring" within the society, whose aim was the improvement of the Nordic race. As an unpublished pamphlet entitled *"Unser Weg"* (Our way) points out, Ploetz and his sympathizers in the Nordic Ring harbored plans for a "Nordic-Germanic race hygiene"—if only as a part of a much broader eugenics program—that would direct its attention to saving the Nordic elements in Western civilization (Ploetz 1911b, p. 2). In addition, these same individuals helped establish other similar, though not secret, *völkisch* organizations, including the little-known and insignificant Bogenklub (1912) and, after the war, the Deutsche Widar-Bund (1919) (Doeleke 1975, p. 46).

The tensions over the race question, while visible during the empire, were exacerbated by the increased political polarization of Germany following the German revolution and the founding of the republic. Formerly the seat of Prussian conservatism, Berlin acquired a decidedly pinkish hue during Weimar. Indeed, of all the German states, Prussia became the republic's staunchest defender. On the other hand, Bavaria, and in particular Munich, became a hotbed of political reaction following the collapse of the feared and hated Munich Soviet republic of 1919. Not only was it the major center of right-wing paramilitary organizations like the Freikorps, but it was also home to the then tiny National Socialist party. In general the political divisions between Berlin and Munich were reflected in the two major chapters of the Deutsche Gesellschaft für Rassenhygiene. During the republic the Berlin chapter was the largest in the society and was sympathetic to the new order; although it undoubtedly had conservative members—some even with a *völkisch* outlook—on the whole it maintained a predominantly centrist–social democratic orientation and rejected as unscientific and politically dangerous any notion of an Aryan eugenics. The Munich chapter tended to be both more right-wing and more open to racist views, although there is no reason to believe that all of the members shared Lenz's blatantly pro-Nordic position.

Yet despite these political and intellectual differences, the Deutsche Gesellschaft never split during Weimar, as has been wrongly suggested by one article on the history of German eugenics (Lilienthal 1979, p. 117). As a result of a general meeting of the society held in Munich in 1922, the headquarters of the Deutsche Gesellschaft was transferred from Munich to Berlin. During the next few years, the decentralized society continued on as before—as a collection of relatively autonomous local chapters. As under the empire, those wishing to join the society during the Weimar years had to join a particular chapter; one could not simply be a member at large. Sometime between 1922 and 1924, the Berlin-based Association of German Registry Officials, a group of six thousand civil servants involved in registering births, marriages, and deaths, became interested in the ideals popularized by the society (Krohne 1925–1926, p. 144). Apparently at the instigation of the association, the Deutscher Bund für Volksaufartung und Erbkunde, an organization dedicated to spreading eugenic ideas to all Germans, including the working class, was formed in Berlin in 1926 (Behr-Pinnow 1927, p. 57). The neutral term *Volksaufartung* (national regeneration) was chosen to make plain its nonracist stance. With approximately fifteen hundred members, the Bund saw itself as a daughter organization of the society (Krohne 1925–26, p. 144), although its left-of-center political orientation and its opposition to racist sentiment made it much more sympathetic to the Berlin chapter than to the one in Munich. The chairman of the Bund was the physician, eugenicist, and high-ranking government official Karl von Behr-Pinnow.

It is not clear exactly why the Bund was formed. While the pro-Aryan sympathies that alienated nonracists in the society may have also contributed to the creation of the Bund, there was undoubtedly a more important reason for its foundation. As was mentioned earlier, the Deutsche Gesellschaft made virtually no effort to reach out to all classes in society. Its rhetoric notwithstanding, the society had gone little beyond attracting a relatively small number of medical professionals and academics to the movement; indeed, during the early 1920s it probably did not have more than one thousand members. It seems as if the leadership of the society did not quite know how to draw large numbers to their fold without compromising their "scientific integrity." The civil servants who formed the Bund, on the other hand, wanted first and foremost to popularize eugenics—to bring the problem of degeneration and the possibility of "national regeneration" to the largest possible number of people (Lenz 1925, p. 349). Although not without class prejudices, those involved in the association were at once less elitist in their view of the hereditary fitness of the working classes and more willing to write in a style that all Germans could understand. This is especially evident in the association's journal, the *Zeitschrift für Volksaufartung und Erbkunde* (1926–1927) and its successor publication, *Volksaufartung, Erblehre, Eheberatung* (1928–1930). Edited by a high-ranking public health official in the Prussian Ministry of Welfare, Artur Ostermann, the two journals published short, nontechnical articles that were decidedly different in style and tone from those found in the *Archiv*. Besides its popular style, the Bund had something else that the society lacked: real influence in government circles. Members of the society had direct links to the Association of German Registry Officials and close ties with the German Ministry of Welfare, the German Ministry of the Interior, and the Prussian Ministry of Welfare. All of these agencies contributed money to the alliance and its journals (Behr-Pinnow 1927, p. 58).

Though lacking any technical, scientific articles, the *Zeitschrift* and *Volksaufartung* voiced many of the same concerns as the older and more established *Archiv*. Both publications continued to warn about the dangers of birthrate decline and the tendency of the fitter classes to have fewer children; both also lamented the slow progress made in bringing genetics and eugenics into the high school classroom (Lenz 1927, 1929; Muckermann 1930; Brüggemann 1927; Isch 1927; Burgdörfer 1928; Spilger 1927; Behr-Pinnow 1928). Yet if there was an overlap, there were also important differences in the two eugenic journals. While carrying racist articles only infrequently, the *Archiv* devoted space to reviews (often written by Lenz) of blatantly racist publications; the two Bund publications were free of any pro-Aryan and anti-Semitic sentiment. Perhaps more importantly, however, the Berlin journals saw as one of their major missions the popularization of Prussia's recently instituted *Eheberatungstellen* (marriage counseling centers). Created by a 1926 decree of the Prussian Ministry of Welfare, these centers may have been a concession to those eugenicists and government health officials who had pleaded, without success, for a compulsory exchange of health certificates for couples prior to marriage. Although undertaken with good intentions, the more than one hundred *Eheberatungstellen* were plagued from the beginning by a shortage of funding and a lack of a unified purpose. Established primarily for genetic counseling, they were not heavily frequented by prospective couples. The Dresden marriage counseling center, the oldest one in Germany, only had sixty-four customers between 1911 and 1915 (Ostermann 1928, p. 295; Scheumann 1928, p. 22).

The attempt to popularize eugenics during the Weimar period was also accompanied by substantial institutional expansion. Before 1920 Germany lacked any institutional center for eugenics and could boast only a few isolated university courses in race hygiene. In 1923 a university chair for race hygiene was founded in Munich (held by Lenz), and by 1932 over forty eugenics lecture courses were given at various German universities—many, if not all, in faculties of medicine (Weiss 1983, p. 304). Two research centers were also established. The German Research Institute for Psychiatry was founded in Munich in 1918, and, with funding and aid from the Rockefeller Foundation, was made a Kaiser Wilhelm Institute (KWI) in 1924. It was directed by Rüdin after 1931. The Kaiser Wilhelm Institute for Anthropology, Human Heredity, and Eugenics in Berlin was founded in 1927 and directed by Eugen Fischer ("Ein deutsches Forschungsintitut" 1936, 1: 131–32).

In addition, the movement was becoming increasingly visible both at home and abroad. In 1926, at the Great Exhibition for Health Care, Social Welfare, and Physical Training held in Düsseldorf, several members of the executive committee of the Deutsche Gesellschaft chose the exhibits for health care. Two years later Munich was to host the International Alliance of Eugenic Organizations, at which time the German eugenicists' foreign colleagues were given an opportunity to visit Ploetz's private research laboratory at Herrsching and were given a guided tour of the KWI for Psychiatry headed by Rüdin (Kröner 1980, pp. 84–85).

Throughout the 1920s and early 1930s, the society also continued to grow, reaching nearly eleven hundred members by 1931 (Muckermann 1932a, pp. 94–95). At a national meeting of the Deutsche Gesellschaft in Munich on 18 September 1931, it merged with the Deutscher Bund für Volksaufartung. The name was changed to the Deutsche Gesellschaft für Rassenhygiene (Eugenik)—the word *Eugenik* was included to demonstrate that the term *Rassenhygiene* was merely its German equivalent. The executive committee was strengthened and given more power, a change that resulted in the Deutsche Gesellschaft becoming more centralized than it had previously been. In addition, members could now join at large (Muckermann 1932a, pp. 95–104; Kröner 1980, p. 87).

The net effect of these changes was to create a larger, more popular, and more influential society that, as Hermann Muckermann put it, was true to "the historical line" of the movement (1931b, p. 48). For Muckermann, a former Jesuit active in the Berlin chapter of the Deutsche Gesellschaft and a zealous popularizer of race hygiene during Weimar, that meant a nonracist eugenics movement. Although Lenz, Rüdin, Ploetz, and a handful of others certainly never gave up their pro-Aryan sentiment, they were willing to put that in the background in the interest of a unified movement. Ironically, their influence over the movement was never weaker than it was at the end of Weimar—a time when the Nazis were gaining strength daily. Despite the fact that the Munich chapter contained many of the prominent leaders of the Deutsche Gesellschaft, it did not grow to the same extent as the Berlin Chapter; indeed, by 1931 even the Stuttgart chapter was larger. The numerous new local chapters that sprung up during the late Weimar years nearly all employed the term *Eugenik* rather than *Rassenhygiene* in their names, largely owing to the influence of Muckermann (Muckermann 1931b, p. 47). Had the Nazis not forced a drastic change in course in 1933, there is every reason to believe that the movement would have become even more similar to its counterpart in Britain.

Depression and sterilization

The depression that got under way in 1929 not only eventually made more than six million people unemployed, but also forced a reexamination of the continued expansion of the welfare state. Calls were heard from industrial circles to trim Germany's welfare budget; "social policy must be limited by the productivity of the economy," it was argued. Although such cries lamenting Germany's economic inefficiency and high welfare costs were not new, they were taken quite seriously by the half-dictatorial, half-parliamentary Brüning government (1930–1932). By 1931 Germany's *Sozialpolitik* had become, at least in the eyes of some, too high "an insurance premium against Bolshevism" (Abraham 1981, pp. 84, 91).

The critique of burgeoning social costs and the desire, even on the part of left-wing politicians, to allocate Germany's dwindling resources in the most cost-effective manner possible did not go unnoticed by race hygienists. This is clearly visible in the more substantial journal *Eugenik,* which superseded the earlier Bund publications in 1930. Edited, like its predecessors, by Ostermann and boasting a circulation of over five thousand, it nonetheless included both racist society members (Lenz and Rüdin) and nonracists (Muckermann) on its editorial board. It was not formally affiliated with the Bund (Lenz 1931a; Muckermann 1932a, p. 99). Although never as well known internationally as the *Archiv, Eugenik* expressed the trends of the movement during Weimar's financially and politically troubled final years.

One concern mirrored in the journal was the problem of crime (Finke 1930; Lange 1931; Muckermann 1932b). If much of Germany's growing crime problem was a manifestation of bad germ plasm, then the millions spent yearly to detain criminals could be saved through an active race hygiene policy. In addition, *Eugenik* carried numerous articles that sought to demonstrate that eugenics was one of the best ways of eliminating waste in the welfare budget. According to one report entitled "Marriage Counseling and Social Insurance," if more people had used the marriage counseling centers, Germany's hereditary defectives—allegedly accounting for between 8 and 10 percent of all those between the ages of sixteen and forty-five—would not constitute such a "heavy burden on our expenditures" ("Eheberatung und Sozialversicherung" 1930, p. 182). More explicit was a state-ment made by Muckermann in an article on welfare and eugenics. Complaining that 3.45 marks was needed daily to support one institutionalized mental defec-tive, and that this saddled Germany with a financial burden of over 185 million marks a year at a time when there was barely enough money to keep healthy indi-viduals from starving, Muckermann presented his readers with a sensible solution to the problem of the Reich's overtaxed social net:

> If one compares the money given out for defectives with the amount which a healthy family has at its disposal, one quickly comes to the conclusion that in the future everything must be done to reduce the number of hereditarily dis-eased individuals—a task that can be achieved by means of eugenics. Besides that, a clear differentiation must be made in the entire welfare system such that the means available are first appropriated for preventive care, and only then given out to people who cannot be brought back to work and life. (1931c, p. 42)

Of course, Muckermann never suggested that Germany's nonproductive elements be treated in an inhumane fashion, but in hard economic times they would become second-class citizens who should receive from the state only the mini-mum amount required to maintain their existence.

Muckermann's cost-benefit analysis reflected Weimar Germany's preoccupa-tion with rationalization and economic efficiency. During the 1920s industrialists sought ways to make Germany competitive on the world market—ways that included the elimination of inefficient facilities, the introduction of better meth-ods of cost accounting and administration, the reorganization of factory work along the lines advocated by Henry Ford and Frederick Taylor, and the amalga-mation of operations and firms into more efficient corporations and cartels (Brady 1933). Although not connected to the industrialists introducing such innovations, eugenicists nonetheless saw the relationship between race hygiene and the various forms of rationalization. As one eugenics supporter succinctly put it:

> We can protect our position in the world and ensure a high level of culture for our people only through a wise human economy [*weise Menschenökonomie*]. Its goal must be an increase in those capabilities of the people who create a larger living space—that is, we must strengthen with respect to procreation, education and employment all those who achieve high quality manual and intellectual work. . . . At the same time it is absolutely essential . . . to limit the number of those who consume more than they produce, who make the struggle for survival of our people difficult, and who depress their [the people's] standard of living. (Winkler 1928, p. 173)

Thus, people became a manipulable resource to be administered in the interest of a healthy and culturally productive nation.

Perhaps nowhere, however, was the true nature of race hygiene better depicted than in the preface to *Eugenik:*

> Civilization has eliminated natural selection. Public welfare and social assis-tance contribute, as an undesired side effect of a necessary duty, to the preser-

vation and further reproduction of hereditarily diseased individuals. A crushing and ever-growing burden of useless individuals unworthy of life is maintained and taken care of in institutions at the expense of the healthy—of whom a hundred thousand are today without their own place to live and millions of whom starve from lack of work. Does not today's predicament cry out strongly enough for a "planned economy," i.e., eugenics, in health policy? ("Geleitwort" 1930)

The devastating financial crisis of the late Weimar period only brought to the fore the logic implicit in eugenics from its very inception: it was a strategy to rationally manage national efficiency in order to preserve Germany's and the West's political and cultural hegemony.

The need to cut welfare costs, together with the constant pressure exerted by Ostermann, Muckermann, and others with influence, finally forced the Prussian government to take action. On 20 January 1932 the Prussian upper house received and approved a resolution by one of its representatives, a Dr. Struve, to recognize eugenics and popularize it in every way possible and to decrease immediately the amount of money given out for the care of the defective to "a level that can be supported by a completely impoverished people" (1932, *Eugenik* 2: 109). On 2 July the Committee for Population Policy and Eugenics of the Prussian Health Council heard talks by Muckermann and three others on the topic "Eugenics in the Service of National Welfare" and consequently adopted several eugenic proposals, including a draft for a sterilization law ("Eugenische Tagung" 1931–1932, pp. 187–89).

The drafting of a sterilization law in Germany was a long time coming. Prominent eugenicists had carefully monitored events in the United States, where sterilization was legally practiced since 1907. During the empire leading members of the Deutsche Gesellschaft did not push even for the voluntary sterilization of hereditary defectives largely because they were certain that the country would find such a practice abhorrent. By the early Weimar years, however, their attitude had changed. Although by and large still opposed to mandatory sterilization, most members of the Deutsche Gesellschaft were open to voluntary sterilization, but they still seemed to place more emphasis on institutionalization and work colonies as a means of preventing the unfit from reproducing ("Aus der rassenhygienischen Bewegung" 1922, p. 374). During the 1920s a few obscure physicians did exploit the ambiguities in paragraph 224 of the Reich's legal code in order to carry out sterilizations on eugenic grounds. One man, a Dr. Gerhart Boeters, not only bragged about the sixty-three sterilizations he performed, but also sought to encourage other physicians, as custodians of the nation's health, to do likewise (Müller 1978, pp. 14–16). Although Boeters was later to lose his civil service position as district physician in Saxony as a result of his boldness, his pleas published in many of Germany's leading medical journals ensured that the issue would be discussed. After prolonged debate among members in the Deutsche Gesellschaft and the medical community, a sterilization law was drafted in 1932 by the Prussian Health Council, which permitted the *voluntary* sterilization of certain classes of hereditarily defective individuals and required that proof be given that the defective traits were in fact genetic. There was no mention of sterilization on either racial or social grounds. In addition, the committee that proposed the bill rejected out of hand the use of euthanasia for eugenic purposes ("Eugenische Tagung" 1931–1932, p. 187).

These proposals were embraced by several medical organizations both inside and outside of Prussia just weeks before the National Socialist takeover in 1933. In general, physicians responded positively to the proposed law. Even in Protestant church circles, the bill had its supporters. By contrast, only the Catholic church, following the 1930 papal encyclical, "Casti Conubii," condemned the practice of sterilization (Lilienthal 1979, p. 120; Baader 1984, p. 869; Nowak 1984). However, owing to the political chaos following the deposition of the Prussian government by the Reich in July 1932, the sterilization draft never became law under the republic, although it would later serve as the basis of the Nazi mandatory sterilization law of July 1933.

Thus, throughout the Weimar years, as during the empire, the movement was concerned first and foremost with boosting Germany's national efficiency and cultural productivity. Despite ideological differences among its members, race hygiene appealed to all its advocates, racist and nonracist alike, as a scientific means of solving social problems. Especially during the last troubled years of the republic, more and more people of all political persuasions turned to the new discipline as one of the only effective ways of reducing the welfare budget and ensuring that Germany maintain its rightful position among the "cultured nations." Late Weimar eugenics expressed even more clearly the managerial logic implicit in German eugenics from its earliest days: population could and should be scientifically manipulated in the interest of power.

Eugenics under the Swastika, 1933–1945

Although the Weimar years witnessed the gradual adoption of a "eugenic outlook" on the part of certain government officials, prior to 1933 the cause of race hygiene was advanced by a relatively small group of intellectuals, primarily medically trained professionals, within the confines of the Deutsche Gesellschaft. The Nazi seizure of power changed this drastically. Now heading the Reich was a man for whom race hygiene represented a key element in a much larger "biological" and racial worldview—a worldview to which the entire nation would be pledged and ultimately sacrificed. Hitler's maniacal obsession with the Aryan race as the motive force of world history assured that anything useful to the preservation of "Nordic blood" would become a cornerstone of national policy and the subject of intense government propaganda. Because much of National Socialist ideology, as one Bavarian Nazi succinctly put it, was in some sense little more than "applied biology" (Proctor 1982, p. 37), it becomes extremely difficult, after 1933, to separate the goals and activities of "professional eugenicists" from the rhetoric and racial policies of Hitler and high-ranking Nazi party members. For our purposes, however, Nazi "race hygiene" will be defined as the activities of professional eugenicists, the Deutsche Gesellschaft, and the two major eugenics institutes during the twelve-year dictatorship. But any examination of eugenics in the

Third Reich cannot neglect the legacy of the pre-1933 movement, nor can it ignore the connection between race hygiene and such Nazi racial policies as the "euthanasia program," the extermination of Europe's gypsy population, and the "final solution." Although none of the latter were viewed by Germany's professional eugenicists as belonging to the province of race hygiene, in at least some instances there were both personal and ideological ties between the two.

Gleichschaltung and change

The new political leadership imposed significant changes upon the race hygiene movement. Not long following the triumph of the new order, the Deutsche Gesellschaft, like all other organizations in the Reich, was *gleichgeschaltet* (coordinated) and subjected to the "Führer principle." This meant, first of all, that the society was no longer an independent organization. It was placed under a special Reich Commission for National Health Service, which, in turn, was directly subordinate to the Reich Ministry of Interior. Accordingly, the society was expected to "support the government in the fulfillment of its race hygienic goals." In addition to becoming a de facto government body, the society lost all semblance of democratic control. In November 1933 Rüdin, director of the KWI for Psychiatry in Munich, was appointed *Reichkommissar* of the society by Nazi Minister of Interior Wilhelm Frick. He, in turn, was in charge of appointing the business manager, as well as the leaders of all the local chapters of the society. Final authority, however, remained in the hands of Frick. The minister of interior could veto all appointments, had to approve any changes in the society's bylaws, and could remove anyone from office at will ("Deutsche Gesellschaft für Rassenhygiene, e. V." 1934, pp. 104–8; Kröner 1980, pp. 92–93).

Even before the new statutes were drawn up early in 1934, Rüdin eliminated the word *Eugenik* from the society's official name and reinstated the one used before the compromise of 1931, the Deutsche Gesellschaft für Rassenhygiene ("Notizen" 1933, p. 467). Unlike the previous name change, this one had more than merely symbolic significance. The Nazi seizure of power eliminated the possibility of a nonracist race hygiene in Germany. Since the *Deutsche Gesellschaft* was now virtually a government organ, and since race hygiene was central to the new order, there could be little if any deviation from the official line. This meant the end of the "Berlin interpretation" of eugenics—an interpretation that appeared to have won the upper hand at the end of the Weimar period. Two of the most influential nonracist eugenicists, Ostermann and Muckermann (the Social Democrat Grotjahn had died in 1931), were removed from their offices and forced into retirement. After 1937 Muckermann was prohibited from writing anything on the subject of eugenics (Lilienthal 1979, p. 123; Ebert 1976, p. 35). With the removal of Muckermann, the movement lost its best popularizer as well as the director of the Eugenics Division of the Berlin KWI for Anthropology, Human Heredity, and Eugenics. These two men were not the only sacrifices of the new regime. Although membership lists for the Deutsche Gesellschaft are not available for the late Weimar and early Nazi years, it can be safely assumed that many of the less prominent nonracists left the society on their own accord or were urged to do so by the newly appointed local chapter leaders. It goes without saying that Jewish members, such as the geneticist Richard Goldschmidt, were forced out of the organization; according to the new 1934 statues, membership in the Deutsche Gesellschaft was restricted to "Germans of Aryan ancestry" (Weindling 1985, pp. 309, 315; "Deutsche Gesellschaft für Rassenhygiene e. V." 1934, p. 107).

Hand in hand with the elimination of nonracist eugenics and its supporters from the newly "coordinated" Deutsche Gesellschaft came a greater preoccupation with race. In the past even Aryan sympathizers like Lenz, Ploetz, and Rüdin had not made *Aufnorderung* (nordification) a cornerstone of their eugenics policy, nor had they publicly suggested that the preservation and racial purification of the "Aryan" population of the Reich should become a primary focus of their attention. After 1933 race hygiene combined *Rassenpflege* (racial care) and *Erbpflege* (genetic care) (Verschuer 1941, p. 125; Gütt 1934a, p. 118). The latter component was equivalent to the old nonracist meritocratic eugenics concerned with the rational management of those mental and physical traits of the population seen as favorable to a more culturally and economically productive Reich. *Rassenpflege* (the management of a population's racial traits) was something new, although Lenz, as well as "racial scientists" (racial anthropologists) such as Hans F. K. Günther, had earlier suggested that high economic and cultural achievement was a product of certain superior races.

This new blatantly racist (and sometimes explicitly anti-Semitic) line was given clear expression by Germany's race hygienists both in public speeches and in their writings. In an address presented at a special meeting of the Deutsche Gesellschaft in 1934, Rüdin stressed the cultural importance of "the Nordic race in world history and especially German history," and concluded that as such it "urgently deserves to be preserved and protected." Although he denied that the goal of preserving and protecting Germany's Nordic and closely related stock meant a devaluation of other races, Rüdin rejected out of hand the crossing of "dissimilar races" (Rüdin 1934a, p. 232; Kröner 1980, p. 94). In his influential *Leitfaden der Rassenghyiene* (Textbook of race hygiene) (1941), Professor Otmar Freiherr von Verschuer (1896–1969), director of the Frankfurt University Institute for Heredity and Race Hygiene, and later director of the KWI for Anthropology, Human Heredity, and Eugenics, discussed the necessity of preserving the "racial peculiarities of the *Volk*" by combating "the penetration of foreign races" (Verschuer 1941, p. 115). Similar statements are found in the writings of other Nazi race hygienists such as Theodor Mollison, Otto Reche, and Martin Staemmler.

Continuities

Yet despite the important changes that the eugenics movement underwent during the Third Reich, there was at least as much continuity as discontinuity. The new preoccupation with race after 1933 in no way lessened the attention devoted to the more traditional concerns of race hygiene (e.g., increasing the birthrate in the "fitter" classes of society; reducing the number of nonproductive elements). Indeed, judging from the plethora of books on the subject, the obsession with reducing the number of the unfit and boosting the ranks of the productive classes through the implementation of a vigorous race hygiene program was far greater than that which existed even in the Weimar years. Popular works such as Otto Helmut's *Volk in Gefahr* (Nation in danger) (over twenty-six thousand copies sold) and Friedrich Burgdörfer's *Völker am Abgrund* (Peoples at the abyss) did not focus their attention on the Jewish menace, but rather used a large number of graphs and diagrams to reiterate such long-standing eugenic concerns as the "hereditary defectives' burden on the German people," "the threat of the sub-humans [criminals]," and "the decline of the fit, the increase of the unfit" (Burgdörfer 1936; Helmut 1934, pp. 26, 28, 30). In one diagram entitled "Fertility and Race," Helmut did not compare Aryans and Jews, but rather tried to demonstrate the alleged Slavic threat facing Germany—the same fear articulated by Schallmayer and others before and during World War I (Helmut 1934, p. 34).

This continuation of earlier themes can be found in statements from those outside the movement as well. In an address given in 1933 to the newly instituted Expert Committee for Population and Racial Policy, a committee set up by the Nazi government to deal with various "racial questions," Interior Minister Frick asserted that "in order to raise the number of genetically healthy progeny, we must first lower the money spent on asocial individuals, the unfit, and the hopelessly hereditarily diseased, and we must prevent the procreation of severely hereditarily defective people" ("Ansprache des Herrn Reichministers" 1933, p. 416). In a short article published the same year, the physician Friedrich Maier urged his readership to replace the system of "welfare, which generally served only the weakest and asocial individuals," with one emphasizing the "management of the health of those portions of the German nation still racially intact in order both to prevent genetically diseased offspring, and to encourage the hereditarily fit individuals in all segments of the population" (1934, p. 56). Thus, although Schallmayer and other nonracists would have viewed the racist and anti-Semitic side of eugenics during the Third Reich as absolutely deplorable and "unscientific," they would not have found all parts of Nazi race hygiene objectionable; its logic and many of its aims were too similar to their own.

Of all the various strategies and programs implemented by the Nazis in the interest of improving the racial substrate of the Reich, none reveals the continuity between pre- and post-1933 race hygiene better than the sterilization law. Formally enacted on 14 July 1933, the *Gesetz zur Verhütung erbkranken Nachwuchses* (Law for the prevention of genetically diseased offspring) was based on the 1932 Prussian proposal initiated by Muckermann, Ostermann, and others, including the director of the Berlin-based KWI for Biology, Richard Goldschmidt (Bock 1986, pp. 80–84; Lilienthal 1979, p. 124; Müller-Hill 1984, p. 32). Unlike the failed Prussian proposal of 1932, however, the Nazi law allowed the *mandatory* sterilization of those individuals who, in the opinion of an *Erbgesundheitsgericht* (genetics health court), were afflicted with (1) congenital feeblemindedness; (2) schizophrenia; (3) manic-depressive insanity; (4) genetic epilepsy; (5) Huntington's chorea; (6) genetic blindness; or (7) genetic deafness. In addition, those suffering from "serious alcoholism" could also be sterilized against their will. The *Gesetz* made no provisions for sterilization based on racial grounds.

Although Ernst Rüdin collaborated in the law's well-publicized interpretative commentary, it is not clear what role, if any, Germany's professional eugenicists had in drafting it. The initial impetus for the *Gesetz* was given by the director of the Commission of National Health Service, Dr. Arthur Gütt (Bock 1986, p. 84). As members of the expert committee, Lenz, Ploetz, and Rüdin merely may have enjoyed the function of "rubber-stamping" a proposal originated by the Ministry of Interior. Nonetheless, they wholeheartedly approved the new measure. Like other members of the medical community, they had good reason to do so. The statutes called for the establishment of genetic health courts and supreme genetic health courts to adjudicate the *Gesetz*, all of which were presided over by a lawyer and two doctors. The *Gesetz* stipulated that one physician be an expert in the field of heredity and that the second be employed by the state. Moreover, since physicians were required to report all individuals afflicted by any genetic illness mentioned above, they were the ones most often responsible for bringing cases to the attention of the courts. Hence, most of Germany's physicians were now afforded ample opportunity to fulfill their obligation as custodians of the nation's health either directly, through their invovlement in the courts, or indirectly, by ensuring that the genetically ill were registered with the courts ("Gesetz zur Verhütung erbkranken Nachwuchses" 1933, pp. 420–23; Proctor 1982, p. 47).

Although initially some seventeen hundred genetic health courts were envisioned by the Nazis (one in each large city and in each county), probably not more than two to three hundred were ever established. It thus proved to be impossible to extend the *Gesetz* to cover an even broader group of "defectives"—something that at least some eugenicists desired. Hence, Rüdin's plea that all "burdensome lives, ethically defective and socially unfit psychopaths, and the huge army of confirmed hereditary criminals" come under the surgeon's knife went largely unanswered (Proctor 1982, p. 47; Müller-Hill 1984, p. 35).

Lenz also believed that the *Gesetz* was too narrow. He spoke of the desirability of sterilizing 1 million feebleminded, 1 million mentally ill, and 170,000 idiots in "the social interest." He at least half-seriously suggested that it would be better if the bottom one-third of the entire population did not reproduce (Projektgruppe "Volk und Gesundheit" 1982, p. 167n). Nonetheless, between 1934 and 1939, estimates on the number of people sterilized range from 200,000 to 350,000 (Baader 1984, p. 865) to 350,000 to 400,000 (Müller-Hill 1984, p. 37; Bock 1986, pp. 237–38). All had passed through the genetic courts, and the overwhelming majority of them were then sterilized against their will. It is estimated that slightly more than half of all operations were performed on the so-called feebleminded. During the first three years of the *Gesetz*, at least 367 women and 70 men died owing to complications following the procedure. The number of related deaths throughout the six-year period probably was much higher (Müller-Hill 1984, p. 36).

During the Nazi period the research conducted in Germany's academic institutes associated with eugenics was similar to investigations carried out during the Weimar period. In the German Research Institute for Psychiatry in Munich, the heavy emphasis placed on twin studies as a means of investigating the inheritance of mental disorders continued during the Third Reich much as it did during the Weimar years. Formed in the hope that the research undertaken would one day help reduce the enormous financial cost of caring for the army of mental defectives, by 1938 the institute had at least eleven researchers working on material collected from over nine thousand identical and fraternal twins (Planck 1936, 1:131–32; Rüdin 1938, p. 195). Although we do not know whether these researchers, like their counterparts in the Berlin institute, were also involved in providing genealogies for individuals whose pure "Aryan lineage" was in question, the major task of Rüdin's institute was to provide the hard evidence for the inheritance of pathological mental traits to aid the government's effort to sterilize the "unfit." However, the institute also made a contribution to positive eugenics by studying genealogies of talented individuals, including shop foremen (and their spouses)—the latter undoubtedly seen as a group of elite workers rising to low-level managerial positions owing to their genetic abilities (Rüdin 1938, p. 198). The backgrounds and exact number of researchers in the institute and the precise nature of the investigations carried out between 1933 and 1945 remain unknown.

Far eclipsing the Munich institute in importance, Berlin's KWI for Anthropology, Human Heredity, and Eugenics remained the institutional center of German race hygiene research throughout the Nazi period. It officially opened on 15 September 1927—a date chosen to coincide with Germany's hosting of the Fifth International Congress of Genetics in Berlin. The institute represented, from its very inception, "the wish of German anthropologists and race hygienists for a central research institute for their disciplines in the Reich" (Kroll 1983, p. 161; Müller-Hill 1984, p. 78). Eugen Fischer, a prominent Freiburg racist eugenicist whose anthropological investigations into the "Rehoboth bastards" (mulattos) of Southwest Africa in 1908 launched his academic career, was chosen to head the Berlin institute as well as its anthropological division. In 1933 he was also appointed rector of the University of Berlin, apparently against the wishes of the Nazis (Müller-Hill 1984, p. 78). Heading the other two original divisions—human heredity and eugenics—were Verschuer and Muckermann, respectively. Muckermann's connections to influential Catholic industrialists were in no small measure responsible for part of the institute's financial backing. In 1933 he was dismissed despite Fischer's efforts to retain him (Müller-Hill 1984, p. 78), and Lenz took over as director of the eugenics division. He remained at his post until 1945 while simultaneously holding a position as professor of eugenics in the faculty of medicine of the University of Berlin. Fischer managed to retain Verschuer as head of the division of human heredity until 1935, despite Nazi suspicions that "he could not be integrated" into the new order because of his "liberal" outlook (Müller-Hill 1984, p. 78). In that year Verschuer received a position at the University of Frankfurt; he did not return to Berlin until 1942, when he was chosen director of the entire institute upon Fischer's retirement.

Owing to the willful destruction of documents toward the end of the Second World War, it is impossible to detail the services and research activities of the institute with any degree of certainty (Müller-Hill 1984, pp. 24–25). The surviving documents, as well as the publications of institute workers for the years 1927 to 1945, suggest that their research activities did not fundamentally change after 1933, although admittedly this evidence is unlikely to tell the entire story. In the divisions of human heredity and eugenics, the focus of investigation during the

Third Reich did not seem to reflect an obsession with either Aryan supremacy or Jewish inferiority. Verschuer and those who came to work under him studied the inheritance of "normal" morphological and physiological traits as well as the inheritance of disease, intelligence, and behavior. Like Rüdin, Verschuer engaged in twin studies. In the eugenics division, the primary concern both before and after 1933 seems to have been differential birthrates of various social groups. In 1930, for example, Muckermann examined the differential birthrates of 3,947 families of German university professors; six years later Ilse Schmidt, a researcher in the eugenics division, studied the relationship between intelligence and urbanization. Another area under investigation in the eugenics division was radiation genetics (Verschuer 1964, pp. 156–58, 160–61). In Fischer's anthropological division the primary focus both before and after 1933 was the genetic analysis of racial crossing. At least according to a later report by Verschuer, virtually every crossing was studied except that between Jews and "Aryans" (1964, pp. 129–36, 159).

While those race hygienists holding research positions seem to have continued with "business as usual" after 1933, their institutional affiliation did obligate them, willingly or unwillingly, to serve the needs of Nazi racial policy. Both institutes, especially the one in Berlin, were expected to aid the government in its effort to improve the German race. What this meant in practice, as revealed in several memos and reports, is that members of the institutes were called upon (1) to teach eugenics, genetics, and anthropology courses to state-employed physicians and SS doctors; (2) to help carry out the sterilization law by providing *Gutachten* (expert testimony) in cases coming before the genetic health courts; and (3) to compose racial testimonials and genealogies for the Ministry of Interior after the passage of the Nuremberg Laws. By 1935, for example, over 1,100 physicians had already taken one of the above-mentioned courses; between 50 and 185 doctors participated in a yearlong continuation course in "genetic and racial care" (Kaiser-Wilhelm-Institut 1935). The writing of *Gutachten* for the genetic health courts was considered so important that Minister of Interior Frick secured money for a total of five assistances for Fischer, Verschuer, and Rüdin just to help them handle the large caseload. Verschuer and Fischer also became members of the Berlin Genetic Health Court and the Berlin Supreme Genetic Health court, respectively. The composition of racial genealogies, however, seems to have been somewhat more unpleasant and "time-consuming" for the particular eugenicists involved (Kaiser-Wilhelm-Institut 1935; Müller-Hill 1984, p. 39). Nonetheless, insofar as Germany's race hygienists were willing to deliver a verdict on the "racial ancestry" of individuals, they were, at least after 1941, indirectly involved in sending Jews to their deaths.

Responsibility and legacy

The eugenicists' willingness to participate in the construction of such racial genealogies raises the question of their connection to other criminal Nazi racial policies. A case in point is the sterilization of the "Rhineland bastards"—the children of German mothers and French African occupation troops stationed in the Rhineland after World War I. Lenz and Rüdin were indeed asked, as members of the Expert Committee for Population and Racial Policy, to give their opinion on what should be done with them. Interestingly enough, neither Lenz nor Rüdin was in favor of mandatory sterilization. Although hardly commendable as a solution to the "problem," Lenz suggested that the children be "exported"; Rüdin opted for their "voluntary" sterilization on pain of deportation (Müller-Hill 1984, pp. 34–35; Pommerin 1979, p. 75). The actual decision to proceed with the forced sterilization of these children, however, was made in 1937 in the Reich Chancellery without further consultation with the eugenicists. Only Fischer and Verschuer were even indirectly involved in this action; both were called upon to write the requisite anthropological testimonials needed to document the childrens' racial ancestry prior to sterilization (Pommerin 1979, p. 78). Whether they willingly prepared the genealogies that resulted in the sterilization of 385 "colored" children remains unknown.

German eugenicists also bore at most only indirect responsibility for the "euthanasia action." Officially, about one hundred thousand so-called useless eaters (mentally ill or retarded patients) were exterminated in Germany between 1939 and 1941. However, recent evidence has demonstrated that the killings began much earlier, did not end until the end of the war, and were not limited to German victims: "useless eaters" in Poland and the Soviet Union, many of them Jews, were also exterminated under the program. Since the history of the destruction of "lives not worth living" is well documented elsewhere, it is not necessary to give a full account of it here (Klee 1983). Suffice it to say that euthanasia was never considered a race hygiene measure by any eugenicist. Only Lenz was in any way involved in an official committee designed to formulate a law permitting euthanasia—a law that apparently never saw the light of day since the action always remained officially secret (Müller-Hill 1984, p. 18). Despite the fact that euthanasia was never seen as a eugenics measure, the action was known and at least halfheartedly accepted by most active race hygiene practitioners. It was, after all, the logical outgrowth of the cost-benefit analysis at the heart of race hygiene. Nonetheless, Germany's race hygiene practitioners were neither in charge of the program nor directly involved in sending any individuals to their deaths.

Perhaps the most commonly held assumption about German eugenics and its practitioners is that they are intricately bound to the activities of the death camps, where a large percentage of Europe's Jewish and gypsy population was exterminated. While there are ideological ties between race hygiene and the destruction of unwanted "racial groups," it would be inaccurate to assume that individual German eugenicists or German race hygiene as a whole was directly responsible for the Holocaust.

Although those particular eugenicists most active during the Nazi period were undeniably anti-Semitic, their socially acceptable brand of anti-Semitism was typical of the German conservative academic mandarins as a whole; these were not people who wanted to see Jews gassed. Lenz provides a typical case in point. During the Weimar years Lenz refused to change his allegedly "objective" position regarding Jews just to please Germany's anti-Semitic movement. He bemoaned the fact that so much energy was being converted into such a "useless racket." Not surprisingly, he never seemed to recognize his own anti-Semitic prejudices and hence continued to talk about anti-Semites as if they were a group to which he in no way belonged. During the Nazi period, however, Lenz was willing to support a somewhat more blatant anti-Semitic position, as evidenced in the change in his description of the Jews between the third edition of the first volume of the *Grundriss* (1927) and the fourth edition of the same volume (1936) (Müller-Hill 1984, pp. 37–38).

However, even after 1933, when it would have been politically expedient, the writings of Lenz and other eugenicists did not emphasize anti-Semitism. Had they been rabid anti-Semites, they could have published such views in any number of journals both before and after 1933. Moreover, none of the eugenicists were involved in any piece of anti-Semitic legislation. Even assuming that many of the eugenicists actually welcomed "early measures" designed to separate and isolate the Jews—an assumption that is by no means firmly established—they had little real influence over any piece of Nazi legislation, let alone legislation relating to the Jews. The Nuremberg Laws forbidding marriages or extramarital relations between Jews and Aryans were composed without the aid of a single "professional race hygienist." Finally, the eugenicists did not take part in the infamous Wannsee Conference of 1942, where plans for the "final solution to the Jewish question" were confirmed by Hitler, Himmler, and other leading Nazi officials.

Absolving the eugenicists of any *direct* responsibility for the "final solution" is, of course, not to excuse or condone their behavior and actions throughout the Nazi period. Ultimately, it was not their anti-Semitism that linked them, however indirectly, to the death camps: in terms of any indirect personal responsibility for the Holocaust, their crimes, like those of large sections of the German population, were largely crimes of omission. By 1933 race hygiene had become an established discipline in Germany, and eugenicists had a vested interest in the continued funding of the field and the institutes to which they belonged. When asked in an interview why Ernst Rüdin wrote an article praising the Nazis, his daughter Edith replied, "He would have sold himself to the devil in order to obtain money for his institute and his research" (Müller-Hill 1984, pp. 130–32). Needless to say, the only way Germany's eugenicists could preserve their positions and secure financial backing for their work was by "playing ball" with Nazi officials. This often meant paying lip service to Nazi programs and joining the party—the latter requested as evidence of loyalty to the regime. Lenz, Fischer, Verschuer, and Rüdin all became party members, but only after 1937 (Müller-Hill 1984, pp. 79, 125, 133).

But, perhaps more important, they expected that their dream of a meritocratic eugenics-based society would be realized in the Third Reich. From a statement made by Lenz in 1931, it is obvious that he welcomed the National Socialists as the only political party willing to take the "eugenics outlook" seriously (1931b, pp. 300–308). Frustrated by the lack of progress in realizing eugenic ideas during the Weimar years, eugenicists active during the Nazi period expected their plans to be realized under Hitler. Even after it became clear to them that Hitler's ideas of race hygiene were not precisely the same as their own, and even after they realized that they were unlikely to be able to exercise any kind of "positive" or moderating influence on Nazi racial policy, Germany's eugenicists showed few qualms about their positions as scientific legitimizers of the kind of racism that sent millions to their deaths. Throughout the Third Reich they simply continued to insist that *their* understanding of eugenics was the scientific one, while attempting to resist taking a rabid anti-Semitic line whenever possible. They sought to hide behind the cloak of "objective science." Fischer, for example, could not be persuaded to say that *all* Jews were inferior. Science, he undoubtedly felt, would not allow such a statement. Hence, for Fischer, Jews were not necessarily always inferior; they were merely "different" (Müller-Hill 1984, p. 78). Yet the eugenicists' attempt to preserve their moral and scientific integrity did not prevent them from using material shipped back from the death camps to the Berlin institute to further their own research. If the evidence presented by Müller-Hill is at all accurate, blood samples and organs extracted from twins and dwarfs were transported from Auschwitz to the Berlin institute in order to allow researchers to continue to advance scientific knowledge (1984, pp. 73–74). However, the eugenicists' crime was not so much their specific theories or their "respectable" anti-Semitism as their willingness to continue on as if their work were totally unrelated to the bestialities carried out in the name of race hygiene by their masters.

What then, if anything, is the legacy of pre-1933 eugenics for the extermination programs of the Third Reich? Can one rightfully speak of an ideological connection between the kind of eugenics articulated by nonracists such as Schallmayer, Muckermann, and Grotjahn and the atrocities carried out in the name of race hygiene by Nazi officials? Throughout its history, race hygiene was a strategy aimed at boosting national efficiency through the rational management of population. Whereas before the Third Reich fitness had generally been understood in purely meritocratic terms, without emphasizing race, after 1933 race and productivity became the two criteria defining fitness. It is not difficult to see the usefulness of race hygiene as a means of creating a stronger Nazi *völkisch* state. From the standpoint of efficiency, a racial policy such as the "euthanasia program," the destruction of "unproductive lives," is not without its logic, as morally perverse as that logic may appear.

But what about the Holocaust? Although the extermination of millions of European Jews cannot really be viewed as a measure designed to boost national efficiency, the interpretation of the Jews as an unfit, surplus, and disposable group is not unrelated to the emphasis implicit in German race hygiene regarding "valuable" and "valueless" people. For the eugenicists, human beings were in some sense variables—objects easily managed or manipulated for some abstract "good." In one of humankind's most barbaric acts to date, there is more than a hint of where the desire to be rid of a "valueless" population can lead. Thus, whatever the intentions of even nonracist eugenicists before 1933, the very logic of eugenics—the rational management of a population for some "higher end"— was a logic readily amenable to other, far more sinister projects than those envisaged by Schallmayer, Muckermann, and Grotjahn. Hence, when all is said and done, it is the *logic* of eugenics far more than its racism that proved to be the most unfortunate legacy of the German race hygiene movement for the Third Reich.

Bibliography

There is relatively little archival material available for the history of the Pre-1933 race hygiene movement in the Federal Republic of Germany. Other than the files in the Bundesarchiv Koblenz and the Hauptstaatsarchiv München (here abbreviated as HSAM), the only sources of significance are the materials on the Berlin institute contained in the Archiv der Max-Planck-Gesellschaft in Berlin-Dahlem, and the completely unsorted materials in the Ploetz family archive at the Ploetz estate in Herrsching (near Munich). There may be some material in the possession of Rüdin's daughter in Munich. Apparently there are also useful holdings in both Potsdam and Merseburg in the German Democratic Republic. In one case in the references cited above, a note contained in *Eugenik* had neither an author nor a title. It was cited merely with the year and the volume and page number. The following abbreviations are used for German eugenics journals: *ARGB* for *Archiv für Rassen- und Gesellschafts-Biologie; VEE* for *Volksaufartung, Erblehre, Eheberatung;* and *ZVE* for *Zeitschrift für Volksaufartung und Erbkunde.*

Abraham, David. 1981. Corporatist Compromise and the Re-emergence of the Labor/Capital Conflict in Weimar Germany. *Political Power and Social Theory* 2: 59–109.

Ackerknecht, Erwin H. 1932. Beiträge zur Geschichte der Medizinalreform von 1848 [Contributions to the history of the 1848 medical reform movement]. *Archiv für Geschichte der Medizin* 25: 51–183.

_____. 1968. *A Short History of Psychiatry.* 2d ed. Trans. S. Wolff. New York and London: Hafner.

Altner, Günter. 1968. *Weltanschauliche Hintergründe der Rassenhygiene des Dritten Reiches* [The ideological background of race hygiene in the Third Reich]. Zurich: EVZ.

Ansprache des Herrn Reichsministers des Innen Dr. Wilhelm Frick auf der ersten Sitzung des Sachverständigenbeirats für Bevölkerungs- und Rassenpolitik [Speech of the Reich Minister of Interior Dr. Wilhelm Frick at the first sitting of the Expert Committee for Population and Racial Policy]. 1933. *ARGB* 27: 412–19.

Aronson, Oscar. 1894. *Ueber Heredität bei Epilepsie* [On heredity in epilepsy]. Berlin: Wilhelm Axt.

Aus der rassenhygienischen Bewegung [From the race hygiene movement]. 1922. *ARGB* 14: 374.

Baader, Gerhard. 1984. Das "Gesetz zur Verhütung erbkranken Nachwuchses"—Versuch einer kritischen Deutung [The "law for the prevention of genetically diseased offspring"—an attempt at a critical interpretation]. In *Zusammenhang. Festschrift für Marielene Putscher* [Coherence. Festschrift for Marielene Putscher], ed. Otto Baur and Otto Glandien, 865–75. Cologne: Wienand.

Baader, Gerhard, and Ulrich Schultz, eds. 1980. *Medizin und Nationalsozialismus: Tabuisierte Vergangenheit—Ungebrochene Tradition?* [Medicine and national socialism: Tabooed past–unbroken tradition?] Dokumentation des Gesundheitstages Berlin 1980, I. Berlin (West): Verlagsgesellschaft Gesundheit mbH.

Baitsch, Helmut. 1970. Das eugenische Konzept einst und jetzt [The eugenics concept then and now]. In *Genetik und Gesellschaft* [Genetics and society], ed. G. Wendt, 59–71. Stuttgart: Wissenschaftliche Verlagsgesellschaft mbH.

Barraclough, Geoffrey. 1981. *An Introduction to Contemporary History.* New York: Penguin.

Baur, Erwin, Eugen Fischer, and Fritz Lenz. 1923. *Grundriss der Menschliche Erblichkeitslehre und Rassenhygiene* [Principles of human heredity and race hygiene]. 2d ed. 2 vols. Munich: J. F. Lehmann.

_____. 1936/1932. *Grundriss der Menschliche Erblichkeitslehre und Rassenhygiene* [Principles of human heredity and race hygiene]. 4th ed. 2 vols. Munich: J. F. Lehmann.

Behr-Pinnow, Karl von. 1913. *Geburtenrückgang und Bekämpfung der Sauglingssterblichkeit* [Birthrate decline and the fight against infant mortality]. Berlin: J. Springer.

_____. 1927. Jahresversammlung des Bundes für Volksaufartung und Erbkunde am 7. Mai [Annual meeting of the Alliance for National Regeneration and the Study of Heredity]. *ZVE* 2: 57–58.

———. 1928. Vererbungslehre und Eugenik in den Schulen [Heredity and eugenics in the schools]. *VEE* 3: 73–80.

———. 1932. Eugenik und Strafrecht [Eugenics and penal law]. *ARGB* 26: 36–56.

Binswanger, Otto. 1888. Geistesstörung und Verbrechen [Mental illness and crime]. *Deutsche Rundschau* 57: 419–40.

———. 1896. *Die Pathologie und Therapie der Neurasthenie: Vorlesungen für Studierende und Aerzte* [The pathology and therapy of neurasthenia: Lectures for students and doctors]. Jena: Gustav Fischer.

Blaschko, Alfred. 1901. Hygiene der Prostitution und venerische Krankheiten [The hygiene of prostitution and venereal disease]. In *Handbuch der Hygiene* [Handbook of hygiene], ed. T. Weyl, X. Jena: Gustav Fischer.

———. 1914. *Geburtenrückgang und Geschlechtskrankheiten* [Birthrate decline and venereal diseases]. Leipzig: Johann Ambrosius Barth.

Bluhm, Agnes. 1907–1908. Die Stillungsnot, ihre Ursachen und die Vorschläge zu ihrer Bekämpfung [The inability to nurse, its causes and suggestions to combat it]. *Zeitschrift für Soziale Medizin* 3: 72–78, 160–72, 261–70, 357–87.

———. 1912. Zur Frage nach der generativen Tüchtigkeit der deutschen Frauen und der rassenhygieneischen Bedeutung der ärztlichen Geburtshilfe [Concerning the question of the reproductive fitness of German women and the eugenic impact of medical obstetrics]. *ARGB* 9: 330–46, 454–74.

———. 1916. Die soziale Versicherung im Lichte der Rassenhygiene [Social insurance in the light of race hygiene]. *ARGB* 12: 15–42.

———. 1917. Rassenhygiene [Race hygiene]. *Concordia: Zeitschrift der Centralstelle für Arbeiterwohlfahrtseinrichtungen* 24: 287.

———. 1930. *Zum Problem Alkohol und Nachkommenschaft* [On the problem of alcohol and offspring]. Munich: J. F. Lehmann.

———. 1934. *Die Alkoholfrage in der Erbforschung* [The alcohol question in hereditary research]. Berlin: Neuland.

———. 1936. *Die rassenhygienischen Aufgaben des weiblichen Arztes* [The eugenic tasks of the female physician]. Berlin: Metzner.

Bock, Gisela. 1983. Racism and Sexism in Nazi Germany: Motherhood, Compulsory Sterilization, and the State. *Signs: Journal of Women in Culture and Society* 8: 400–421.

———. 1986. *Zwangsterilsation und Nationalisozialsmus: Studien zur Rassenpolitik und Frauenpolitik* [Compulsory sterilization and national socialism: Studies pertaining to racial and sexual policy]. Schriften des Zentralinstituts für sozialwissenschaftliche Forschung der Freien Universität Berlin, Band 48. Opladen: Westdeutscher Verlag.

Bolle, Fritz. 1962. Darwinismus und Zeitgeist [Darwinism and the spirit of the age]. *Zeitschrift für Religions- und Geistesgeschichte* 14: 143–78.

Bonhoeffer, K. 1949. Ein Rückblick auf die Auswirkungen und die Handhabung des nationalsozialistischen Sterilisierungsgesetzes [A look back at the consequences and administration of the national socialist sterilization law]. *Der Nervenarzt* 20: 1–5.

Bornträger, Jean. 1913. *Der Geburtenrückgang in Deutschland: Seine Bewertung und Bekämpfung* [The birthrate decline in Germany: Its measurement and elimination]. Würzburg: Curt Kabitzsch.

Brady, Robert A. 1933. *The Rationalization Movement in German Industry.* Berkeley, Calif.: University of California Press.

Braencker, Wilhelm. 1917. *Die Entstehung der Eugenik in England* [The origins of eugenics in England]. Hildburghausen: L. Nonnes.

Breiling, Rupert. 1971. *Die nationalsozialistische Rassenlehre: Entstehung, Ausbreitung, Nutzen und Schaden einer politischen Ideologie* [The National Socialist race theory: Origins, diffusion, uses and harm of a political ideology]. Meisenheim am Glan: Anton Hain.

Brentano, Lujo. 1908–1909. Die Malthusische Lehre und die Bevölkerungsbewegung der letzten Dezennien [The Malthusian teachings and the population fluctuation of the last decades]. *Abhandlungen der Akademie der Wissenschaften* 567–625.

Fircks, Arthur von. 1898. *Bevölkerung und Bevölkerungspolitik* [Population and population policy]. Leipzig: C. L. Hirschfeld.

Fischer, Alfons. 1933. *Geschichte des deutschen Gesundheitswesens* [History of the German health care system]. 2 vols. Berlin: F. A. Herbig.

Fischer, Eugen. 1912–1913. Sozialanthopologie [Social anthropology]. In *Handwörterbuch der Naturwissenschaften* [Handbook of the natural sciences], 9: 172–88. Jena: Gustav Fischer.

———. 1913. *Die Rehobother Bastards und das Bastardisierungsproblem beim Menschen* [The Rehoboth bastards and the bastardization problem among humans.]. Jena: Gustav Fischer.

———. 1914. *Das Problem der Rassenkreuzung beim Menschen* [The problem of racial crossing among humans]. Freiburg i. B.: Speyer und Kaerner.

———. 1927a. Das Preisausschreiben für den besten nordischen Rassenkopf [The competition for the best nordic racial head]. *Volk und Rasse,* no. 1: 1–11.

———. 1927b. *Rasse und Rassenentstehung beim Menschen* [Race and the origin of races among humans]. Berlin: Ullstein.

———. 1930. Aus der Geschichte der deutschen Gesellschaft für Rassenhygiene [From the history of the German Society for Race Hygiene]. *ARGB* 24: 1–5.

———. 1933. *Der völkische Staat, biologisch gesehen* [The folkish state, from a biological viewpoint]. Berlin: Junker und Dunnhaupt.

Forel, Auguste. 1909. *Die sexuelle Frage* [The sexual question]. 9th ed. Munich: Ernst Reinhardt.

Foucalt, Michel. 1978. *The History of Sexuality. Volume 1: An Introduction.* New York: Pantheon.

Fuld, E. 1885. Das rückfälliger Verbrechertum [Repeat offenders]. *Deutsche Zeit- und Streit-Fragen* 453–84.

Gasman, Daniel. 1971. *The Scientific Origins of National Socialism: Social Darwinsim in Ernst Haeckel and the German Monist League.* New York: American Elsevier.

Gaupp, Robert. 1904. Über den heutigen Stand der Lehre von "geborenen Verbrecher" [On the current position of the theory of the "born criminal"]. *Monatsschrift für Kriminalpsychologie und Strafrechtsreform* 25–42.

———. 1925. *Die Unfruchtbarmachung geistig und sittlich Kranker und Minderwertiger. Erweitertes Referat, erstattet auf der Jahresversammlung des Vereins für Psychiatrie am 2. September 1925 in Kassel* [The sterilization of the mentally and morally ill and inferior. Expanded lecture given at the annual meeting of the Association for Psychiatry on 2 September 1925 in Kassel]. Berlin: J. Springer.

Geiger, Theodor. 1934. *Erbpflege: Grundlagen, Planung, Grenzen* [Genetic care: Basis, planning, limits]. Stuttgart: Enke.

Geleitwort [Preface]. 1930. *Eugenik* 1.

Gerlada, Alfred. 1973–74. Untersuchungen zur Geschichte des *Archivs für Rassen- und Gesellschaftsbiologie* [Investigations into the history of the *Archiv für Rassen- und Gesellschaftsbiologie*]. Unpublished paper, Johannes Gutenberg-Universität, Mainz.

Gerhardt, J. P. 1904. *Zur Geschichte und Literatur des Idiotenwesens in Deutschland* [On the history and literature of mental retardation in Germany]. Hamburg: privately published.

Germany. 1907. Kaiserliches Statistisches Amt. *Statistisches Jahrbuch für das Deutsche Reich* [Statistical yearbook of the German Reich]. Berlin: Heymann.

———. 1931. Reichskommission, Internationale Hygiene-Ausstellung, Dresden, 1930–1931. *Die Entwicklung des Deutschen Gesundheitswesens* [The development of the German health care system]. Berlin: Arbeitsgemeinschaft sozialhygienischer Reichsfachverbände.

Gerstenhauer, M. R. 1912. Rassenverfall durch Zivilisation und seine Bekämpfung [Racial decline through civilization and its prevention]. *Deutsche Tageszeitung,* 20 May.

Gesetz zur Verhütung erbkranken Nachwuchses [Law for the prevention of genetically diseased offspring]. 1933. *ARGB* 27: 420–23.

Gladen, Albin. 1974. *Geschichte der Sozialpolitik in Deutschland* [History of social policy in Germany]. Weisbaden: Steiner.

Glass, Bentley. 1981. A Hidden Chapter of German Eugenics between the Two World Wars. *Proceedings of the American Philosophical Society* 125: 357–67.

Gottstein, Adolf. 1909. Die Entwicklung der Hygiene im letzten Vierteljahrhundert [The development of hygiene in the last quarter century]. *Zeitschrift für Sozialwissenschaft* 12: 65–82.

Graham, Loren R. 1977. Science and Values: The Eugenics Movement in Germany and Russia in the 1920s. *American Historical Review* 82: 1133–64.

Grassmann, K. 1896. Kritische Ueberblick über die gegenwärtige Lehre von der Erblichkeit der Psychosen [Critical overview of the contemporary theory of the development of heredit-ability of psychoses]. *Allgemeine Zeitschrift für Psychiatrie* 52: 960–1022.

Grotjahn, Alfred. 1898. *Der Alkoholismus nach Wesen, Wirkung und Verbreitung* [Alcoholism in its essence, effects and propagation]. Leipzig: G. H. Wigand.

––––––. 1904. Soziale Hygiene und Entartungsproblem [Social hygiene and the problem of degeneration]. In *Handbuch der Hygiene* [Handbook of hygiene], ed. Th. Weyl, 4th Suppl. Vol., 727–89. Jena:Gustav Fischer.

––––––. 1912. *Soziale Pathologie: Versuch einer Lehre von sozialen Beziehungen der menschlichen Krankheiten als Grundlage der sozialen Medizin und der sozialen Hygiene* [Social pathology: An attempt at a theory of the social conditions of human diseases as a basis for social medicine and social hygiene]. Berlin: A. Hirschwald.

––––––. 1914. *Geburtenrückgang und Geburtenregelung* [Birthrate decline and family planning]. Berlin: Louis Marcus.

––––––. 1926. *Die Hygiene der menschlichen Fortpflanzung: Versuch einer praktischen Eugenik* [The hygiene of human reproduction: An attempt at a practical eugenics]. Berlin: Urban und Schwarzenberg.

Grotjahn, Alfred, and I. Kaup, eds. 1912. *Handwörterbuch der sozialen Hygiene* [Handbook of social hygiene]. Leipzig: F.C.W. Vogel.

Gruber, Max von. 1903. Führt die Hygiene zur Entartung der Rasse? [Does hygiene lead to racial degeneration?]. *Münchener Medizinische Wochenschrift* 50: 1713–18, 1781–85.

––––––. 1909. Vererbung, Auslese und Hygiene [Heredity, selection and hygiene]. *Deutsche Medizinische Wochenschrift* 46: 1993–96, 2049–53.

––––––. 1914. *Ursachen und Bekämpfung des Geburtenrückganges im Deutschen Reich* [The causes and prevention of birthrate decline in the German Reich]. Munich: J. F. Lehmann.

––––––. 1918. Deutsche Gesundheitspflege [German health care]. *Deutsche Revue* 43: 64–78.

––––––. 1922. "Wilhelm Schallmayer." *ARGB* 14: 52–56.

Gruber, Max von, and Ernst Rüdin. 1911. *Fortpflanzung, Vererbung, Rassenhygiene* [Reproduction, heredity, race hygiene]. Munich: J. F. Lehmann.

Günther, Maria. 1982. *Über die Konstituierung der Rassenhygiene an den deutschen Universitäten vor 1933* [The establishment of race hygiene in the German universities before 1933]. Medical dissertation, Mainz. Mainz: privately printed.

Güse, Hans-Georg, and Norbert Schmacke. 1976. *Psychiatrie zwischen bürgerlicher Revolution und Faschismus* [Psychiatry between bourgeois revolution and fascism]. 2 vols. Hamburg: Athenäum.

Gütt, Arthur. 1934a. Ausmerze und Lebensauslese in ihre Bedeutung für Erbgesundheit und Rassenpflege [Elimination and selection in their meaning for genetic health and racial care]. In *Erblehre und Rassenhygiene im völkischen Staat* [Genetics and race hygiene in the folkish state], ed. Ernst Rüdin, 104–19. Munich: J. F. Lehmann.

––––––. 1934b. *Dienst an der Rasse als Aufgabe der Staatspolitik* [Service for the race as a task of state policy]. Berlin: Juncker und Dünnhaupt.

Haas, G. 1936. Über Sterilisierungsoperationen: Ein Bericht aus der Frauenklinik Tübingen über die Jahre 1918–1930 [On sterilization operations: A report from the gynecological clinic at Tübingen in the years 1918–1930]. Medical dissertation, Tübingen.

Haeckel, Ernst. 1876. *The History of Creation.* Trans. and revised by E. Ray Lankester. Vol. 1. London: Henry S. King and Co.

Hafner, Karl Heinz, and Rolf Winau, 1974. "Die Freigabe der Vernichtung lebensunwerten Lebens": Eine Untersuchung zu der Schrift von Karl Binding und Alfred Hoche ["The right to eliminate lives not worth living": An investigation of the article of Karl Binding and Alfred Hoche]. *Medizinhistorisches Journal* 9: 227–54.

Hegar, A. 1896. Brüste und Stillen [Breast and nursing]. *Deutsche Medizinische Wochenschrift* 22: 539–41.

Helmut, Otto. 1934. *Volk in Gefahr: Der Geburtenrückgang und seine Folgen für Deutschlands Zukunft* [Nation in danger: The birthrate decline and its consequences for Germany's future]. Munich: J. F. Lehmann.

Hentig, Hans von. 1914. *Strafrecht und Auslese* [Penal law and selection]. Berlin: J. Springer.

Hertwig, Oscar. 1922. *Zur Abwehr des ethischen, des sozialen, des politischen Darwinismus* [On the warding off of ethical, social, and political Darwinism]. 2d ed. Jena: Gustav Fischer.

Herz, Friedrich. 1915. *Rasse und Kultur* [Race and culture]. 2d ed. of *Moderne Rassentheorien* [Modern race theories]. Leipzig: Alfred Kröner.

Hillel, Marc, and Clarissa Henry. 1978. *Of Pure Blood*. Trans. Eric Mossbacher. New York: Pocket Books, 1978.

Hoffmann, Geza von. 1913. *Die Rassenhygiene in den Vereinigten Staaten von Nordamerika* [Race hygiene in the United States of America]. Munich: J. F. Lehmann.

———. 1914. Eugenics in Germany: Society of Race Hygiene Adopts Resolution Calling for Extensive Program of Positive Measures to Check Decline in Birth-Rate. *Journal of Heredity* 5: 435–36.

———. 1916. *Krieg und Rassenhygiene* [War and race hygiene]. Munich: J. F. Lehmann.

Hoffman, H. 1970. Erhebungen über die im Rahmen des Gesetzes zur Verhütung erbkranken Nachwuchses in den Jahren 1934–1945 durchgeführten Sterilisationen im Raume Nürnberg-Fürth-Erlangen. 2. Beitrag [Investigations of the sterilizations carried out under the law for the prevention of genetically diseased offspring in the years 1934–1945 in the Nuremburg-Fürth-Erlangen. 2d contribution]. Medical dissertation, Erlangen-Nürnberg.

Hoffmann, Walther G. 1963. The Take-Off in Germany. In *The Economics of Take-Off into Sustained Growth*, ed. W. W. Rostow. London: Macmillan, pp. 95–118.

Hohorst, Gerd, Jürgen Kocka, and Gerhard A. Ritter. 1974. *Sozialgeschichtliches Arbeitsbuch: Materialien zur Statistik des Kaiserreiches 1870–1914*. Munich: C. H. Beck.

Horn, D. 1972. Erhebungen über die im Rahmen des Gesetzes zur Verhütung erbkranken Nachwuchses in den Jahren 1934–1945 durchgeführten Sterilisationen im Raume Nürnberg-Fürth-Erlangen. 3. Beitrag [Investigations of the sterilizations carried out under the law for the prevention of genetically diseased offspring in the years 1934–1945 in the Nuremberg-Fürth-Erlangen. 3d contribution]. Medical dissertation, Erlangen-Nürnberg.

Huerkamp, Claudia. 1980. Ärzte und Professionalisierung in Deutschland [Doctors and professionalization in Germany]. *Geschichte und Gesellschaft* 6: 349–82.

Internationale Gesellschaft für Rassenhygiene. 1907. 2. Bericht der Internationalen Gesellschaft für Rassenhygiene. Jahresbericht 1907 [2nd report of the International Society for Race Hygiene. Annual report 1907]. Unpublished, Ploetz family archive.

Isch. 1927. Ist Geburtensteigerung bei den Intellektuellen möglich? [Is the increase of the birthrate among intellectuals possible?]. *ZVE* 2: 141–42.

Jörger, J. 1905. Die Familie Zero [The family Zero]. *ARGB* 2: 494–559.

Just, Günter, ed. 1932. *Eugenik und Weltanschauung* [Eugenics and world-view]. Berlin and Munich: A. Metzner.

Kahn, Ernst. 1930. *Der internationale Geburtenstreik: Umfang, Ursachen, Wirkungen, Gegenmassnahmen* [The international birth strike: Extent, causes, effects, remedies]. Frankfurt am Main: Societäts-Verlag.

Kaiser-Wilhelm-Institut für Anthropologie, menschliche Erblehre und Eugenik. 1935. Tätigkeitsbericht von Anfang Juli 1933 bis 1. April 1935 [Activity report from the beginning of July 1933 to 1 April 1935]. Unpublished, Archiv der Max-Planck-Gesellschaft, Folio 2401, doc. 49.

Die Kallikaks unserer Zeit [The Kallikaks of our time]. 1932. *Eugenik* 2: 202–8.

Kater, Michael H. 1983. Die "Gesundheitsführung" des Deutschen Volkes [The "health leadership" of the German people]. *Medizinhistorisches Journal* 18: 349–75.

Kaup, I. 1913. Was kosten die minderwertigen Elemente dem Staat und der Gesellschaft? [What do the inferior elements cost the state and society?]. *ARGB* 10: 723–48.

Kelly, Alfred. 1981. *The Descent of Darwin: The Popularization of Darwinism in Germany, 1860–1914*. Chapel Hill, N.C.: University of North Carolina Press.

Klee, Ernst. 1983. *"Euthanasie" im NS-Staat: Die "Vernichtung lebensunwerten Lebens"* ["Euthanasia" in the National Socialist state: The "elimination of lives not worth living"]. Frankfurt am Main: S. Fischer.

Knieke, H. 1903–1904. Die Verstaatlichung des Aerztewesens [The nationalization of the medical profession]. *Politisch-anthropologische Revue* 2: 402–9.

Knodel, John E. 1974. *The Decline of Fertility in Germany 1871–1939*. Princeton, N.J.: Princeton University Press.

Köllmann, Wolfgang. 1969. The Process of Urbanization in Germany at the Height of the Industrialization Period. *Journal of Contemporary History* 4: 59–76.

Kopp, Walter. 1934. *Gesetzliche Unfruchtbarmachung* [Legal sterilization]. Leipzig: Lipsius und Fischer.

Kraepelin, Emil. 1886. Bernhard von Gudden. *Münchener Medizinische Wochenschrift* 33: 577–607.

Krafft-Ebing, Richard Freiherr von. 1885. *Über gesunde und kranke Nerven* [On healthy and diseased nerves]. Tübingen: Laupp'sche Buchhandlung.

———. 1890. *Lehrbuch der Psychiatrie* [Textbook of Psychiatry]. 4th ed. Stuttgart: Ferdinand Enke.

———. 1899. Nervosität und neurasthenische Zustände [Nervousness and neurasthenic conditions]. In *Specielle Pathologie und Therapie* [Special pathology and therapy], ed. Hermann Nothnagel, vol. 2, pt. 2, 1–210. Vienna: Alfred Holder.

Kresse, Oskar. 1912. *Der Geburtenrückgang in Deutschland, seine Ursachen und die Mittel zu seiner Beseitigung* [Birthrate decline in Germany, its causes and the means toward its elimination]. Berlin: John Schwerin.

Krohne, Otto. 1925/26. Position of Eugenics in Germany. *Eugenics Review* 17: 143–46.

Kroll, Jürgen. 1983. *Zur Entstehung und Institutionalisierung einer naturwissenschaftlichen und sozialpolitischen Bewegung: Die Entwicklung der Eugenik/Rassenhygiene bis zum Jahre 1933* [On the origins and institutionalization of a scientific and socio-political movement: the development of eugenics/race hygiene up to 1933]. Sozialwiss. Dissertation, Tübingen. Tübingen: privately printed.

Kröner, Hans-Peter. 1980. Die Eugenik in Deutschland von 1891 bis 1934 [Eugenics in Germany from 1891 to 1934]. Medical dissertation, Münster.

Krügel, Rainer. 1984. *Friedrich Martius und der konstitutionelle Gedanke* [Friedrich Martius and the constitutional idea]. Marburger Schriften zur Medizingeschichte, Band 11. Frankfurt am Main: Peter Lang.

Kruse, Walter. 1903. Entartung [Degeneration]. *Zeitschrift für Sozialwissenschaft* 6: 359–76.

———. 1898. Physische Degeneration und Wehrfähigkeit bei europäischen Völkern [Physical degeneration and military fitness among European peoples]. *Zentralblatt für allgemeine Gesundheitspflege* 17: 457–73.

Kudlien, Friedolf. 1982. Max v. Gruber und die frühe Hitlerbewegung [Max v. Gruber and early Hitler movement]. *Medizinhistorisches Journal* 17: 373–89.

Kurella, Hans. 1893. *Naturgeschichte des Verbrechers: Grundzüge der criminellen Anthropologie und Criminalpsychologie* [The natural history of the criminal: The characteristics of criminal anthropology and criminal psychology]. Stuttgart: Ferdinand Enke.

Lange, Johannes. 1931. Verbrechen und Vererbung [Crime and heredity]. *Eugenik* 1: 165–73.

Lapouge, Georges Vacher de. 1909. Ueber die natürliche Minderwertigkeit der modernen Bevölkerungsklassen [On the natural inferiority of modern social classes]. *Politisch-anthropologische Revue* 8: 393–409, 454–64.

Ledbetter, Rosanna. 1976. *A History of the Malthusian League 1877–1927*. Columbus, Ohio: Ohio State University Press.

Leibfried, Stephan, and Florian Tennstedt. 1981. *Berufsverbote und Sozialpolitik 1933: Die Auswirkungen der nationalsozialistischen Machtergreifung auf die Krankenkassenverwaltung und die Kassenärzte* [Occupational restrictions and social policy 1933: The impact of the National Socialist seizure of power on the administration of health insurance and

insurance fund doctors]. Arbeitspapier des Forschungsschwerpunktes Reproduktionsrisi-
ken, sozial Bewegungen und Sozialpolitik. 3d ed. No. 2. Bremen: Universität Bremen.
Leitsätze der Deutschen Gesellschaft für Rassenhygiene zur Geburtenfrage [Principles of the German Society for Race Hygiene on the birthrate question]. 1914. *ARGB* 11: 134–35.

Lenz, Fritz. 1910. Über die Verbreitung der Lues, spezielle in Berlin, und ihre Bedeutung als
Faktor des Rassentodes [On the diffusion of syphilis, especially in Berlin, and its importance in racial death]. *ARGB* 7: 306–27.

_____. 1915a. Deutsche Gesellschaft für Bevölkerungspolitik [German Society for Population
Policy]. *ARGB* 11: 555–57.

_____. 1915b. Zum Begriff der Rassenhygiene und seine Benennung [On the concept of race
hygiene and its naming]. *ARGB* 11: 445–48.

_____. 1916. Rassenhygienische Bevölkerungspolitik [Eugenic population policy]. *Deutsche Pol-
itik* 1: 1658–68.

_____. 1917a. Bevölkerungspolitik und "Mutterschutz" [Population policy and "protection of
mothers"]. *ARGB* 12: 345–48.

_____. 1917b. Überblick über die Rassenhygiene [Overview of race hygiene]. *Jahreskurse für
ärztliche Fortbildung* 8: 16–50.

_____. 1917c. Zur Erneuerung der Ethik [Toward the renewal of ethics]. *Deutschlands Erneu-
erung* 35–56.

_____. 1918. Vorschläge zur Bevölkerungspolitik mit besonderer Berücksichtigung der Wirt-
schaftslage nach dem Kriege [Suggestions for population policy with special attention to
the economic conditions after the war]. *ARGB* 12: 440–68.

_____. 1919. Wilhelm Schallmayer. *Münchener Medizinische Wochenschrift* 66: 1294–96.

_____. 1923. Review of *Handbuch der Judenfrage* [Handbook of the Jewish question]. *ARGB*
15: 428–32.

_____. 1924. Eugenics in Germany. *Journal of Heredity* 15: 223–31.

_____. 1925. Ein "Deutscher Bund für Volksaufartung und Erbkunde" [A "German Alliance for
National Regeneration and the Study of Heredity"]. *ARGB* 17: 349–50.

_____. 1927. Kinderreichtum und Rassenhygiene [Abundance of children and race hygiene].
ZVE 2: 104–5.

_____. 1929. Die bevölkerungspolitische Lage und das Gebot der Stunde [The population policy
situation and the demand of the hour]. *ARGB* 21: 241–53.

_____. 1930. Alfred Ploetz zum 70. Geburtstag [Alfred Ploetz on his 70th birthday]. *ARGB* 24:
viii–xv.

_____. 1931a. "Eugenik, Erblehre, Erbpflege" ["Eugenics, Genetics, Genetic Care"]. *ARGB* 23:
451–52.

_____. 1931b. Die Stellung des Nationalsozialismus zur Rassenhygiene [The position of
National Socialism on race hygiene]. *ARGB* 25: 300–308.

_____. 1956. Über die Grenzen praktischen Eugenik [On the limits of practical eugenics]. *Acta
Genetica,* no. 6: 13–24.

_____. 1960. Die soziologische Bedeutung der Selektion [The sociological importance of
selection]. In *Hundert Jahre Evolutionsforschung: Das wissenschaftliche Vermächtnis
Charles Darwins* [One hundred years of evolutionary research: The scientific legacy of
Charles Darwin], ed. Gerhard Heberer, 368–96. Stuttgart: Gustav Fischer.

Lilienthal, Georg. 1979. Rassenhygiene im Dritten Reich: Krise und Wende. [Race hygiene in
the Third Reich: Crisis and turning point]. *Medizinhistorisches Journal* 14: 114–33.

_____. 1980. "Rheinlandbastarde". Rassenhygiene und das Problem des rassenideologische
Kontinuität. Zur Untersuchung von Reiner Pommerin: "Sterilisierung der Rheinlandbastarde" ["The Rhineland bastards." Race hygiene and the problem of the continuity of the
ideology of race. On the investigation of Reiner Pommerin: "Sterilization of the Rhineland bastards"]. *Medizinhistorisches Journal* 15: 426–36.

_____. 1984. Zum Anteil der Anthropologie an der NS-Rassenpolitik [The part played by
anthropology in National Socialist racial policy]. *Medizinhistorisches Journal* 19: 148–60.

Löwenfeld, Leopold. 1910. Über medizinische Schutzmassnahmen gegen Verbrechen [On medical preventive measures against crime]. *Sexualprobleme* 6: 300–327.

Lutzhöft, Hans-Jürgen. 1981. *Der Nordische Gedanke in Deutschland 1920 bis 1940* [The nordic idea in Germany 1920 to 1940]. Kieler Historische Studien, Band 14. Stuttgart: Klett-Cotta.

McHale, Vincent E., and Eric A. Johnson. 1976–1977. Urbanization, Industrialization and Crime in Imperial Germany. *Social Science History* 1: 45–78, 210–47.

Maier, Friedrich. 1934. Die neue Staatsmedizinische Akademie in München [The new state medical academy in Munich]. *ARGB* 28: 56–57.

Mann, Gunter. 1969. Medizinisch-biologische Ideen und Modelle in der Gesellschaftslehre des 19. Jahrhunderts [Bio-medical ideas and models in the social theory of the 19th century]. *Medizinhistorisches Journal* 4: 1–23.

———. 1973. Rassenhygiene—Sozialdarwinismus [Race hygiene—social Darwinism]. In *Biologismus im 19. Jahrhundert in Deutschland: Vorträge eines Symposiums vom 30. bis 31.10.1970 in Frankfurt am Main* [Biologism in the 19th century in Germany: Lectures at a symposium from 30–31 October 1970 in Frankfurt am Main], ed. Gunter Mann. Stuttgart: Ferdinand Enke.

———. 1975. Biologie und Geschichte: Ansätze und Versuche zur biologischen Theorie der Geschichte im 19. und beginnenden 20. Jahrhundert [Biology and history: Beginnings and attempts at a biological theory of history in the 19th and early 20th centuries]. *Medizinhistorisches Journal* 10: 281–306.

———. 1977. Biologie und der "neue Mensch" [Biology and the "new man"]. In *Medizin, Naturwissenschaft, Technik und das zweite Kaiserreich* [Medicine, natural science, technology and the second empire], ed. Gunter Mann and Rolf Winau, 172–88. Göttingen: Vandenhoeck und Ruprecht.

———. 1978. Neue Wissenschaft im Rezeptionsbereich des Darwinismus: Eugenik—Rassenhygiene [A new science in the field of Darwinism: Eugenics—race hygiene]. *Berichte zur Wissenschaftsgeschichte* 1: 101–11.

———. 1983. Sozialbiologie auf dem Wege zur unmenschlichen Medizin des Dritten Reiches [Social biology on the road to the inhuman medicine of the Third Reich]. In *Unmenschliche Medizin. Geschichtliche Erfahrungen. Gegenwärtige Probleme und Ausblick auf die zukünftige Entwicklung. Seminar* [Inhuman medicine. Historical experiences. Present problems and the outlook for future developments]. Bad Nauheimer Gespräche der Landesärztekammer Hessen. Mainz: Kirchheim.

Marcuse, Max. 1907. Gesetzliche Eheverbote für Kranke und Minderwertige [Legal marriage restrictions for the diseased and inferior]. *Soziale Medizin und Hygiene* 2: 96–108, 163–75.

———. 1908. Die Verhütung der Geisteskrankheiten durch Eheverbote [The prevention of mental illnesses through marriage restrictions]. *Allgemeine Zeitung* (Munich), 13 June.

Marten, Heinz-Georg. 1983. *Sozialbiologismus: Biologische Grundpositionen der politischen Ideengeschichte* [Social biologism: Biological theories underlying the history of political ideas]. Frankfurt am Main and New York: Campus.

Martius, Friedrich. 1901. Das Vererbungsproblem in der Pathologie [The heredity problem in pathology]. *Berliner Klinische Wochenschrift* 38: 781–83, 814–18.

———. 1909. *Neurasthenische Entartung einst und jetzt* [Neurasthenic degeneration then and now]. Leipzig and Vienna: Franz Deuticke.

———. 1910. Die Bedeutung der Vererbung für Krankheitsentstehung und Rassenerhaltung [The importance of heredity for the etiology of disease and racial preservation]. *ARGB* 7: 470–89.

———. 1914. *Konstitution und Vererbung in ihren Beziehungen zur Pathologie* [Constitution and heredity in their relations to pathology]. Berlin: J. Springer.

Mersmann, Ingrid. 1978. *Medizinische Ausbildung im Dritten Reich* [Medical education in the Third Reich]. Medical dissertation, Munich. Munich: privately printed.

Meyer, Ludwig. 1885. Die Zunahme der Geisteskranken [The increase in the mentally ill]. *Deutsche Rundschau* 78–94.

Mitschelich, Alexander, and Fred Mielke, eds. 1978. *Medizin ohne Menschlichkeit: Dokumente des Nürnberger Ärzteprozesses* [Medicine without humanity: Documents of the Nuremberg doctors' trial]. Frankfurt am Main: Fischer.

Möbius, Paul Julius. 1900. *Ueber Entartung* [On degeneration]. Wiesbaden: J. F. Bergmann.

Mock, Wolfgang. 1981. Manipulation von oben oder Selbstorganisation an der Basis? Einige neuere Ansätze in der englischen Historiographie zur Geschichte des deutschen Kaiserreichs [Manipulation from above or self-organization from below? Some new beginnings in the English historiography of the Germany empire]. *Historische Zeitschrift* 232: 358–75.

Monheim, Maria. 1928. *Rationalisierung der Menschenvermehrung* [Rationalization of human reproduction]. Jena: Gustav Fischer.

Mosse, George L. 1964. *The Crisis of German Ideology: Intellectual Origins of the Third Reich.* New York: Grosset and Dunlap.

———. 1978. *Toward the Final Solution: A History of European Racism.* London: J. M. Dent and Sons.

Muckermann, Hermann. 1906. *Attitudes of Catholics towards Darwinism and Evolution.* St. Louis: Herder.

———. 1920. *Kind und Volk: Der biologische Wert der Treue zu den Lebensgesetzen beim Aufbau der Familie* [Child and nation: The biological value of loyalty to the laws of life in the growth of the family]. 3d ed. Freiburg i. B.: Herder.

———. 1922. *Vererbung und Auslese* [Heredity and selection]. Freiburg i. B.: Herder.

———. 1930. Differenzierte Fortpflanzung [Differential reproduction]. *ARGB* 24: 269–89.

———. 1931a. Alfred Ploetz und sein Werk [Alfred Ploetz and his work]. *Eugenik* 1: 261–65.

———. 1931b. Hauptversammlung der Deutschen Gesellschaft für Rassenhygiene (Eugenik), München, 18. September 1931 [Main meeting of the German Society for Race Hygiene (Eugenics), Munich, 18 September 1931]. *Eugenik* 2: 47–48.

———. 1931c. Illustrationen zur Frage: Wohlfahrtspflege und Eugenik [Illustrations to the question: Welfare and eugenics]. *Eugenik* 2: 41–42.

———. 1932a. Aus der Hauptversammlung der Deutschen Gesellschaft für Rassenhygiene (Eugenik) zu München am 18. September 1931 [From the main meeting of the German Society for Race Hygiene (Eugenics) at Munich on 18 September 1931]. *ARGB* 26: 94–101.

———. 1932b. Eugenik und Strafrecht [Eugenics and penal law]. *Eugenik* 2: 104–9.

———. 1932c. *Rassenforschung und Volk der Zukunft: Ein Beitrag zur Einführung in die Frage vom biologischen Werden der Menschheit* [Racial research and the nation of the future: An introduction to the question of the biological development of humanity]. Berlin: Metzner.

———. 1933. *Volkstum, Staat und Nation, eugenisch gesehen* [People, state and nation from a eugenic standpoint]. Essen: Fredebeul und Koenen.

Muckermann, Hermann, and O. Freiherr von Verschuer. 1931. *Eugenische Eheberatung* [Eugenic marriage counseling]. Berlin: Dümmler.

Mühlen, Patrik von zur. 1977. *Rassenideologien: Geschichte und Hintergründe* [Racial ideologies: History and background]. Berlin and Bad Godesberg: J.H.W. Dietz.

Müller, Joachim. 1978. Sterilisation und Gesetzgebung bis 1933 [Sterilization and legislation up to 1933]. Unpublished paper delivered at the Institut für die Geschichte der Medizin in Mainz, 7 November 1978.

Müller-Hegemann, Dietrich. 1959. Über den Einfluss der Eugenik auf die deutsche Medizin [On the influence of eugenics on German medicine]. *Deutsche Gesundheitswesen* 14: 429–36.

Müller-Hill, Benno. 1984. *Tödliche Wissenschaft: Die Aussonderung von Juden, Zigeunern und Geisteskranken 1933–1945* [Deadly science: The elimination of Jews, gypsies and mentally ill 1933–1945]. Hamburg: Rowohlt.

Müssigang, Albert. 1968. *Die soziale Frage in historischen Schule der deutschen Nationalökon-*

omie [The social question in the historical school of German economics]. Tübingen: J.C.B. Mohr.

Näcke, Paul. 1899. Kastration bei gewissen Klassen von Degenerierten als ein wirksamer sozialer Schutz [Castration among certain classes of degenerates as an effective social defense]. *Archiv für Kriminalanthopologie* 1: 58–84.

Nemitz, Kurt. 1981. Die Bemühungen zur Schaffung eines Reichsgesundheitsministeriums in der ersten Phase der Weimarer Republik 1918–1922 [The efforts to create a Reich health ministry in the first phase of the Weimar Republic]. *Medizinhistorisches Journal* 16: 424–45.

Notizen [Notes]. 1933. *ARGB* 27: 467.

Nowacki, Bernd. 1983. *Der Bund für Mutterschutz (1905–1933)* [The League for the Protection of Mothers (1905–1933)]. Abhandlungen zur Geschichte der Medizin und der Naturwissenschaften, Heft 48. Husum: Matthiesen.

Nowak, Kurt. 1984. *"Euthanasie" und Steriliserung im Dritten Reich: Die Konfrontation der evangelischen und katholischen Kirche mit dem "Gesetz zur Verhütung erbkranken Nachwuchses" und der "Euthanasie"-Aktion* ["Euthanasia" and sterilization in the Third Reich: The confrontation of the Protestant and Catholic churches with the "law for the prevention of genetically diseased offspring" and the "euthanasia" action]. 3d ed. Göttingen: Vandenhoeck und Ruprecht.

Oettingen, Alexander von. 1882. *Die Moralstatistik in ihrer Bedeutung für eine Sozialethik* [Moral statistics in their importance for a social ethics]. 3d ed. Erlangen: A. Deichert.

Oettinger, W. 1912. Selektion und Hygiene [Selection and hygiene]. *Deutsche Vierteljahrsschrift für öffentliche Gesundheitspflege* 44: 606–26.

Olberg, Oda. 1906. Bemerkungen über Rassenhygiene und Sozialismus [Remarks on race hygiene and socialism]. *Die neue Zeit* 25: 725–33.

_____. 1907. Rassenhygiene und Sozialismus [Race hygiene and socialism]. *Die neue Zeit* 26: 882–87.

Ostermann, Artur. 1928. Eheberatungstellen [Marriage counseling centers]. *VEE* 3: 293–98.

Peltzer, otto. 1925–1926. Das Verhältnis der Sozialpolitik zur Rassenhygiene: Eine kritische Studie über die Auswirkung biologischer Erkenntnisse auf die Lösungsversuche sozialer Probleme [The relationship of social policy to race hygiene: A critical study on the impact of biological knowledge on the attempts to solve social problems]. Dissertation, Munich.

Planck, Max, ed. 1936. *25 Jahre Kaiser Wilhelm-Gesellschaft zur Förderung der Wissenschaften* [25 years of the Kaiser Wilhelm Society for the Advancement of the Sciences]. Berlin: J. Springer.

Plate, Ludwig. 1900. *Die Bedeutung und Tragweite des Darwinischen Selektionsprinzipes* [The meaning and significance of the Darwinian selection principle]. Leipzig: W. Engelmann.

Platen-Hallermund, Alice. 1948. *Die Tötung Geisteskranken in Deutschland* [The killing of the mentally ill in Germany]. Frankfurt am Main: Frankfurter Hefte.

Ploetz, Alfred. 1894. "Rassentüchtigkeit und Sozialismus." [Racial fitness and socialism]. *Neue Deutsche Rundschau* 5: 989–97.

_____. 1895a. Ableitung einer Rassenhygiene und ihrer Beziehung zur Ethik [The construction of a race hygiene and its relationship to ethics]. *Vierteljahresschrift für Wissenschaftliche Philosophie* 19: 368–77.

_____. 1895b. *Die Tüchtigkeit unsrer Rasse und der Schutz der Schwachen: Ein Versuch über Rassenhygiene und ihr Verhältnis zu den humanen Idealen, besonders zum Sozialismus* [The fitness of our race and the protection of the weak: An attempt to create a race hygiene and its relationship to humane ideals, especially socialism]. Berlin: Gustav Fischer.

_____. 1902. Sozialpolitik und Rassenhygiene in ihrem prinzipiellen Verhältnis [Social policy and race hygiene in their principle relationship]. *Archiv für Soziale Gesetzgebung und Statistik* 17: 393–420.

_____. 1903. Der Alkohol im Lebensprozess der Rasse [Alcohol in the life-history of the race]. *Internationale Monatsschrift zur Erforschung des Alkoholismus und Bekämpfung der Trunksitten* 239–89.

_____. 1904a. Die Begriffe Rasse und Gesellschaft und die davon abgeleiteten Disciplinen [The concepts of race and society and the disciplines derived from them]. *ARGB* 1: 2–26.

_____. 1904b. Kritik über Max von Gruber: "Führt die Hygiene zur Entartung der Rasse" [Criticism of Max von Gruber: "Does hygiene lead to the degeneration of the race"]. *ARGB* 1: 157.

_____. 1906a. Ableitung einer Gesellschaftshygiene und ihrer Beziehung zur Ethik [Construction of a hygiene of society and its relationship to ethics]. *ARGB* 3: 253–60.

_____. 1906b. Zur Abgrenzung und Einleitung des Begriffs Rassenhygiene. [On the definition and introduction of the concept of race hygiene]. *ARGB* 3: 864–66.

_____. 1907. Denkschrift über die Gründung der Internationalen Gesellschaft für Rassen-Hygiene [Memorandum on the founding of the International Society for Race Hygiene]. Unpublished, Ploetz family archive.

_____. 1909. Gesellschaften mit rassenhygienischen Zwecken [Societies with race hygiene purposes]. *ARGB* 6: 277–81.

_____. 1911a. Die Begriffe Rasse und Gesellschaft und einige damit zusammenhängende Probleme [The concepts of race and society and some related problems]. *Schriften der Deutschen Gesellschaft für Soziologie* 1: 111–36.

_____. 1911b. Unser Weg [Our way]. Unpublished, Ploetz family archive.

_____. 1911c. Ziele und Aufgaben der Rassenhygiene [Goals and tasks of race hygiene]. *Vierteljahrsschrift für Öffentliche Gesundheitspflege* 43: 164–191.

_____. 1913. Neo-Malthusianismus und Rassenhygiene [Neo-Malthusianism and race hygiene]. *ARGB* 10: 166–72.

_____. 1932. Dr. Agnes Bluhm 70 Jahre am 9. Januar 1932 [Dr. Agnes Bluhm 70 years old on 9 January 1932]. *ARGB* 26: 63.

_____. 1935. Trostworte an einem naturwissenschaftlichen Hamlet [Condolences to a scientific Hamlet]. Reprint of an article in the *New Yorker Volkszeitung,* 6 November 1892. *ARGB* 29: 88–89.

Ploetz, Alfred, Anastasius Nordenholz, and Ludwig Plate. 1904. Vorwort [Foreword]. *ARGB* 1: iv.

Poliakov, Leon. 1977. *The Aryan Myth: A History of Racist and Nationalist Ideas in Europe.* Trans. E. Howard. New York: New American Library.

Pommerin, Reiner. 1979. *Sterilisierung der Rheinlandbastarde: Das Schicksal einer farbigen deutschen Minderheit 1918–1937* [Sterilization of the Rhineland bastards: The fate of a colored German minority 1918–1937]. Düsseldorf: Droste.

Proctor, Robert N. 1982. Pawns or Pioneers? The Role of Doctors in the Origins of Nazi Racial Science. Unpublished manuscript, Harvard University.

Projektgruppe "Volk und Gesundheit," ed. 1982. *Volk und Gesundheit: Heilen und Vernichten im Nationalsozialismus* [Nation and health: Healing and extermination in national socialism]. Tübingen: Tübinger Vereinigung für Volkskunde.

Rehse, Helga. 1969. Euthanasie, Vernichtung unwerten Lebens und Rassenhygiene in Programmschriften vor dem Ersten Weltkrieg [Euthanasia, elimination of lives not worth living and race hygiene in programmatic documents before the First World War]. Medical dissertation, Heidelberg.

Reibmayr, Albert. 1906–1907. Die biologische Gefahren der heutigen Frauenemanzipation [The biological dangers of modern women's emancipation]. *Politisch-anthropologische Revue* 5: 445–68.

Reichel, Heinrich. 1931. Alfred Ploetz und die rassenhygienische Bewegung der Gegenwart [Alfred Ploetz and the contemporary race hygiene movement]. *Wienischer Klinische Wochenschrift* 44: 1–9.

Ribbert, Hugo. 1910. *Rassenhygiene: Eine gemeinverständliche Darstellung* [Race hygiene: A popular presentation]. Bonn: Friedrich Cohen.

Ringer, Fritz. 1969. *The Decline of German Mandarins: The German Academic Community, 1890–1933.* Cambridge, Mass.: Harvard University Press.

Risson, Renate. 1983. *Fritz Lenz und die Rassenhygiene.* [Fritz Lenz and race hygiene]. Medical dissertation, Mainz, 1982. Husum: Matthiesen.

Roberts, James S. 1980. Der Alkoholkonsum deutscher Arbeiter im 19. Jahrhundert [The alcohol consumption of German workers in the 19th century]. *Geschichte und Gesellschaft* 6: 220–42.

Rosen, George. 1958. *A History of Public Health*. New York: M. D. Publications.

―――. 1975. Die Entwicklung der sozialen Medizin [The development of social medicine]. In *Seminar: Medizin, Gesellschaft, Geschichte* [Seminar: Medicine, society, history], ed. H. -U. Deppe and M. Regus. Frankfurt am Main: Suhrkamp.

Roth, Guenther. 1963. *The Social Democrats in Imperial Germany*. Reprint. New York: Arno Press, 1979.

Rüdin, Ernst. 1904. Zur Rolle der Homosexuellen im Lebensprozess der Rasse [On the role of homosexuals in the life-history of the race]. *ARGB* 1: 99–109.

―――. 1910. Über den Zusammenhang zwischen Geisteskrankheit und Kultur [On the relationship between mental illness and culture]. *ARGB* 7: 722–48.

―――. 1934a. Aufgaben und Ziele der Deutschen Gesellschaft für Rassenhygiene [Tasks and goals of the German Society for Race Hygiene]. *ARGB* 28: 228–33.

―――, ed. 1934b. *Erblehre und Rassenhygiene im völkischen Staat* [Genetics and race hygiene in the folkish state]. Munich: J. F. Lehmann.

―――. 1938. 20 Jahre menschliche Erbforschung an der Deutschen Forschungsanstalt für Psychiatrie in München, Kaiser-Wilhelm Institut [20 years of human heredity research at the German Research Institute for Psychiatry, Kaiser Wilhelm Institute]. *ARGB* 32: 193–203.

Rutgers, Johannes. 1908. *Rassenverbesserung, Malthusianismus und Neumalthusianismus* [Racial improvement, Malthusianism and neo-Malthusianism]. Trans. Martina G. Kramers. Dresden and Leipzig: Heinrich Minden.

Saller, Karl. 1932. Über Intelligenzunterschiede der Rassen Deutschlands. [On intelligence differences in the races of Germany]. *Eugenik* 2: 220–25.

―――. 1961. *Die Rassenlehre des Nationalsozialismus in Wissenschaft und Propaganda* [The race theory of National Socialism in science and propaganda]. Darmstadt: Progress.

Samson-Himmelstjerna, Hermann von. 1902. *Die gelbe Gefahr als Moralproblem* [The yellow peril as a moral problem]. Berlin: Deutscher Kolonial Verlag.

Schade, Heinrich. 1936. Ausländische Stimmen zur deutschen Erb- und Rassenpflege [Foreign voices on German genetic and racial care]. *Der Erbarzt* 3: 60–61.

Schäfter, Cäsar. 1934. *Volk und Vererbung: Eine Einführung in die Erbforschung, Familienkunde, Rassenlehre, Rassenpflege und Bevölkerungspolitik* [Nation and heredity: An introduction to genetic research, family studies, racial theory, racial care and population policy]. Leipzig: G. B. Teubner.

Schallmayer, Wilhelm. 1895. *Über die drohende physische Entartung der Culturvölker* [On the threatening physical degeneration of civilized nations]. 2d ed. Berlin: Heuser.

―――. 1903. *Vererbung und Auslese im Lebenslauf der Völker: Eine staatswissenschaftliche Studie auf Grund der neueren Biologie* [Heredity and selection in the life-process of nations: A social scientific study on the basis of the newest biology]. Jena: Gustav Fischer.

―――. 1904. Zum Einbruch der Naturwissenschaften in das Gebiet der Geisteswissenschaften [On the invasion of the humanities by the natural sciences]. *ARGB* 1: 586–97.

―――. 1905. *Beiträge zu einer Nationalbiologie* [Contributions to a national biology]. Jena: Hermann Costenoble.

―――. 1907. Rassehygiene und Sozialismus [Race hygiene and socialism]. *Die neue Zeit* 25: 731–40.

―――. 1908a. Der Krieg als Züchter [War as selector]. *ARGB* 5: 364–400.

―――. 1908b. Die Politik der Fruchtbarkeitsbeschränkungen [The policy of restricting fertility]. *Zeitschrift für Politik* 2: 391–439.

―――. 1909. Was ist von unserem sozialen Versicherungswesen für die Erbqualitäten der Bevölkerung zu erwarten? [What is to be expected from our social insurance system for the hereditary quality of the population?] *Archiv für Soziale Hygiene und Demographie* 3: 27–65.

―――. 1910a. Gobineaus Rassenwerk und die moderne Gobineauschule [Gobineau's race theory and the modern Gobineau school]. *Zeitschrift für Sozialwissenschaft*, N.F., 1: 553–72.

———. 1910b. *Vererbung und Auslese in ihrer soziologischen und politischen Bedeutung* [Heredity and selection in their sociological and political importance]. 2d ed. Jena: Gustav Fischer.

———. 1911a. Rassedienst [Racial service]. *Sexualprobleme* 7: 433–43, 534–47.

———. 1911b. Sozialistische Entwicklungs- und Bevölkerungslehre [Socialist development and population theory]. *Zeitschrift für Sozialwissenschaft*, N.F., 2: 511–30.

———. 1913. Soziale Massnahmen zur Verbesserung der Fortpflanzungsauslese [Social measures for the betterment of reproductive selection]. In *Krankheit und soziale Lage* [Illness and social situation], ed. M. Mosse and G. Tugendreich, 841–59. Munich: J. F. Lehmann.

———. 1914a. Ernst Haeckel und die Eugenik [Ernst Haeckel and eugenics]. In *Was Wir Ernst Haeckel Verdanken* [What we owe Ernst Haeckel], ed. Heinrich Schmidt, vol. 2, 367–72. Leipzig: Unesma.

———. 1914b. Sozialhygiene und Eugenik [Social hygiene and eugenics]. *Zeitschrift für Sozialwissenschaft*, N.F., 5: 329–39, 397–408, 505–13.

———. 1914–1915. Zur Bevölkerungspolitik gegenüber dem durch den Krieg verursachten Frauenüberschuss [On population policy in regard to the excess of women caused by the war]. *ARGB* 11: 713–37.

———. 1917. Einführung in die Rassehygiene [Introduction to race hygiene]. In *Ergebnisse der Hygiene, Bakteriologie, Immunitätsforschung und experimentalen Therapie* [Results of hygiene, bacteriology, immunity research and experimental therapy], ed. W. Weichards, vol. 2, 433–532. Berlin: J. Springer.

———. 1918a. Kriegswirkungen am Volkskörper und ihre Heilung [The effects of the war on the national body and their remedy]. *Die Umschau* 22: 1–24.

———. 1918b. *Vererbung und Auslese: Grundriss der Gesellschaftsbiologie und der Lehre vom Rassedienst* [Heredity and selection: Principles of social biology and the theory of racial service]. 3d ed. Jena: Gustav Fischer.

———. 1919a. Neue Aufgaben und neue Organisation der Gesundheitspolitik [New tasks and new organization of health policy]. *Archiv für Soziale Hygiene und Demographie* 13: 225–70.

———. 1919b. Sicherung des Volksnachwuchses und Sozialisierung der Nachwuchskosten [Securing the national progeny and the socialization of the cost of the progeny]. *Die Umschau* 23: 497–500, 517–20.

Scheumann, F. K. 1928. Sinn und Wesen der Eheberatung [The meaning and essence of marriage counseling]. *VEE* 3: 19–22.

Schopol, Dr., ed. 1932. *Die Eugenik im Dienste der Volkswohlfahrt: Bericht über die Verhandlungen eines zusammengesetzten Ausschusses des Preussischen Landesgesundheitsrats vom 2. Juli 1932* [Eugenics in the service of national welfare: Report on the proceedings of the constituted committee of the Prussian Health Council of 2 July 1932]. Veröffentlichungen aus dem Gebiete der Medizinalverwaltung, XXXVIII. Band—5. Heft. Berlin: Richard Schoetz.

Schraepler, Ernst, ed. 1964. *Quellen zur Geschichte der sozialen Frage in Deutschland* [Sources for the history of the social question in Germany]. 2d ed. 2 vols. Göttingen: Musterschmidt.

Schulte, W. 1965. "Euthanasie" und Sterilisation im Dritten Reich [Euthanasia and sterilization in the Third Reich]. In *Deutsches Geistesleben und Nationalsozialismus: Eine Vortragsreihe an der Universität Tübingen* [German intellectual life and national socialism: A lecture series at the University of Tübingen], ed. A. Flitner, 73–89. Tübingen: Rainer Wunderlich.

Schwarz, Hanns. 1950. *Ein Gutachten über die ärztliche Tätigkeit im sogenannten Erbgesundheitsverfahren* [A testimonial on the activity of doctors in the so-called genetic health system]. Halle: Marhold.

Seeberg, Reinhold. 1913. *Der Geburtenrückgang in Deutschland, eine sozialethische Studie* [Birthrate decline in Germany, a study in social ethics]. Leipzig: A Deichert.

Seidler, Eduard. 1977. Der politische Standort des Arztes im zweiten Kaiserreich [The political position of physicians in the second empire]. In *Medizin, Naturwissenschaft, Technik und das zweite Kaiserreich* [Medicine, natural science, technology and the second empire], ed. Gunter Mann and Rolf Winau, 87–102. Göttingen: Vandenhoeck und Ruprecht.

Seidler, Horst, and Andreas Rett. 1982. *Das Reichssippenamt entscheidet: Rassenbiologie im Nationalsozialismus* [The Reich Genealogical Office decides: Racial biology in national socialism]. Vienna and Munich: Jugend- und Volk-Verlagsgesellschaft.

Siemens, Hermann W. 1916. Die Proletarisierung unseres Nachwuchses, eine Gefahr unrassenhygienischer Bevölkerungspolitik [The proletarianization of our offspring, a danger of dysgenic population policy]. *ARGB* 12: 43–55.

_____. 1917a. *Die biologischen Grundlagen der Rassenhygiene und der Bevölkerungspolitik* [The biological basis of race hygiene and population policy]. Munich: J. F. Lehmann.

_____. 1917b. Kritik der rassenhygienischen und bevölkerungspolitischen Bestrebungen [Criticism of race hygiene and population policy efforts]. *Politisch-anthropologische Monatsschrift* 15: 547–51.

Sigerist, Henry E. 1930. Das Bild des Menschen in der modernen Medizin [The view of humans in modern medicine]. *Neue Blätter für den Sozialismus* 1: 97–106.

Soloway, Richard Allen. 1982. *Birth Control and the Population Question in England, 1877–1930*. Chapel Hill, N.C.: University of North Carolina Press.

Sombart, Werner. 1897. Ideale der Sozialpolitik [Ideals of social policy]. *Archiv für Soziale Gesetzgebung und Staat* 10: 1–48.

Spilger, Dr. 1927. Vererbungslehre und Rassenhygiene im biologischen Unterricht der höheren Schulen [Genetics and race hygiene in the biology instruction of the higher schools]. *ARGB* 19: 63–69.

Staemmler, Martin. 1933a. *Rassenpflege im völkischen Staat* [Racial care in the folkish state]. Munich: J. F.Lehmann.

_____. 1933b. *Rassenpflege und Schule* [Racial care and the schools]. Langensalza: H. Beyer.

_____. 1933c. Die Sterilisierung Minderwertiger vom Standpunkt des Nationalsozialismus [The sterilization of the inferior from the standpoint of National Socialism]. *Eugenik* 3: 97–110.

Steinmetz, Sebald R. 1902. Die erbliche Rassen- und Volkscharakter [The hereditary racial and national character]. *Vierteljahrsschrift für wissenschaftliche Philosophie und Soziologie* 26: 77–126.

_____. 1904a. Feminismus und Rasse [Feminism and race]. *Zeitschrift für Sozialwissenschaft* 7: 751–68.

_____. 1904b. Der Nachwuchs der Begabten [The offspring of the talented]. *Zeitschrift für Sozialwissenschaft* 7: 1–25.

Stern, Fritz. 1975. The Political Consequences of the Unpolitical German. In *The Failure of Illiberalism: Essays on the Political Culture of Modern Germany*. Chicago: University of Chicago Press.

Stroothenke, Wolfgang. 1940. *Erbpflege und Christentum* [Genetic care and Christianity]. Leipzig: Leopold Klotz.

Ternon, Yves, and Socrate Helman. 1969. *Histoire de la médecine SS: ou, Le mythe du racisme biologique* [History of SS medicine: or, The myth of biological racism]. Tournai: Casterman.

Thom, Achim, and Horst Spaar, eds. 1983. *Medizin im Faschismus. Symposium über das Schicksal der Medizin in der Zeit das Faschismus im Deutschland 1933–1945. Protokoll* [Medicine in fascism. Symposium on the fate of medicine in the fascist period in Germany 1933–1945. Protocol]. Berlin (East): Akademie der Ärztliche Fortbildung der DDR.

Thomann, Klaus-Dieter. 1979. Die Zusammenarbeit der Sozialhygieniker Alfred Grotjahn und Alfons Fischer [The cooperative work of the social hygienists Alfred Grotjahn and Alfons Fischer]. *Medizinhistorisches Journal* 14: 251–74.

_____. 1983. Das Reichsgesundheitsamt und die Rassenhygiene: Eine Rückblick anlässlich der Verabschiedung des "Gesetzes zur Verhütung erbkranken Nachwuchses" vor 50 Jahren

[The Reich Health Office and race hygiene: A look back on the occasion of the passage of the "law for the prevention of genetically diseased offspring" 50 years ago]. *Bundesgesundheitsblatt* 26: 206–13.

Thompson, Larry V. 1971. *Lebensborn* and the Eugenics Policy of the *Reichsführer-SS. Central European History* 4: 57–71.

Tönnies, Ferdinand. 1903. Eugenik [Eugenics]. *Schmollers Jahrbuch für Gesetzgebung und Verwaltung* 29: 273–90.

———. 1905. Zur naturwissenschaftlichen Gesellschaftslehre [On the scientific theory of society]. *Schmollers Jahrbuch für Gesetzgebung und Verwaltung* 31: 27–102, 1283–1321.

———. 1907. Zur naturwissenschaftlichen Gesellschaftslehre: Eine Replik [On the scientific theory of society: A reply]. *Schmollers Jahrbuch für Gesetzgebung und Verwaltung* 33: 49–114.

Tutzke, Dietrich. 1979. *Alfred Grotjahn.* Biographien hervorragender Naturwissenschaftler, Techniker und Mediziner, Band 36. Leipzig: BSB B. G. Teubner.

Über die Vererbung von Geisteskrankheiten nach Beobachtung in preussischen Irrenanstalten [On the heritability of mental illnesses according to observations in Prussian mental asylums]. 1879. *Jahrbuch für Psychiatrie und Neurologie* 1: 65–66.

Verhandlungen des Ersten Deutschen Soziologentages [Proceedings of the first German sociological conference]. 1911. Tübingen: J.C.B. Mohr.

Verschuer, Otmar Freiherr von. 1930. Vom Umfang der erblichen Belastung im deutschen Volke [The extent of hereditary taint in the German people]. *ARGB* 24: 238–68.

———. 1934. *Erbpathologie: Ein Lehrbuch für Ärzte* [Genetic pathology: A textbook for physicians]. Dresden: T. Steinkopff.

———. 1938. Rassenbiologie der Juden [Racial biology of the Jews]. In *Forschungen zur Judenfrage* [Researches on the Jewish question], Band 3. Hamburg: Hanseatische Verlagsanstalt.

———. 1941. *Leitfaden der Rassenhygiene* [Textbook of race hygiene]. Leipzig: Georg Thieme.

———. 1944. *Erbanlage als Schicksal und Aufgabe* [Genes as fate and task]. Berlin: W. de Gruyter.

———. 1964. Das ehemalige Kaiser-Wilhelm Institut für Anthropologie, menschliche Erblehre und Eugenik [The former Kaiser Wilhelm Institute for Anthropology, Human Heredity and Eugenics]. *Zeitschrift für Morphologische Anthropologie* 55: 127–74.

Vierkandt, A. 1906. Ein Einbruch der Naturwissenschaften in die Geisteswissenschaften? [An invasion of the humanities by the natural sciences?] *Zeitschrift für Philosophie und philosophische Kritik* 128: 168–77.

Vondung, Klaus, ed. 1976. *Das Wilhelminische Bildungsbürgertum: Zur Sozialgeschichte seiner Ideen* [The Wilhelmine educated middle classes: On the social history of its ideas]. Göttingen: Vandenhoeck und Ruprecht.

Wagner, Adolf. 1872. *Rede über die soziale Frage* [Speech on the social question]. Berlin: Wiegandt and Grieben.

Wahl, M. 1885. Über den gegenwärtigen Stand der Erblichkeitsfrage in der Lehre von der Tuberculose [On the present sitatuion of the heredity question in the theory of tuberculosis]. *Deutsche Medizinische Wochenschrift,* no. 1: 3–5; no. 3: 36–38; no. 4: 34–36; no. 5: 69–71; no. 6: 88–90.

Waite, Robert G. L. 1969. *Vanguard of Nazism: The Free Corps Movement in Postwar Germany 1918–1923.* New York: Norton Library.

Waldinger, Robert J. 1973. The High Priests of Nature: Medicine in Germany, 1883–1933. B.A. thesis, Harvard University.

Wehler, Hans-Ulrich. 1973. Sozialdarwinismus im expandierenden Industriestaat [Social Darwinism in an expanding industrial state]. In *Deutschland in der Weltpolitik des 19. und 20. Jahrhunderts: Festschrift für Fritz Fischer* [Germany in the world affairs of the 19th and 20 centuries: Festschrift for Fritz Fischer]. Düsseldorf: Bertelsmann.

———. 1977. *Das Deutsche Kaiserreich* [The German empire]. Göttingen: Vandenhoeck und Ruprecht.

Weindling, Paul. 1981. Theories of the Cell State in Imperial Germany. In *Biology, Medicine and*

Society 1840-1940, ed. Charles Webster, 99–155. Cambridge: Cambridge University Press.

_____. 1983. The Medical Profession, Social Hygiene and the Birth Rate in Germany, 1914–1918. Unpublished manuscript.

_____. 1984. Die Preussische Medizinalverwaltung und die "Rassenhygiene": Anmerkungen zur Gesundheitspolitik der Jahr 1905-1933 [The Prussian medical administration: Remarks on health policy during the years 1905 to 1933]. *Zeitschrift für Sozialreform* 30: 675–87.

_____. 1985a. Weimar Eugenics: The Kaiser Wilhelm Institute for Anthropology, Human Heredity and Eugenics in a Social Context. *Annals of Science* 303–18.

_____. 1985b. Race, Blood and Politics. *The Times Higher Education Supplement,* 19 July, 13.

Weinreich, Max. 1946. *Hitler's Professors: The Part of Scholarship in Germany's Crimes against the Jewish People.* New York: Yiddish Scientific Institute-YIVO.

Weismann, August. 1889. On Heredity. In *Essays on Heredity and Kindred Problems,* ed. Edward Poulton, Selmar Schönland, and Arthur Shipley. Oxford: Clarendon Press.

Weiss, Sheila F. 1983. Race Hygiene and the Rational Management of National Efficiency: Wilhelm Schallmayer and the Origins of German Eugenics, 1890-1920. Ph.D. dissertation, The Johns Hopkins University.

_____. 1986. Wilhelm Schallmayer and the Logic of German Eugenics. *Isis* 77: 33–46.

Wertham, Frederic. 1973. *A Sign for Cain.* New York: Warner Paperback Library.

Winau, Rolf. 1983. Natur und Staat. Oder: Was lernen wir aus den Prinzipien der Deszendenztheorie in Beziehung auf die innerpolitische Entwicklung und Gesetzgebung der Staaten? [Nature and state. Or: what can we learn from the theory of evolution about internal political development and state legislation?]. *Berichte zur Wissenschaftsgeschichte* 6: 123–32.

Winkler, W. F. 1928. Bevölkerungspolitische Zukunftsfragen Europas [Population policy questions concerning the future of Europe]. *VEE* 3: 169–173.

Wolf, Julius. 1912. *Der Geburtenrückgang* [The birthrate decline]. Jena: Gustav Fischer.

Woltmann, Ludwig. 1903. *Politische Anthropologie: Eine Untersuchung über den Einfluss der Descendenztheorie auf der Lehre von der politischen Entwicklung der Völker* [Political anthropology: An investigation of the influence of the theory of evolution on the theory of the political development of nations]. Leipzig and Eisenach: Anst.

Wuttke-Groneberg, Walter, ed. 1982. *Medizin im Nationalsozialismus: Ein Arbeitsbuch* [Medicine in National Socialism: A workbook]. 2d ed. Rottenburg: Schwäbische Verlagsgesellschaft.

Yeboa, Joseph. 1968. Zum sozialen und eugenischen Darwinismus am Ausgang des 19. Jahrhunderts [On social and eugenic Darwinsim at the end of the 19th century]. Medical dissertation, Heidelberg.

Young, E. J. 1968. *Gobineau und der Rassismus: Eine Kritik der anthropologischen Geschichtstheorie* [Gobineau and racism: A critique of anthropological history theory]. Meisenheim am Glan: Anton Hain.

Zapp, Albert. 1979. *Untersuchungen zum Nationalsozialistischen Deutschen Ärztebund (NSDÄB)* [Investigations into the National Socialist Doctor's Association (NSDÄB)]. Medical dissertation, Kiel. Kiel: privately printed.

Ziegler, Heinrich E. 1893. *Die Naturwissenschaft und die socialdemokratische Theorie: ihr Verhältniss dargelegt auf Grund der Werke von Darwin und Bebel* [Natural science and social democratic theory: Their relationship as presented through the works of Darwin and Bebel]. Stuttgart: Enke.

_____. 1899. Das Verhältnis der Sozialdemokratie zum Darwinismus [The relationship of social democracy to Darwinism]. *Zeitschrift für Sozialwissenschaft* 2: 424–32.

_____. 1902. *Über den derzeitigen Stand der Descendenztheorie in der Zoologie* [On the present state of the theory of evolution in zoology]. Jena: Gustav Fischer.

_____. 1903. Einleitung zu dem Sammelwerke *Natur und Staat* [Introduction to the series *Nature and State*]. In *Natur und Staat: Beiträge zur naturwissenschaftliche Gesellschaf-*

tslehre [Nature and state: Contribution to a scientific theory of society]. Jena: Gustav
    Fischer.

_____. 1918. *Die Vererbungslehre in der Biologie und in der Soziologie* [The theory of heredity
    in biology and sociology]. Jena: Gustav Fischer.

_____. 1922. Review of *Zur Abwehr des ethischen, des sozialen, des politischen Darwinismus*
    [On the warding off of ethical, social, and political Darwinism] by Oskar Hertwig. *ARGB*
    14: 212–18.

Zmarzlik, Hans G. 1963. Sozialdarwinismus in Deutschland als geschichtliches Problem [Social
    Darwinism in Germany as a historical problem]. *Vierteljahreshefte für Zeitgeschichte* 11:
    246–73.

CHAPTER 3

The Eugenics Movement in France
1890–1940

William H. Schneider

Eugenics movements in non-English-speaking countries have yet to be studied enough to produce a significant secondary literature, and one result of this lack of information has been the suspicion, if not presumption, that eugenics was a peculiarly Anglo-Saxon affair. This, in turn, has permitted the development of a very narrow definition of eugenics based on some peculiarly English and American circumstances, such as an early acceptance of Mendelian heredity, or strong race and class prejudice. Historians of biology have further narrowed the perspective by viewing eugenics as an infertile offshoot of an emerging genetics with which it shared common origins.

The French case demonstrates the advantage of considering eugenics more broadly: that is, as a widespread phenomenon found at the turn of the nineteenth century in most Western industrial societies. Accordingly, its roots lay in the social class differentiation and conflict, economic cycles, and increased growth of government, as well as the scientific view of the universe, that were some of the most obvious features of the new modern society. Eugenics was a reaction to the perception that society was in a state of decline and degeneration. Its novelty was the self-proclaimed scientific means it proposed to resolve the problem, but this reaction was common to a long list of countries in the world of 1900. In the international context eugenics was less a pseudoscientific, failed branch of applied human genetics than a biologically based movement for social reform. Accordingly, it was influenced by many cultural and social crosscurrents in each of the

Research for this work was made possible by grants from the Penrose Fund of the American Philosophical Society, the National Endowment for the Humanities, and the National Science Foundation under Grant SES-821852. In addition to the other authors in this book, I would like to thank Robert A. Nye and Richard Soloway for their comments.

countries where it developed. Studies of eugenics in non-English-speaking countries may even cause us to revise existing interpretations of the American and
British experience, taking into account such previously neglected elements as positive eugenics and the persistence of neo-Lamarckian hereditary ideas.

The French Setting

The French approach to eugenics grew out of a response to perceptions of decline
at the end of the nineteenth century that were shared with many other countries
(Nye 1984; Soloway 1982b, 1982c; Jones 1971; Nordau 1968). These perceptions
had roots deep in the antibourgeois prejudices of the traditional aristocratic elites,
while others attributed a general cultural decline to the new art and literature
being created for the middle- and even working-class audiences of the day. There
were also complaints about the growth of cities and the decline of the simpler life
of the countryside. Even in France rural society was rapidly disappearing by 1900
(Jones 1971, pp. 127–51; Weber 1976). Perceptions of biological decline appeared
relatively late in the century and were reflected in a rising concern over the health
and physical stature of the population. After midcentury, statistics on army
recruits and school-age children suggested real declines in measurable physical
traits (Broca 1866–1867). Of more immediate concern by the end of the century
was the apparent rise in the number of criminals, alcoholics, and those afflicted
with tuberculosis and venereal disease. The novels of Zola and plays of Eugène
Brieux, whose *Damaged Goods* inspired such English-speaking authors as George
Bernard Shaw and Upton Sinclair, took these issues out of the restricted deliberations of scientists and placed them squarely before the general public (Schiefley
1917, pp. 384–85; Shaw 1911). None of these developments was unique to France,
but another problem was perceived both earlier and to a more alarming extent by
the French: a declining birthrate.
    Although a decline in birthrate is now recognized as a common feature of all
industrializing societies, when the phenomenon first became apparent in the mid-
nineteenth century, its manifestation was greeted with surprise and shock. To be
sure, prior to 1850 the population problem, especially of the cities, had been seen
as one of too rapid growth (Chevalier [1958] 1973), but as the birthrate slowed,
the population actually showed signs of leveling off, thus prompting some to
sound the warning of "depopulation." Fueling this concern was the growing military rivalry in Europe since armies drew their manpower from mass conscrip-
tion. On both these counts France appeared to be the country most seriously
affected. The Franco-Prussian War of 1870–1871 had demolished the reputation
of French military supremacy on the Continent, and the population of the new
German Empire was larger and growing faster than that of France. Although the
German birthrate would soon follow the pattern of the French, at the time it
appeared that France was headed for depopulation or at least a diminished base
for military recruitment (Lewis 1962).
    One might expect that the French would have produced some all-embracing,
logical proposals to solve these social ills: historically, it was the birthplace of the

Enlightenment, the French Revolution, and, by extension, the major ideologies that had shaped social reform throughout the nineteenth century. In most cases, however, these ideologies were ill prepared to deal with questions of biological decline. Conservatives and the Catholic church were hard-pressed to admit, let alone explain, these political, economic, and demographic changes. By the end of the century, however, some conservative nationalists had found common cause with the Catholic opposition to birth control in an effort to increase the size of the population and the strength of the French army. The result was the founding and financing of natalist organizations in the 1890s, which became active in the first decades of the twentieth century (Talmy 1962). Liberalism traditionally had opposed any attempt by government to direct the biological or other destiny of any individual, and the French revolutionary tradition was often cited by eugenicists as a reason for their failure to make more progress. One exasperated proponent noted, "In our country of 'liberté' it is easier to overthrow a Bastille or Ministers than established customs" (Baudet 1927, p. 579). Socialists naturally had no such qualms about remaking society, but their interests were so narrowly focused on economic and political questions at the turn of the century that questions of health and demography were ignored.

Others did come forth with proposals for biological regeneration to counter these perceptions of decline in both the quality and quantity of the French population. Some, like the various leagues against alcoholism, tuberculosis, and venereal disease, may seem, at first, far removed from eugenics, but they did view the problems in terms of a threat to all society and called accordingly for government actions to meet the threat. Other proposals were much closer to the eugenics of Galton and Davenport. One of the earliest of these was by Clémence Royer (1830–1902), who called in her preface to the 1862 French translation of Darwin's *Origin of Species* for the elimination of "the weak, the infirm, the incurable, the wicked themselves and all the disgraces of nature . . . which they perpetuate and multiply indefinitely" (p. lxvi). Beginning in the 1880s, the recluse librarian and amateur anthropologist Georges Vacher de Lapouge (1854–1936) outlined an even more elaborate program of "anthroposociology," which included a plan for artificial insemination using "a very small number of males of absolute perfection . . . to inseminate all the females worthy of perpetuating the race" (1896, pp. 472–73). While natalist organizations called the decline in the birthrate a national peril, a controversial but surprisingly strong neo-Malthusian group welcomed it as a sign of "conscientious procreation" that could only improve the quality of the population (Talmy 1962; Ronsin 1980).

Into this welter of proposals for biological regeneration stepped the founders of the French eugenics movement: a group of doctors, scientists, and statisticians whose theoretical framework could tie these diverse elements together and whose social stature gave them a visibility and legitimacy that made them attractive as spokesmen for the many groups concerned with the question of decline. Members of the French Establishment had little reason to suspect them of radically undermining social stability: their emphasis was on the "conservation" of the species. At the same time they also appealed to those wishing to use science for the improvement of the species—regardless of the entrenched traditions that might

be challenged. Finally, they had an institutional base that individuals like
Lapouge had never enjoyed, thus providing them legitimacy, meeting places,
offices, and positions to be filled by their followers.

The doctors and statisticians who founded French eugenics occupied chairs
and offices at the Faculty of Medicine in Paris and the new Statistique générale
of France. What more trusting and established figure was there than the baby doctor concerned with the health and welfare of the newborn? What more sober and
objective authority than the census takers who gathered data on births, deaths,
and occupational characteristics of the French population? The new idea that they
used to draw the diverse groups together was the concept of puericulture, defined
by its champion Adolphe Pinard (1844–1934) as "knowledge relative to the reproduction, the conservation and the amelioration of the human species" (Pinard
1899). His chair of obstetrics at the Faculty of Medicine gave him an additional
institutional base because the appointment also made him head of the Baudel-
ocque Clinic, where for a quarter century before World War I he and his students
worked on problems of pregnancy, prenatal care, infant mortality, and especially
those problems attributed to the effects of alcoholism, tuberculosis, and venereal
disease. Hence, both the natalists and the social hygienists could rightly count
Pinard as one of their own. Along with Lucien March (1859–1933), head of the
Statistique générale of France, Pinard sat on government commissions on
depopulation, wrote articles, and lectured widely to the French public on puericulture (Schneider 1986). When Pinard called for "conscious and responsible procreation" as the solution to many of the problems causing the degeneration of the
French population, it was difficult to find anyone who disagreed. In its May 1912
issue, the editor of the radical journal *Le Malthusien* noted that Pinard's ideas
"are the same ideas that we constantly propose"; he mockingly praised "the courage of Dr. Pinard who has no fear of proclaiming this in the press and in the halls
of the Academy of Medicine" (p. 332).

Producing healthy babies is as uncontroversial a goal as can be imagined, so
the rapid and widespread acceptance of puericulture is hardly surprising. Indeed,
puericulture would have been little more than a call for prenatal care and breast-
feeding without its hereditarian underpinnings. It was soon clear, however, that
Pinard envisioned the well-being of the infant not as simply deriving from the
health of the pregnant mother, but as intertwined with the health of previous generations and those yet unborn. Pinard's appreciation of heredity was evident as
early as 1899, when he advanced the idea of "puericulture before procreation," a
phrase to describe the workings of heredity and "the dominant influence of the
procreators" (p. 144). In addition, Pinard believed in the hereditary transmission
of acquired characteristics, which meant that the heredity of newborns was subject to all sorts of environmental influences, both past and present. The neo-
Lamarckian heredity of puericulture was attractive to a wide variety of specialists
concerned with infant health because it gave their work an importance not only
for the present, but for subsequent generations as well.

Neo-Lamarckism profoundly influenced the nature of eugenics in France.
Throughout the entire history of the movement, there were eugenicists concerned
about the effect of degenerative environmental and social influences because they

would be inherited; it also made them presume that the inherited quality of the population could be improved by removing such influences. This made cooperation possible with such other groups in France as the influential natalist movement and the various leagues against the "social plagues." The crucial role of neo-Lamarckism was no accident, for, in a larger sense, French society as well as French eugenics got in neo-Lamarckism precisely the hereditary theory it desired because its optimistic justification for health and social reform was very compatible with the political and social philosophy of the French Third Republic (Hayward 1959, 1961; Clark 1984, pp. 67–75). With reinforcement from the broader intellectual and political community, neo-Lamarckism became that much harder to challenge within the scientific community.

The French Eugenics Society

In the short run their neo-Lamarckism placed Pinard, March, and others in touch with a wide variety of people, all sharing the common goal of improving the human race. Hence, when the English Eugenics Education Society issued a call for an international congress to be held in London in July 1912, the French responded with an organizing committee of over forty members, the largest outside of Great Britain (Schneider 1982). Nineteen Frenchmen attended the congress and four gave papers. Pinard was too ill to attend, but his paper was read for him. It was entitled "Considerations générales sur la 'puericulture avant la procreation'" (Pinard 1912), but the English translated the latter phrase as "education before procreation," thus indicating their unfamiliarity with the term puericulture. It was not just the French scientific community whose attention was attracted by the London eugenics congress. Parisian papers gave it daily coverage while it was in session, and at least one newspaper sent a special correspondent to report on the meetings with an eye to educating the public about the new ideas. "The Eugenics Congress Studies the Laws of Heredity" and "What Is Eugenics?" were two of the headlines in *Le Journal,* but other coverage by the *Journal des Débats* appealed to French national pride by highlighting French participants at the congress. This included an account of the address by Senator Paul Doumer, a natalist leader and future president of the republic, who expressed his hope that the next international congress would be held in Paris.

The London congress was the direct inspiration for the founding of a formal eugenics organization. The French Eugenics Society was created in December 1912 by Pinard, March, and over a hundred other founding members. Its first president was Edmond Perrier (1844–1921), the head of the Museum of Natural History and one of the few biologists active in the movement. Yves Delage (1854–1920), the founder of *Année Biologique* who was freindly with Charles Davenport, lent his name as an honorary president of the French organizing committee for the London congress, but neither he nor any active research biologists participated in the eugenics society. Although Lucien Cuénot (1886–1951) and Jean Rostand (1894–1977) wrote on eugenics, neither was a member of the organized movement. Cuénot, who was virtually the only French biologist doing genetic research before the thirties, presented a paper at the 1921 international eugenics congress in New York entitled "Genetics and Adaptation" (Cuénot 1923), but his primary goal remained the resolution of the Lamarckian-Mendelian dichotomy rather than any direct eugenic applications of hereditary theory. Rostand, in addition to his other writings, translated Morgan's drosophila work, as well as H. J. Muller's *Out of the Night,* into French, but he remained noncommittal as to the viability of eugenics and only protested late against the extreme Nazi racism attached to it (Rostand 1939). Perrier's earlier career made him one of the foremost spokesmen for neo-Lamarckism in France (Conry 1974; Stebbin 1974). He had not hesitated to draw rather far-reaching political conclusions from his observation of "association" and "solidarism" in living organisms that were contrary to the assumptions of competition in nature that social Darwinists used to support their ideas (Perrier 1881). But by the time he assumed the presidency of the eugenics society, he had moved away from research, having served as administrative head of the natural history museum since 1900.

Anthropologists were more in evidence among eugenicists, especially criminal anthropologists like Georges Papillaut of the Ecole d'anthropologie, who became a vice-president of the eugenics society, and Georges Paul-Boncour, also a professor at the Ecole d'anthropologie and director of the Institut médico-pédagogique at Vitry. However, the majority of physical anthropologists at the Ecole d'anthropologie kept their distance from the eugenics movement, largely because of an earlier dispute with Vacher de Lapouge and his ostracism from French anthropology (Manouvrier 1899). Likewise, statisticians did not show a great interest in the society after it was formally organized, with the important exception of Lucien March, who became the secretary-treasurer.

By far the most active eugenicists came from medicine and physiology, fields that accounted for over half of the founding members of the society. This included Pinard, who succeeded Perrier as president of the French Eugenics Society in 1921 and was probably the most respected obstetrician in France during the first decades of the century (Dumont and Morel 1968, pp. 74–75). Thanks to that reputation Pinard ran successfully for public office after his retirement from the Faculty of Medicine, serving as a deputy in the French National Assembly from 1918 to 1928. Pinard's successor as president was Eugène Apert (1868–1940), also president of the French Pediatric Society. Charles Richet (1850–1935), a vice-president of the society, was a 1913 Nobel laureate and holder of the chair of physiology at the Faculty of Medicine, as well as being a popularizer of science through his journals, the *Revue Scientifique* and the *Revue des Deux Mondes.*

In choosing to call themselves the French "eugenics" society, the organizers made clear their identification with the ideas and talks they had heard in London. Yet, in the founding statutes of the organization published in the first issue of its journal, *Eugénique,* one can see the goals of Pinard's puericulture virtually unchanged: "reproduction, preservation, and improvement of the species," with particular attention to "questions of heredity and selection in their application to the human species, and questions relative to the influence of environment, economic status, legislation and customs on the quality of successive generations and on their physical, intellectual, and moral aptitudes" (1912). Thus, neo-Lamarckism was an operational presumption in the society's statutes.

The French Eugenics Society was very active in the short time it functioned before the First World War. Monthly meetings were held for a year and a half at the Faculty of Medicine. Its journal, *Eugénique,* appeared monthly and published papers given at meetings, reviews of the literature, and summaries of the work of eugenicists in other countries. A membership drive was begun, members wrote articles for a wide variety of other publications, and guest speakers were brought in to address the society, including the neo-Lamarckian C.-W. Saleeby from the English eugenics society.

At the outset, then, the French Eugenics Society developed organizationally like the comparable societies in England, Germany, and the United States. Moreover, although the French society could be distinguished from those in Great Britain and America by its emphasis on puericulture and neo-Lamarckism, nonetheless, at least in the period before World War I, it was very similar to them in the broad range of ideas and concerns it encompassed. For example, natalist eugenicists such as the deputy Adolphe Landry definitely stressed the need to improve health and environmental conditions to "preserve and improve the species," but men like Jean Laumonnier and Charles Richet stressed the importance of diminishing the number of undesirable elements of the population. In an article for the *Revue Bleue* that asked "What Is Eugenics?" Landry stated that sterilization and marriage restriction for individuals with defects were "repugnant to our need for liberty and our delicate individualism." Instead, he argued for an emphasis on the positive goal of trying to improve these unhealthy elements as well as "fortifying elements of mediocre quality and preserving from evil those that are healthy" (Landry 1913, p. 782).

In contrast, Laumonnier wrote an article at approximately the same time for the *Larousse Mensuel* that placed first emphasis on eliminating the undesirables. Although he agreed that the healthy elements should be encouraged to reproduce through programs that Landry and Pinard championed—puericulture before and after birth, the work of temperance societies, and construction of better housing—he placed most of his emphasis on negative measures that, "like military service, quarantines or mandatory vaccinations entail certain restrictions on individual liberty" (Laumonnier 1912, p. 455). Among these he included control of immigration, marriage restriction, and sterilization. This negative position was spelled out even more extensively in Charles Richet's *Sélection humaine,* written before the war but not published until 1919. The tone of the book was clear from the outset and in accord with the harshest of English or American rhetoric: "The fact of nature is the crushing of the weak. The fact of society is the protection of the weak. Thus, the social state vitiates the grand law of selection which is essentially the survival of the strong" (Richet 1919, p. 17). Richet was no less frank in describing the various measures necessary to bring society in line with the laws of nature, including segregation of the races and ending care for the "mentally deficient." He also proposed the use of sterilization to prevent such people from procreating, but as a practical expedient Richet admitted a willingness to accept marriage regulation until such time as public sentiment found sterilization more acceptable (1919, pp. 89–93, 164–65).

The Twenties

World War I had a profound effect on French eugenics. With the outbreak of war, the French Eugenics Society suspended its operations, and although work resumed in 1919, in many ways the organization was never the same again. Meetings were held only twice a year, the society's journal appeared less frequently, and overall membership dropped. Perhaps even more important, the naive spirit of optimism about improving the human race was gone. Something of the intellectual shock produced by the war is reflected in the following remarks made at one of the first postwar meetings of the eugenics society by Georges Papillaut and reprinted in *Eugénique:*

> The concrete, living world does not obey the fantasies of reformers. It has strict and even harsh laws upon whose discovery we must try to focus our efforts. The truth is, it is the only way to be useful to our country and the generations that follow. Do not begin in another form the humanitarian dreams of 1914. They are too costly! (1914–1922, p. 247)

From its beginning, however, eugenics had been a reaction to perceptions of decline and fall, so that in another sense the war made the work of eugenicists appear more necessary than ever. In France the most obvious effect of the war had been the loss of life, with over 1.3 million French soldiers killed. When added to the prewar fear of depopulation, the result was an even greater emphasis on positive eugenics and the abandonment of proposals to eliminate the birth of "dysgenics." The new watchword of the French Eugenics Society was "social hygiene," a concept that had a long prewar history dating back at least to the last quarter of the nineteenth century (Sicard 1927a, p. 44). It was vague, however, and was applied in many different contexts. In fact, the most frequent definitions were phrased simply in terms of the diseases it sought to combat, especially the "social plagues" of tuberculosis, alcoholism, and venereal disease. For example, the 1904 founding statutes of the Alliance d'hygiène sociale used precisely this definition, and as late as 1930 the Larousse dictionary defined social hygiene as "the most appropriate methods to limit the ravages of social diseases such as tuberculosis, cancer and syphilis and to combat plagues such as alcoholism." One of the earliest to lend scientific substance to the idea of social hygiene was Emile Duclaux, director of the Pasteur Institute, who argued in a 1902 book that there was a broader theoretical dimension implied by the term, which "envisages illnesses not in themselves but from the social viewpoint; that is, from the point of view of their repercussions on society and the ability of society more or less to preserve itself and fight them" (p. 5). Here was the medical prescription for the overall health of society. Neo-Lamarckism already provided a theoretical link between social hygiene and eugenics to the extent that diseases were seen as hereditary. The First World War made such an association even more logical by its emphasis on the subservience of the individual to society as a whole.

The first organized activity of the French Eugenics Society was a conference entitled "The Eugenic Effects of the War." The talks given by key members of the society demonstrated the new emphasis on social hygiene. The long-standing Lamarckian Edmond Perrier specifically questioned the effectiveness of negative measures championed by eugenicists in other countries. In their place he urged that

> the environment itself must cease to be a cause of degradation. National vices like alcoholism, lack of personal care which propagates contagion, and overindulgence of all kinds, must be unmercifully proscribed. Homes and cities, all that touches communal life must be the object of carefully coordinated attention. (1922, p. 20)

Eugène Apert's address, "Eugenics and National Health," noted that there had been a rise in the incidence of syphilis and tuberculosis during the war and concluded, "The population after the war is in such condition as to make more necessary than ever the health measures that had already been called for before the war" (1922, p. 60).

The conference was also indicative of a new organizational strategy of the French Eugenics Society that aimed at reaching the intellectual, scientific, and political decision makers of France. This was a change from its prewar work, which most closely resembled that of a learned society. Postwar activities were more outwardly directed toward reaching the educated public through conferences, lectures, and public university courses. The eugenics society also helped support or start organizations with similar but more limited goals. For example, in 1922 André Honorrat, a senator and former minister of education, approached the French Eugenics Society with a proposal to create a Committee of Union against the Venereal Peril. It was accepted, and a joint board of directors was created for the new organization. Support was also given to a series of public lectures on social hygiene at the Sorbonne, presented every spring by the public health physician Just Sicard de Plauzole. In the aftermath of the war, anthropologists at the Ecole d'anthropologie founded the International Institute of Anthropology, in order to foster international cooperation. One of its sections was devoted to eugenics, and the local national committee for France was virtually identical with the French Eugenics Society. Although it met infrequently, this organization served as a formal link between eugenicists and the Ecole d'anthropologie that would later prove to be important.

Of more immediate concern to the new agenda of the eugenics society was social hygiene, which quickly became a subject of great public attention. Although eugenicists could not claim credit for single-handedly making it such an important issue, they were certainly in line with a broad current of postwar sentiment in France. With the creation of a National Office of Social Hygiene in 1924, however, the direction of the social hygiene movement was beyond the control of eugenicists. Of course, many members of the organizational council of the new office were members of the eugenics society, including Honnorat (who was named president); Pinard and his son-in-law Raymond Couvelaire who succeeded him to the chair of obstretrics at the Faculty of Medicine; Edouard Jeanselme; and Sicard de Plauzole. However, the budget of the office was quite large and its constituency very broad—ranging from the long-established Venereal Disease Prevention Service and several visiting nursing schools to a new Colonial Social Hygiene Service and organizations working on cancer, mental health, and typhoid fever. As a result, the long-range goals of eugenicists were quickly lost in the effort to meet the immediate needs of particular groups supported by the office (Schneider 1982, pp. 282–83).

With their major issue of social hygiene preempted by other organizations, the members of the French Eugenics Society faced a crisis of identification in the mid-twenties. This coincided with an organizational crisis that was partly the result of a generational shift in its membership. Many of the founders of the French Eugenics Society either died or retired from active professional life during this period. For example, Louis Landouzy, the first vice-president of the French Eugenics Society and dean of the Faculty of Medicine, died in 1917, the same year that Frederic Houssay, another society officer, died. Perrier, the first president of the eugenics society, died in 1921. Charles Richet retired from the medical faculty in 1925, and Lucien March, secretary-treasurer of the society, also retired from the direction of the Statistique générale of France in the early twenties. Pinard, already retired from the Faculty of Medicine since 1914, ended his political career in 1928. Although many in the new group rising to prominence in French eugenics—such as the pediatricians Eugène Apert and Georges Schreiber, who became president and secretary-treasurer in the twenties—shared the training and outlook of the founding generation, others like the public health doctors Réné Martial and Just Sicard de Plauzole only began their careers at the turn of the century and had very different backgrounds and views from the earlier generation.

Related to this change in personnel was a change in the status of the French Eugenics Society itself that had far-reaching consequences in the following decade. As noted before, the society, through its founders, had close ties with the Faculty of Medicine in Paris where it held its regular meetings. In addition, there was sufficient membership in the organization to support the printing and distribution of a journal. After 1926, however, the cost of the journal exceeded income from membership subscriptions, and it was decided to publish the articles and minutes of the society in *Revue Anthropologique,* the journal of the Ecole d'anthropologie in Paris (*Revue anthropologique* 1927, pp. 225–26). This was possible because of the cooperation of the eugenics society with the International Institute of Anthropology, which also used the school's journal as its official organ. In addition, the French Eugenics Society merged with the eugenics section of the French national committee of the International Institute of Anthropology, but meetings continued to be held on an irregular basis, only once or twice a year. This left the *Revue Anthropologique* as the only continuing organizational foundation for eugenics in the remainder of the twenties and the thirties. Rather than signaling a decline or even an end to eugenics in France, the late twenties and especially the thirties witnessed an upsurge in writing and discussion of eugenic measures. As will be seen, although the lack of organization hindered the practical implementation of a eugenics program, it allowed all sorts of people to enter the debate with new ideas.

A good example of the continuing influence of eugenics after the reorganization of the French Eugenics Society was the campaign for a premarital examination law beginning in 1926. This campaign was the first concrete proposal for legislation by French eugenicists, and ultimately was enacted into law. Moreover, it marked a shift away from the positive, social hygiene program toward negative measures designed to discourage procreation of the unfit. Both of these trends were to continue and grow in the thirties as the call for other eugenic measures entered the national political debate.

The proposal for a physical examination before marriage came after a conference on the subject sponsored by the French Eugenics Society in May and June of 1926. In his opening remarks to the first meeting, Pinard announced his intention of introducing a bill in the Chamber of Deputies that read, "Every French [male] citizen wishing to marry or remarry can be entered in the civil registry only if he has a medical certificate dated from the day before, establishing that he has contracted no contagious diseases" (*Eugénique* 1922–1926, pp. 267–68). The debate over the proposal was extensive, even among eugenicists who raised such questions as whether it would be sufficient to have only the male examined, whether the other spouse should see the results, and whether marriage should be proscribed depending on the results of the exam. Although the bill was reported on favorably by parliamentary committee in 1927, it was delayed by counterproposals and continuing disagreements among doctors and eugenicists (Schneider 1982).

One important result of the proposal for the premarital examination was that it tested seriously for the first time the broad coalition of support that eugenics had enjoyed in France since its beginnings. Indeed, the very lack of organized opposition up to that point is perhaps the most eloquent testimony to the effectiveness of the milder policies of French eugenicists in securing allies. Of course, it is more than a coincidence that these serious questions were first raised when the eugenics society proposed its first legislative action. The alliance survived the initial questions raised by the premarital examination even among such important groups as the Catholic church and natalist organizations. It took the outside stimulus of the 1930 papal encyclical and calls for more extreme negative eugenic measures within France during the thirties before an open break occurred.

Opposition to Eugenics

When a premarital examination was first proposed by the French Eugenics Society, church and natalist organizations had not opposed it, although some were wary of how the procedure might be implemented. For example, Edouard Jordan, a professor of medieval history at the Sorbonne who had been a prominent member of the postwar natalist congresses and the Association of Christian Marriage, accepted the basic justification of the measure proposed by the French Eugenics

Society. In a 1926 article he pointed out that it would be "unreasonable to think only of numbers and not be concerned about the quality" of the population. The premarital exam, Jordan agreed, appeared to offer a common ground for cooperation in that "everyone could agree that children should be born under the best of circumstances." He did, however, caution his fellow natalists that "one can draw very different conclusions from eugenics, and according to the manner in which it is understood and practiced, it could be a powerful ally or redoubtable adversary in the campaigns that we will be pursuing" (Jordan 1926, p. 1).

As an example of his concerns, Jordan offered an institute in Hamburg, described by Georges Schreiber in his talk at the conference on the premarital exam as a marriage counseling center. Jordan's information, however, was that only about 10 percent of the institute's work was with couples engaged to be married; the other 90 percent came solely for birth control information. Jordan also wondered about what was to be done with couples not passing the examination. But his biggest surprise was Pinard's announcement that he intended to propose a law that would make the premarital exam mandatory. Although Jordan stopped short of opposing it, he noted that all along he had presumed a voluntary exam. A mandatory certificate which had to be presented to government authorities was to him "a rather different hypothesis."

Church leaders also participated in the ongoing debate about the Pinard proposal. Jordan spoke at a conference on the subject sponsored by the prestigious Committee of Social and Political Studies in 1928; at the same time, church officials also joined in offering their ideas concerning eugenics in general and the premarital exam in particular. As late as April 1930, René Brouillard, a Jesuit theologian wrote, "In principle, Catholic morality does not condemn all eugenic science." Differences occurred, he said, when one "passes into the realm of practice and forgets that man the animal is not the total man." Of the two most commonly mentioned eugenic measures, he found sterilization "absolutely repugnant to Catholic morality." On the regulation of marriage, however, he had a more open attitude. A premarital physical examination seemed a good idea to Brouillard in the overall practice of marriage counseling, but making it mandatory raised a question since he felt that a medical exam should not be sufficient grounds for "a legal interdiction of marriage" (1930, pp. 115–16).

Although Brouillard answered Jordan's question about a mandatory exam law negatively, he was not reluctant to discuss eugenics or find a way to incorporate aspects of it into church teachings, and in this respect he was typical of French churchmen. In fact, the following month, May 1930, the Association of Christian Marriage held a national congress in Marseille devoted entirely to the subject of the church and eugenics. The attitude of most at the congress was summarized in a final address by Msgr. Dubourg, the archbishop of Marseille, who concluded:

> If the goal of the new science [eugenics] is, as its name indicates, to assure good offspring, it can only inspire our sympathy and find in Christian morality an auxiliary, even a very precious guide, because we profess that if God commanded man to multiply, He did not wish him to multiply poorly. (p. 224)

To be sure, the congress soundly condemned birth control as a eugenic measure, but in his preface to the published proceedings of the congress, Jordan still welcomed eugenics as "an invitation to reflect upon the responsibilities involved in procreation." The words could not have been better chosen by Pinard. Jordan now even expressed support for an obligatory exam, like the alternative proposed by the deputy Duval-Arnould, which contained provisions for the exchange of results by spouses. This sentiment was echoed by Jean Arnould, former chief of gynecology at the Faculty of Medicine in Marseille who spoke on the premarital exam to the congress on the church and eugenics. Arnould found the proposed mandatory exam law "morally, socially and eugenically" advantageous (p. 124).

Thus, on the eve of the papal encyclical, the French church still expressed a very conciliatory attitude toward eugenics, based largely on an accommodation over the premarital examination. French churchmen lauded the "discretion" of French eugenicists who distinguished their program from the much harsher "Anglo-Saxon eugenics" advocated in England and the United States (Jordan 1926; Pietri 1930, p. 73). Jordan himself summed up this position in an extraordinary work entitled "Eugenics and Morality" published in 1931 but written just before the papal encyclical of December 1930. It was clear to Jordan that the negative eugenic measures of the Americans and others, especially the sterilization laws and advocacy of birth control that were being discussed more and more in Europe, were pushing eugenics in a different direction from the positive program that had been emphasized by the French Eugenics Society since its beginning. In fact, the work almost amounted to a 175-page plea to French eugenicists to return to their original track. "It would continue its legitimate warnings against unfortunate births, but it would concentrate again more on the improvement of the milieu, on the progress of medicine, on general hygiene, urban planning" (Jordan 1931, p. 173).

Jordan was criticizing the Americans and British, to be sure, but his real targets were those Frenchmen who were sympathetic to the Anglo-Saxon ideas. The most complete statement of this position in French (and the one most frequently cited by Jordan) was Charles Richet's *Sélection humaine*. Although the book appeared more than a decade earlier, there could not have been a more inauspicious time for its dramatic negative eugenics program to have been proposed. The early twenties was precisely when the positive program of social hygiene reached its peak in an effort to recover from the serious loss of life in the war. By 1930, however, the depression had changed things considerably, and it was not surprising to see Richet's views cited by many of the new generation of eugenicists.

Jordan found virtually none of Richet's proposals acceptable, even arguing against his compromise position on marriage restriction. Leaving the moral question aside for the moment, Jordan argued on pragmatic grounds, "Which would be a better course of action: forbid marriage by alcoholics or revoke the rights of distillers and limit the number of bars?" Jordan juxtaposed the neo-Lamarckian presumptions of the founders of French eugenics with the existing mood of the times:

> Take a poor family, because it is assumed by many eugenicists that a poor person is a degenerate. Raise the wages, find the family healthy lodging, family subsidies and try to raise the standard of living. Won't their health have a chance to be maintained and improved? But does a society which wants to do nothing effective against alcohol or degradation or slums or other social plagues, have

the right to avenge itself, in a way, on the victims of its own inactions, and to have recourse to the contemptuous and harsh methods such as sterilization, on the pretext that it is simple and final? (1931, p. 172)

It is doubtful that Jordan hoped to dissuade proponents of such negative eugenic measures, but a more realistic hope may have been to plead with the Catholic church hierarchy to leave room for the positive eugenics that had been championed by the French. When the encyclical was published, its contents clearly showed that Jordan had failed.

By most accounts, Pius XI's papal bull of 31 December 1930, "Casti Conubii" [On Christian marriage], was aimed primarily at the Anglican bishops' endorsement of contraception at their Lambeth conference earlier in the year (Noonan 1965, pp. 424–25; Soloway 1982a, p. 254). The church also took advantage of the occasion to condemn several other practices that were increasingly advocated in the name of eugenics. This included the American state sterilization laws that were applied with greater frequency after the U.S. Supreme Court confirmed their constitutionality in 1927. Similar laws were under discussion or newly legislated in European countries as well. The encyclical directly condemned these attempts to sterilize "defectives" by legislation. The effect would be

to deprive these of that natural faculty by medical action despite their unwillingness; and this they do not propose as an infliction of grave punishment under the authority of the state for a crime committed, nor to prevent future crimes by guilty persons, but against every right and good they wish the civil authority to arrogate to itself a power over a faculty which it never had and can never legitimately possess. (Pope Pius XI 1930, pp. 22–23)

The position of the church on sterilization was hardly surprising, and neither was the encyclical's condemnation of abortion on any grounds, "social or eugenic" or even "medical and therapeutic," in the words of the letter, "however much we may pity the mother whose health and even life is gravely imperiled in the performance of the duty allotted to her by nature." It went on, however, to condemn as well those who "put eugenics before aims of a higher order, and by public authority wish to prevent from marrying all those whom, even though naturally fit for marriage, they consider, according to the norms and conjectures of their investigations would, through hereditary transmission, bring forth defective offspring." Thus, the church was also opposing the use of premarital examinations—the very proposal that French Catholics had seen as a possible meeting ground for accommodation with the eugenicists. As if to remove any doubt, the Holy Office issued a decree on 21 March 1931 that "declared false and condemned the theory of eugenics, either positive or negative," and disapproved of the means it proposed "to improve the human race, neglecting the natural, divine, or ecclesiastical laws which concern marriage or the rights of individuals" (Brouillard 1931, p. 441).

Most French churchmen and their allies in natalist organizations took the encyclical as a signal to end the equivocal position they had held on eugenics from the beginning. For example, writing in the Jesuit review *Etudes,* René Brouillard credited the encyclical and subsequent decree as being the catalyst for his response to the increasing publicity given in recent years by the press to "eugenic views, even the most radical, without the most elementary reserve and with a sympathy that is out of place in [such a] publication of high moral principles as the *Débats*" (Brouillard 1931, p. 454). The reference was to a 22 January 1930 article by Henri Varigny in the prestigious newspaper *Journal des Débats*. In his lengthy two-part article, Brouillard welcomed the papal condemnation of these practices—abortion, birth control, sterilization, marriage restriction—which he noted were against church doctrine and too drastic to be justified by the uncertain scientific knowledge of genetics (1931, p. 578–82).

At the end of the second part, however, the author indicated that he was not willing to give up completely the idea of eugenics, at least the overall goal that it sought to achieve. Brouillard attempted to make the case for retaining a notion of eugenics that was different from what he called the "Anglo-Saxon, Galtonian" version. It is a remarkable testimony to the power and attraction of the idea of eugenics that even in the face of the new church edicts, Brouillard still sought to define a "Catholic eugenics": one that was a "eugenics of life" as opposed to the "eugenics of death" as preached in the United States and England (1931, pp. 597–600). Reading between the lines of the encyclical and emphasizing what was not condemned rather than what was, Brouillard spelled out the features of the new eugenics:

> sanctification of marriage and the duties of spouses; attention to morality and health at the time of conception and birth; action by the state, associations and the church against public immorality, social diseases, alcoholism, slums, etc. to develop the economic well-being, general hygiene, puericulture, healthy dwellings, the prosperity of families . . . all of which would constitute a moral, family, social and Christian eugenics. (1931, p. 597)

Thus, Brouillard continued to hope for a version of eugenics in France completely purged of the harsher negative elements with which it had so long coexisted.

Of course, such was not to be the case. Individuals like Sicard de Plauzole and new organizations like the Association of Sexological Studies continued to express their views and declare themselves even more openly in support of the measures condemned by the church. Soon the major natalist organizations joined with church organizations in the attack. In January 1931 Fernand Boverat of the National Alliance against Depopulation wrote the first article in the organization's journal openly attacking eugenics (Boverat 1931). This article was in striking contrast to a 1928 review of the first volume of Marie-Thérèse Nisot's *La question eugénique dans divers pays,* which was called by the natalist journal "an indispensable repertory for all those . . . interested in the present future of the race" (Nisot 1927). When the Association of Sexological Studies adopted a program in 1933 calling for a mandatory premarital exam and the creation of public clinics to give advice on contraception and perform sterilizations and abortions, the reaction of the natalists was equally vigorous in opposition. This was particularly necessary, Boverat noted, because it was proposed by a serious group that "contains among its officers and members a large number of distinguished personalitites belonging especially to the medical world" (Boverat 1933).

The opposition obviously did not silence or overwhelm these proponents of the harsher eugenics of the thirties. However, the papal encyclical and other opposition were indicative of the end of the old coalition that had founded the French eugenics movement before the war. During the thirties French eugenics was to be very different from what it had been in earlier decades.

The Thirties and the Population Question

If World War I was an important turning point in the history of French eugenics, in many ways the thirties saw an equally significant shift in the movement. Although the most obvious reason for the change might appear to be the economic depression, there were also internal changes in personnel and organization involved, as well as the papal encyclical and the work of eugenicists in the United States and Nazi Germany. Regardless of origins, the result was a shift in French eugenics toward a greater emphasis on harsher, usually negative eugenic measures such as the premarital examination, immigration restriction, and birth control. Discussion even began about the use of sterilization. Although most of these ideas had surfaced in France during preceding years, with the exception of the premarital exam they had not been pursued or advocated in an organized fashion by the majority of French eugenicists.

Of all these measures, the one with broadest support was the proposal for a law requiring a physical examination before marriage. A campaign began in November 1926 after Pinard introduced his bill in the Chamber of Deputies to require an attestation of "no appreciable symptoms of contagious diseases" before a couple could receive a marriage license. Although action on Pinard's bill was delayed while eugenicists argued among themselves over the best way to clarify the vague language, it was revived again in the early thirties by Justin Godart, a former health minister, who proposed a modified form of the bill in the French senate. Thereafter, the idea was delayed only because it became entangled with the the the call for a more general *carnet de santé* popularized by Louise Hervieu and others in the midthirties, which proposed that a health card be required for every individual from cradle to grave for presentation at appropriate stages in one's life, including marriage (France. *Journal Officiel* 1938). Support for this health card was not just a sympathetic response to the plight of Hervieu, a well-known painter and writer whose syphilis contracted at childbirth was only manifested later in life. The dramatization of her story in two novels, *Sangs* and *Crime,* became bestsellers during the thirties, but the scientific and governmental support for her campaign owed much to the shift in opinion that the eugenicists had correctly anticipated and exploited in proposing the premarital exam. Success eventually came when a Ministry of Public Health decree of 2 June 1939 made the health card mandatory, although the law defined it as "a strictly personal document which no one could require to be divulged." This latter restriction was removed in November 1942 by the Vichy regime as part of the same decree that established a separate mandatory premarital exam (Schneider 1982, pp. 287–90). De Gaulle's

Provisional Government retained the exam after the war, and it remains the law of France to this day.

Another major eugenic issue in the thirties was immigration. Eugenics society members first voiced concern in the 1920s about the large influx of workers and refugees to France in the wake of postwar political and economic dislocation, but their attention had been focused primarily on the physical condition of those coming to France. In the thirties, however, a full-blown immigration restriction program was advocated, with eugenic warnings of biological decline from inter-mixing of incompatible races. An even more radical change for French eugenics in the thirties was the call to liberalize laws on contraception, which went directly against the previous alliance between French eugenicists and natalists. This fundamental shift also produced the first serious discussion of sterilization of the "unfit," which had previously been dismissed out of hand as being unacceptable to French mores (Landry 1913). In the face of these developments, those supporting the milder eugenics of the twenties did not simply fade away. In fact, they were strengthened in the late thirties when the French Left finally entered the debate on eugenics, supporting a family and public health policy in the tradition of Pinard and the *puericulteurs*. Hence, eugenics managed to become a part of a wide range of political ideologies in France on the eve of World War II.

The single most important reason for the emergence of the harsher negative eugenics program in the thirties was the coming of the Great Depression. It influenced events both within and outside France that prompted specific new eugenic proposals, but, even more important, it also dramatically changed the general climate of opinion and mood of the times, thus making negative eugenics more acceptable to the public. Although for a short time France was spared its effects, by late 1931 the unemployment and economic decline that had been seen in other countries arrived in France. In England or America such developments undercut eugenic arguments that the conditions of lower, poorer classes were the result of biology, for how could their numbers be multiplying faster than their birthrate (Ludmerer 1972, p. 127)? In French eugenics there was no such contradiction. One reason, of course, is that neo-Lamarckians presumed the opposite relationship between poverty and biology. The lower classes were worse off biologically because they were poor. An increase in their numbers only raised the fears of more rapid biological decline because of the effects of deteriorating environment.

The conclusions drawn by many French eugenicists from the depression were Malthusian ones—that is, the problems were the result of demography and economics. In words that would have made the English parson smile (albeit grimly), they described a world with too many mouths to feed and too few resources. A new word entered the French eugenic vocabulary: *overpopulation,* the cause not only of economic woes, but of wars as well. As circumstances would have it, Europeans had a convenient example at hand beginning in 1931 when Japan invaded Manchuria. The image of the teeming "Asian masses" is an old one in Europe, but a book with that title written by Etienne Dennery on the eve of the Manchurian war expressed it in contemporary demographic language that was cited throughout the thirties (Toulouse 1931b; Haury 1934; Richet 1935, pp. 11–22; Sicard 1935). For example, Gaston Bouthol's *Population dans le monde,* written in 1935, criticized the Japanese preoccupation with population in words that could just as easily have applied to the French natalists of the twenties:

> They are intoxicated with the dizziness of figures. "Tomorrow we will be one hundred million," is the theme of exaltation which is found in the Japanese newspapers. No matter that the difficulties and miseries will grow in proportion, the essential thing is that the numbers make them proud. (pp. 238–39)

Bouthol then repeated Dennery's observation of what such a growth of population brings:

> To whoever has traversed these overpopulated countries, it is incontestable that overpopulation is a cause of their malaise, disorder and fundamental weakness. The abundance of the miserable, the unemployed, and those without skills makes a country anemic rather than reinforced. . . . The number of inhabitants does not necessarily increase the power of a country if it diminishes the output of each inhabitant. (Dennery 1930, pp. 87–88)

Bouthoul's own conclusion was an explicit attempt to view demographic questions in a more balanced light, and he warned, "Those who maintain that the amelioration of humanity depends on the uninterrupted growth of the population are as far from the evidence as those who see restriction as the essential remedy of all past and present difficulties" (1935, p. 235).

This new view of the population question in France was possible in part because of new statistics from recent years that showed France's two rivals, Britain and Germany, with a precipitate drop in their birthrates. In fact, by 1932 the French rate of 17.3 births per thousand was actually higher than Britain's 15.3 and Germany's 15.1. This general leveling off of the population in Europe was therefore doubly welcome to French observers concerned with France's relative position in an overcrowded world. The respected economist Charles Gide noted that "the density of the population in Europe appears to have attained almost the maximum compatible with its present-day resources" (Sicard 1935, p. 81). To those who were tempted to see France in a position of actual advantage since it had the lowest population density in western and central Europe, others cautioned, "Populations must be proportional to the resources and not just the surface of the territory" (Toulouse 1931b).

It did not take long for the effects of this new view to be felt by the natalists. In the 1932 elections to the Chamber of Deputies, the natalist group lost two-thirds of its members, including the leader, Adolphe Landry (Talmy 1962, 1:179–214). For the first time, authors wrote critically of the whole technique of projecting long-range future population statistics from limited short-term trends. This, of course, had been one of the major reasons for the wave of fear about depopulation in France at the end of the nineteenth century. Bouthol devoted a whole chapter to "demographic forecasting," in which he criticized such predictions as the common view in 1890 that Germany would have one hundred million inhabitants by 1920 while France would only have thirty million; the actual figures in the midtwenties were forty million for France and sixty million for Germany (Bouthol 1935, p. 234). By 1935 members of the Academy of Medicine were being lectured "On the Pretended 'Depopulation' in France." The premise of the author, Alexandre Roubakine, who had formerly been attached to the Hygiene Section of the League of Nations, was that natalists had erred focusing their attention solely on birthrate. For, although the rate in France and all Europe, for that matter, was dropping, the mortality rate was dropping even faster. In fact, Roubakine prophetically noted:

If there is a decline in the birthrate in Europe, its population is, nevertheless, growing more rapidly than that of Asia. Moreover, since the habitable spaces are much more restricted in Europe than Asia, it is the expansion of the white race of Europe which presents the greatest danger for the world today. (1935, p. 143)

Such a dramatic change in perceptions of population growth had an obvious effect of softening attitudes toward contraceptive measures. By the midthirties, for example, the use of birth control was openly discussed by the French Association of Women Doctors. New organizations like the Association of Sexological Studies called for the repeal of the 1920 ban on the sale and advertisement of contraceptive devices, as did established organizations like the League for the Rights of Man. Such standard medical reference works as the *Encyclopédie médico-chirurgicale* (Laffont and Audit 1934) justified the practice of birth control in cases of women "whose motherhood would be dangerous for themselves or for the future of the race, by the inferior quality of the infants they would bring into the world." Although part of the reason was a desire to diminish the estimated five hundred thousand yearly illegal abortions in France, the authors pointed to growing support for the concept of "motherhood by consent."

There were obvious eugenic implications in these new ideas of birth control and overpopulation that eugenicists sought to turn to their advantage. One of the most articulate and persistent was Just Sicard de Plauzole, president of the French League against Venereal Disease, who became a leading spokesman for the revisionist eugenic view of the population problem in the 1930s.

Sicard de Plauzoles

Sicard represents the new generation of French eugenicists who came to the movement in the twenties and thirties from a background in public health. His family background was perhaps the most aristocratic of all eugenicists. Born in 1872 at Montpellier, he traced his family origins back to Raymond de Plauzoles, made a count in 1230 by the king of Aragon. In the 1700s members of the family began pursuing medicine as a career, with no less than nine ancestors having been physicians when Just was born. At that time his father, Henri Sicard, was a professor of medicine at Montpellier, and shortly thereafter he was named dean of the Faculty of Science and Medicine at Lyon.

Sicard de Plauzoles attended medical school in Paris, where he studied with Pinard and Richet, but it was Louis Landouzy who directed his work toward public health in general and tuberculosis in particular. After graduation he pursued his interest in "public medicine," as he called it, joined Fournier's Society of Moral and Sanitary Prophylaxis and published *Tuberculosis* (1900) and *Motherhood and the National Defense against Depopulation* (1909). He joined the League for the Rights of Man in 1898 and became a member of its central committee shortly thereafter. He also participated in popular *cours libres* on social hygiene at the College Libre des Sciences Sociales and the Sorbonne in the prewar years.

Despite the nature of his interests and his contact with Pinard, Richet, and Landouzy, Sicard did not become associated with the French Eugenics Society until after the war. The manner of his affiliation illustrates the postwar changes in eugenics. In 1919 Sicard de Plauzoles began directing a tuberculosis clinic in Paris, and the following year he became the general secretary of the Venereal Disease Commission of the Ministry of Health. In this capacity he joined the French Eugenics Society in May of 1922, seeking support for a regular series of public talks on social hygiene. In December of that year, the eugenics society created a Committee of Union against the Venereal Peril, which joined the National League against the Venereal Peril headed by Sicard de Plauzoles. Twelve hundred francs were given to the league by the eugenics society, which at the same time agreed to cosponsor Sicard's *cours libres* on social hygiene recently approved by the Paris Faculty of Medicine (*Eugénique* 1922–1926, pp. 53, 169–70). Although Sicard de Plauzoles eventually became the head of the French Society of Moral and Sanitary Prophylaxis and general secretary of the Social Hygiene Council created by the Ministry of Health in 1938 (while retaining his other titles), it was this series of lectures given every year until 1941 that brought him the most notoriety and permitted him the widest latitude in working out his eugenic ideas.

The *cours libres* of fifteen to twenty-five lectures, usually running from January through March, were begun in 1922 and reflected the general interest in social hygiene during the postwar years. There had been some delay when the course was first proposed in 1920 because Léon Bernard, who occupied the chair of hygiene at the Faculty of Medicine, which had to authorize the course, objected that it would duplicate instruction at the school. A compromise was worked out the following year whereby the lectures would be given instead at the Grand Amphitheater of the Sorbonne (France. Archives Nationales 1921). This did not diminish the official sanction given to the course as indicated from the attendance at the opening sessions by members of the French public health and medical establishment—including professors and deans of the medical faculty, senators, deputies, and even ministers of health. Moreover, the list of other institutions acting as cosponsors of the lectures (besides the medical school and eugenics society) included the Ministry of Labor and Health, the National Commission of Defense against Tuberculosis, the Franco-American League against Cancer, the National League against Alcoholism, and the Mental Health League. When Bernard died in 1932, the *cours libres* were moved to the Faculty of Medicine where Louis Tanon, the new occupant of the chair of hygiene, presided over the opening lecture and introduced Sicard de Plauzoles with the admission that the course had always belonged at the medical school (Sicard 1933, pp. 137–38).

Sicard de Plauzoles' ideas on eugenics were implicit in his notion of social hygiene developed in the twenties. In his lectures, he always cited the work of Pinard and Richet, both of whom were obviously not opposed to having their names associated with his ideas since they frequently attended and even spoke at opening lessons of the public hygiene course. From 1927 to 1932 Pinard attended all opening lectures except one; Richet also attended in 1929 and ceremonially opened the lecture series in 1930. By this time Sicard's ideas were already reaching a wider audience thanks to the publication of the book, *Principes d'hygiène* (Sicard 1927a), based on the first five years of the courses, which included a preface by Pinard.

A key concept of the book, which also revealed the influence of Taylorism on Sicard's generation of public and industrial hygienists, was *zootechnie humaine*, which he defined as "the art of procreating, perfecting and utilizing man as a work-producing machine." Eugenics' role in the process, he claimed, was to ensure that the best "human capital" would be produced, while social hygiene would help to ensure the best possible return on this invested capital. Sicard's economic and technical language was phrased in an equation:

$P = n + p + i + a + e + r + m$, where
  P ("prix de revient") = the cost of return,
  n ("naissance") = cost of pregnancy and birth,
  p ("puericulture") = cost of rearing,
  i ("instruction") = cost of education,
  a ("apprentissage") = cost of apprenticeship,
  e ("entretien") = upkeep,
  r ("retraite") = cost of retirement, and
  m ("maladie") = cost of health.

The social value of an individual thus equaled the total productivity of an individual's life minus the total of these "maintenance" costs (Sicard 1927a, pp. 83–89). If this view of humanity seems to be simply an extreme extension of the positivistic, Tayloristic view of humanity, it also followed precedent. Before the turn of the century, social hygienists dramatized the social costs of diseases like tuberculosis by assuming a monetary value of human life (twenty-five thousand francs was a commonly cited prewar figure) and multiplying it by the number of deaths caused by the disease (Rochard 1888; Duclaux 1902; Sicard 1927a, pp. 87–88).

Sicard de Plauzoles' attack on those who were content merely to count the number of births to determine the value of the population was first made in the opening session of the 1932 social hygiene course entitled "The Future and the Preservation of the Race: Eugenics," with Justin Godart and Pinard in attendance. He began with an admonition:

It is infantile to measure the vigor and future of a population by the number of births registered every year. What constitutes the value of a nation is the number of health adults in condition to work, produce, and reproduce healthfully; . . . and what counts is less the number of births than their quality. (Sicard 1932, pp. 199–200)

Having argued in favor of quality over quantity, Sicard then expressed his alarm at the qualitative decline of the French population. This was happening, he insisted, because

> the lower classes, the poorer classes, have a much higher birthrate than the upper, richer classes. . . . Misery, along with alcoholism, syphilis and tuberculosis, is a powerful factor of degeneration . . . and children of poorer classes compared to children of the richer classes show an inferiority of physical, intellectual and moral development . . . caused by fatigue and deprivation of the mother during gestation, by insufficient feeding in early years, by poor housing conditions and by working at an early age. (Sicard 1932, p. 201)

Most important to Sicard was the fact that the inferiority did not disappear because it was transmitted and increased from generation to generation.

Here was the greatest danger of all. For, assuming as Sicard did that lower-class families had an average of five children, while upper-class families had two, in two generations the descendants of the lower half of the population would represent 85 percent of the people and in five generations, 99 percent. From this Sicard concluded:

> The increased swamping of superior classes of society by the lower classes will certainly result in the complete bankruptcy of the nation in gifted, capable and energetic individuals. It cannot take long before the whole of the population is lowered to a level which today is that of the uncultured classes. . . . In summary, as the population grows in number, it diminishes in quality. It is the lower categories that are the most prolific: the defectives multiply; the elites disappear. The result is a progressive bastardization, a degeneration which is more and more pronounced. Anything that can reduce the proliferation of the lower classes, in any country, will be a benefit for humanity. (p. 210)

The major point here is not Sicard's class prejudice nor the fanciful notions about differential birthrates that his fellow eugenicist Lucien March had done much to disprove in his prewar studies. Although Sicard does not cite the source, it appears that his birthrate figures for Paris, London, Vienna, and Berlin were based on Rainer Fetscher's 1924 book *Essentials of Race Hygiene* (Marchand 1933, pp. 48–49). Rather, it is the fact that a serious program of class-based negative eugenics was being proposed that considered birth control, especially for the lower classes, the only solution to the decline of the species. The French government's efforts to encourage larger families were considered by Sicard as a policy that "favors the multiplication of inferior classes and runs directly counter to natural selection and the progress of the species." Hence, he concluded, "birth control is justified as a means of artificial selection to prevent the evils that result from an unhealthy or exaggerated fertility" (Sicard 1932, p. 216).

There was at least one attempt in the thirties to give this new approach an organizational base: the founding in late 1931 of the Association of Sexological Studies, which included as members Sicard de Plauzoles and Justin Godart, as well as Victor Basch, the president of the League for the Rights of Man, several deputies and senators, and a large number of doctors. The chief organizer was Henri Toulouse, a well-respected psychiatrist and head of the French National

Mental Health League (Toulouse 1931a, pp. 598–607; 1931b). Warning against "reverse selection," Victor Basch called for repeal of the 1920 law against birth control as the first order of business for the new association. Although he did not subscribe to all the ideas of Sicard de Plauzoles mentioned above, Basch did make clear that his reasons were based on eugenics, citing Sicard's definition that it

> wants procreation to be no longer the result of blind passion and chance, but on the contrary, something of conscious will and reflection by healthy parents, vigorous in mind and body, wise and prudent, knowing the task they are undertaking, willing and able to carry it through to a good conclusion. (Godart 1933)

Basch's support was significant because his League for the Rights of Man was the most important civil liberties organization in France. It had been founded at the time of the Dreyfus Affair and was the rallying point of Left intellectuals supporting Dreyfus after Zola's publication of "J'accuse" forced the question into public light. After World War I the league increased its activities not just in political matters, but also in many social and health questions that were of interest to eugenicists. For example, the league took a position in favor of the premarital examination and opposed to the 1920 legislation against birth control based on the right of the infant to a healthy life. The man behind both of these league positions was the ubiquitous Sicard de Plauzoles (1927b, p. 103; 1929a, p. 539).

Sicard had been a founding member of the league and quickly moved into the inner circle of its directors. He became a member of its central committee in 1903 and a vice president from 1911 to 1919. In the 1920s his renewed interest in social hygiene prompted Sicard to bring before the league such matters as the mandatory declaration of tuberculosis and venereal disease (justified by the right of others to a healthy life), mandatory declaration of pregnancy and the prohibition of work by expectant mothers just prior to giving birth (justified by the right of the infant to be born healthy), and mandatory breast-feeding in the first ten months of life (justified by the right of a child to its mother's milk). In fact, largely at Sicard's instigation, the question of the conflict between the child's rights and mother's rights was debated by various committees of the league (1922, pp. 447–58; 1923, 150–51; 1928; 1929b, pp. 229–30). In 1927 Sicard obtained the support of the league for the French Eugenics Society's proposed premarital exam law as part of the concept of "protection of the child before procreation and during pregnancy." This was an extension of the concept of children's rights that went back to the earliest days of the league and a concern over questions of access to education (Buisson 1921; Sicard 1936).

The following year Sicard brought up the matter of the 1920 legislation prohibiting publicity in favor of birth control and secured passage by the league's central committee of the following resolution: "That the law of 31 July 1920 be revised; that all provisions contrary to the free expression of opinions be deleted; and that in particular paragraph 2 article 3 aimed at, 'publicity for birth control and against the birth rate' be deleted" (Sicard 1929a, p. 539). In his article reporting the results of the central committee's decision, Sicard was not yet as strident in his criticism of the natalists as he would be six years later, but he did note that eugenics offered a middle position between the populationist doctrine of "go forth and multiply" and the Malthusian claim that increased population only brought "misery and suffering." Eugenics, he stated, concentrates "less on the number than the quality of the products."

After the creation of the Association of Sexological Studies, Basch, in his capacity as president of the League for the Rights of Man, made birth control "the question for October 1932" in the league's journal. In addition to urging the repeal of the law of July 1920, Basch added a call for the creation of birth control counseling centers (1932, pp. 413–14). In 1933 the league backed a bill introduced in the chamber by the Left urging amnesty for those guilty of breaking the law, and the following year the league formally protested the arrest and conviction of Jeanne Humbert for spreading birth control propaganda. In fact, the league was so outspoken on the issue that it published a disclaimer in its journal in 1936, stating, "The league defends the rights of children . . . but it has not created any outside organizations to this effect, and it has no link with any group specializing in the defense of the rights of children" (p. 76). Evidently, Basch's and Sicard's participation in the Association of Sexological Studies did not constitute such a link.

The birth control question was only one part of the association's overall eugenics program that went far beyond calling for a repeal of the 1920 legislation against birth control. In February 1933 the association formally voted to support a six-point program including a mandatory premarital exam and the creation of public clinics to give advice on contraception and perform sterilizations and abortions. The latter two could be voluntary or performed for medical reasons such as those "in the public interest (physiological and mental hereditary defects, impulses of a criminal or sexual order) for which a list would be established according to the advice of competent medical societies" (Godart 1933, p. 10).

The ideas of Sicard de Plauzoles and others in the Association of Sexological Studies illustrate two effects of the depression on eugenics and contraception questions in France during the thirties. First, although the depression did not create the birth control movement, it did change the climate of opinion enough to provide an opportunity for those who favored the use of contraception as a negative eugenic measure to make their case. At the same time, the economic decline and rising unemployment undercut the natalist position that had dominated French eugenic thought for so long.

Immigration and Sterilization

Birth control was not the only negative eugenic measure receiving more attention in the thirties. At least two others—immigration restriction and sterilization— were discussed or advocated in ways that were markedly different from previous years. In the case of sterilization, the simple fact of serious discussion was a dramatic change. Immigration, on the other hand, had been the subject of attention at least since the turn of the century in France, and during the twenties and thirties it became an even more prominent issue in French health and political circles (Mauco 1932; Millet 1938; Bonnet 1976; Cross 1983).

Although the tradition of immigration to France had a long prewar history (especially from countries such as Belgium, Italy, and Poland), it was the increased number of immigrants after the war that attracted the public's attention. France was not alone in receiving large numbers of people pushed by political and economic upheaval in central and eastern Europe after the war, but the labor shortage in the rebuilding of France also made it a particularly attractive country for immigrants. In addition, the migration of people from French colonies such as North Africa continued after the war. Immigration to France was further stimulated when the United States began restricting immigration drastically in 1921, thus eliminating the country that in prewar years had been the world's largest recipient of immigrants. All of these developments increased immigration fourfold in the twenties. Total numbers of foreigners were counted in the millions. In 1931 over 11 percent of the work force was filled by foreigners, and some sectors, such as mining, metallurgy, and construction, contained 30 to 40 percent (Millet 1938, pp. 7–15; Cross 1983, pp. 122–42).

The economic downturn and rising unemployment of the thirties greatly increased sentiment against these foreign workers in France. Despite the delayed effects of the depression—as late as June 1930 there were still complaints of worker shortages in French mining and metallurgical industries—by the end of 1930, organized labor was demanding restrictions on foreign workers. Bills to this effect were introduced in the National Assembly, and in early 1932 some mildly restrictive measures were passed into law (Bonnet 1976, pp. 201–35). But economic decline was not the only explanation behind anti-immigrant views, for there had been a strong xenophobic current in the twenties—that is, even at the height of the labor shortage. This included some questions raised about the new immigrants by members of the French Eugenics Society, based on rather fuzzy notions of racial and cultural affinity whose only common thread, in hindsight, seems to be a correlation of perceived assimilability to the geographical distance from France of the immigrant's country of origin (Richet 1919, pp. 60–80; Apert 1923, pp. 152–57).

In the thirties anti-immigrant sentiments were buttressed by a new eugenic argument warning against the supposed biological danger posed by these foreigners. It was based on the discovery of the existence of hereditary human blood types and differences in the pattern of their distribution among ethnic groups. In the early 1900s medical researchers had determined that there were four distinct blood types in the ABO system, plus the fact that blood types were inherited according to the Mendelian laws of genetics. During the First World War, it was discovered that different peoples had distinctly different proportions of the types among their populations. For example, native Americans had almost 90 to 95 percent type "O"; in subcontinental India almost 50 percent of the population had type "B"; and western European peoples had almost equal proportions (40 to 45 percent) of the "A" and "O" types (Schneider 1983). The response among anthropologists was a rush to sample the blood types of peoples all over the globe in the hopes that a new measure of race had been discovered.

In France René Martial, a public health doctor like Sicard de Plauzole, was one of the leading experts on health and immigration during the twenties. When he heard of the new blood-group anthropology in the early thirties, Martial saw it as the key to screening potential immigrants. At its simplest, Martial urged, "Keep the 'O's' and the 'A's,' eliminate the 'B's,' only keep the 'AB's' if the psychological examination is favorable" (Martial 1934, p. 323). This was directed primarily against Jews and other eastern European migrants to France because studies had shown a higher percentage of type "B" blood among these populations. Martial's writings were picked up by French right-wing racists who used them in the thirties to bolster their call of "France for the French." An ironic result of Martial's use of the blood-group discoveries was that he was one of the few writers to popularize Mendelian genetics in France before World War II (Martial 1933).

The question of sterilization was a new issue in French eugenics in the thirties even more directly stimulated by developments outside France. The inspiration for a discussion of the question came not from the Nazis, as is commonly assumed, but from the Americans after the 1927 *Buck v. Bell* U.S. Supreme Court decision that upheld the sterilization law of Virginia. This was followed shortly by sterilization legislation in Sweden and the French-speaking Swiss canton of Vaud.

In the past, only iconoclastic figures like Charles Richet had seriously proposed sterilization, and even he had realized it was unlikely to be accepted, so he called for the premarital examination as a temporary measure until public opinion changed. Although these initial articles did not wholeheartedly endorse the new American and Swiss legislation, for the first time they seriously examined the question of sterilization. At a minimum they contained substantial descriptions of the new laws, albeit often with abundant warnings of caution or skepticism (Heger-Gilbert 1928, p. 71). Almost invariably, however, the writers admitted either explicitly or implicitly the legitimacy and scope of the problem sterilization proposed to resolve: elimination of, in the words of one author, "the refuse of life, the sickly such as tuberculers, incurable defects, the insane and also those socially dangerous because of nerves, alcoholism and especially the morally pathogenic such as criminals and the socially demented" (Levrat 1930). Even those who thought sterilization to be extreme endorsed at least the more moderate premarital exam as a means of achieving the same end (Hamel 1933).

One of the earliest and most thorough examinations of the question was a 1930 article by Georges Schreiber, vice-president of the French Eugenics Society. His approach was to make sterilization more acceptable by first examining "therapeutic sterilization," and then discussing "eugenic, penal, economic and social" sterilizations. The examples he chose from his experience as a pediatrician were intended to elicit sympathy for the women whose lives were threatened and in some cases even lost because of pregnancies they could not bring to term. For example, he spoke of

> a woman who comes every week to my clinic. She has three young children and expects a fourth. The three babies have rickets, serious rickets. The father is an alcoholic, and the mother probably is too. At home, "There is misery!" says the visiting social nurse who follows them closely. They are piled into a small room, the father barely makes a living. This is a family that one can consider as being

in the worst possible condition. Yet the woman is pregnant again. Do you believe it would be desirable for this woman to bring into the world a fourth child? (Schreiber 1929, p. 265)

Having asked his rhetorical question, Schreiber could offer his own general answer that, "from the practical point of view of daily consultation . . . there are cases where accumulated defects and misery make human sterilization legitimate."

Schreiber was more cautious on the question of penal and economic sterilization. Nevertheless, he cited Louis Heuyer's 1926 talk on the premarital exam identifying the hereditary trait of "instinctive perversion" as the origin of criminality and delinquency (Heuyer 1927, pp. 132–36), and suggested that sterilization would be justified to prevent its transmission. On the other hand, Schreiber was critical of the 1909 California statute requiring castration for certain crimes because it was too broad in its assumption of inherited criminal traits (1929, p. 267). In the end he concluded that the general question of sterilization should at least be studied further in France without "the false sentimentality which risks simply multiplying the number of miserable beings."

Early in 1932 there was a chance to sample a broader cross section of opinion when, as part of preparation for the Third International Eugenics Congress, Charles Davenport sent a letter to the French Eugenics Society requesting "the opinion of the French public on questions of reducing the fertility of the 'socially inadequate,'" by means of sterilization and birth control. In response Henri Vignes, a member of the Ecole d'anthropologie who had earlier surveyed opinion on the premarital exam, sent letters to twenty doctors and sociologists, of whom half replied. Whereas earlier mention of the subject in the French Eugenics Society had prompted immediate disclaimers, only a few of those surveyed in response to Davenport's letter condemned sterilization outright or saw no instances when it was justified. Like Schreiber, most observers of the day saw sterilization as another means, albeit extreme, to a laudable end: the prevention of procreation by undesirables. It was the practical matter of public acceptance of sterilization that was most often cited as a reason for attempting other, less controversial measures to achieve the same ends, such as the premarital exam or Davenport's suggestion of agricultural work colonies segregated by sex to prevent procreation by the "socially inadequate." Significantly, the only respondent urging caution on the scientific grounds that not enough was known yet about human heredity to sanction sterilization measures was the one nonscientist who replied: Georges Inman, a novelist and lawyer. Such was not the case with the physicians (Apert, Briand, Drouet, Schreiber, and Turpin), anthropologists (Papillaut and Paul-Boncour), or psychologists (Biot and Jeudon) who responded.

A final point about the timing of the discussion of the sterilization question is the fact that it was begun well before the Nazis came to power in Germany. Hence, when the Law for the Preservation of the (Aryan) Race was passed in Germany in July 1933, articles in French journals simply added the German law to the list of those in the United States and Switzerland, as well as laws under consideration in England and Scandinavia (Vignes 1934; Piechaud and Marchaud

1934; Swarc 1934). The general pattern of the articles in medical journals was to give a detailed description of the laws, their rationale, and their application, with a short section listing support or objections and occasionally a concluding paragraph or two about the author's moral qualms or skepticism concerning the exactitude of knowledge about heredity.

The Nazi measures did produce a wider divergence of views on sterilization. Georges d'Heucqueville, a doctor for the public insane asylums, considered sterilization justified for alcoholics who

(1) have already given birth to defective children, [and] (2) have already been hospitalized or committed at least two times, for example, in a state of alcoholic intoxication or simply demonstrate a permanent intellectual weakening by their incapacity to accomplish regular tasks. (1935, p. 214)

On the other end of the spectrum of opinion was a Dr. Lowenthal of the Academy of Medicine, who ridiculed the whole notion of sterilization on the Lamarckian grounds that defects were acquired by action of the environment such as venereal disease, alcoholism, and "psychic trauma" (1935, p. 251). The one feature of the Nazi sterilization program that struck observers as significantly different was the numbers involved. Even Georges Schreiber called "audacious" the fact that sixteen thousand sterilizations were reported in the first year after the enactment of the German laws (*Revue anthropologique* 45, p. 87). Yet, as late as 1939, an article in the well-respected *Concours Médical* said of the German legislation, "These laws which appear at first sight to be an affront to individual liberty and consequently to the welfare of the citizen, have as their goal the rational pursuit of that welfare (Tisserand 1939).

On the whole, therefore, one would have to call the French response to the German sterilization program muted. In part, this was because it was seen as only one manifestation of sterilization programs in many countries. Moreover, the Nazis' measures were tied to a wider program of population, eugenic, and race laws passed by Hitler's regime. Some of these, like the anti-Semitic race legislation, were strongly criticized in France not only on moral, but also on scientific grounds. Even right-wing French theorists like Martial and the anthropologist Georges Montandon never subscribed to the possibility, let alone the advantage, of achieving racial purity. Other measures, however, were actually envied. For example, the Nazi repopulation program was especially lauded by the very natalist organizations in France that had come to oppose eugenics because of its new attachment to ideas of birth control. One 1934 article in the bulletin of the National Alliance against Depopulation even reprinted a section of *Mein Kampf* describing how the state should encourage large families, and asked wistfully why no French prime minister spoke or acted like Hitler did (p. 101). The Communists, too, admitted the efficiency of Hitler's program of making state loans to young couples setting up a household (Raymond 1936, p. 108). Even the respected geneticist Lucien Cuénot, who had no love for the Germans (he lived through World War I on the front lines at his university in Nancy), wrote admiringly in a 1936 article of the "great number of measures" passed by the Nazis, some eugenic, others "para-eugenic," and some repopulationist, that had as their goal the "practice of suppressing dysgenics" in the population. Cuénot concluded that, as a result, Germany would be "in twenty years a power that could dare anything"; he warned with a not very subtle sarcasm that "France, headed towards ruin by its absence of a family policy, would make a very nice German colony" (1936, p. 22).

Eugenics and the French Left

The thirties saw the compromise position of the French Eugenics Society come under attack from church and natalist organizations, as well as from those advocating harsher, more extreme measures like sterilization and immigration restriction. But advocates of the older, broad compromise eugenics were hardly silent in the decade. In addition to holding occasional meetings and publishing reports in the *Revue Anthropologique,* eugenicists like Eugène Apert and Georges Schreiber also represented France at international eugenics meetings. They even attempted to create a new international organization as an alternative to the International Eugenics Federation, which was dominated by the Nazis and extreme Americans after the midthirties. The International Latin Federation of Eugenics Societies was founded in Mexico late in 1935 and held a congress in Paris in conjunction with the 1937 World's Fair, with representatives from Latin American countries as well as France, Italy, Romania, Spain, and Switzerland (Turpin 1938). The outbreak of World War II prevented the new federation from doing much else than publishing the proceedings of the congress, but of more lasting significance was the fact that the ideas of the moderate French eugenicists received support from an important new political group in the thirties: the French Left. This meant that some form of eugenic policy was being advocated across the political spectrum from extreme Right to Left, showing how widely accepted eugenics had become in France by the end of the thirties.

The new position of the Left in France was not comparable to developments in other European countries where "left-wing" utopian eugenic programs had developed from the beginning. Around the turn of the century, there was one significant grass-roots neo-Malthusian movement aimed at the working class: Paul Robin's "regeneration" league, which contained many eugenic elements. It was soon preoccupied with a bitter debate that focused attention narrowly on the question of birth control (Ronsin 1980, pp. 164–84; McLaren 1983, pp. 177–89). Robin, therefore, did not have the opportunity to develop a broader program, and few other Socialist or labor leaders supported him even on this one issue. With the exception of occasional debates on issues like alcoholism, there was little mention of health, let alone birth control or eugenics, in Socialist publications in the prewar years (Quillent 1914; Prestwick 1980; Talmy 1962, 2:179–80). In the postwar years these and other policy questions were overshadowed by the doctrinal split between French Communists and Socialists produced by the Russian Revolution.

The first mention of any eugenic topic in the party publications of the new French Communist party came in the early thirties with a discussion of abortion in two articles of the *Cahiers du Bolchevisme* in 1931 and 1932 (Abeau 1931;

Abran 1932). As might be expected, the author of the first article was highly critical of the "repressive" French laws that not only failed to prevent abortions, but
also had the effect of making it possible for only the wives of the bourgeoisie to
pay the high price for safe, clandestine abortions by doctors and midwives.
Women of the working class had to resort to other means. As the author succinctly put it, "Done by people without medical instruction, with crude instru-
ments, under miserable hygienic conditions, these operations present very grave
dangers and result . . . in a very high proportion of deaths and injuries" (Abeau
1931, p. 791). Not surprisingly, she contrasted the French situation with that in
the Soviet Union, where abortion was legal and accessible in clinics, thus making
the death and injury rate almost negligible. It is significant that she also made
special note of the fact that the Soviet population continued to grow at a rate of
over two million inhabitants yearly. Despite the awareness of the wider natalist
issue, this article is most telling by its being an exception, with no references to
previous positions on the issues having been taken by the French Communist or
Socialist parties. The second article, which appeared the following year, openly
called for legalization of abortion, but neither article gave any indication of the
sweeping legislative proposal to be introduced in 1933 by the handful of Communist deputies of the French chamber.

The Law for the Protection of Maternity and Childhood was prompted,
according to its authors, "by the economic crisis which has struck the capitalist
world," leaving fifty million workers unemployed, as well as hundreds of millions
in the Far East and India (France. *Journal Officiel* 1933). Among the effects of this
desperate situation in other countries was a more relaxed attitude toward the use
of contraception, but according to the authors of the legislation, in its shortsighted
approach to the problem of birthrate, the French bourgeoisie had retained the law
of 1920, with the result being an increase in the number of clandestine abortions.
The Communist deputies called for a general revision of existing laws and the
introduction of new legislation creating offices to coordinate existing programs for
pregnant women and new mothers, expanding the number of refuges for expectant mothers, and providing day nurseries for new mothers if other programs
could not allow them to stay home with newborns. The most controversial features of the proposed legislation, however, were the final two sections, which
called for revocation of the Law of 31 July 1920 and a complete revision of the
statutes on abortion. Legal abortions would be permitted when the health of the
pregnant woman was endangered and "for eugenic reasons when necessary to prevent the procreation of defects or insanity" (France. *Journal Officiel* 1933, p. 866).

The deputies admitted in their conclusion that the bill had "no chance to be
supported let alone adopted by the majority of the Chamber," despite the fact that
it complied "with the suggestions and current evidence of eugenics, medical and
surgical sciences," not to mention "the interest of all human society" (France.
Journal Officiel 1933, p. 871). Their reasoning was that the present situation
suited the capitalists' interests: "The presence of hundreds of thousands of unemployed, the presence of an army of momentarily nonproductive reserve workers
constitutes for them an argument and a pretext for lowering salaries and thus aug-
menting their profits."

Two years later, in November 1935, a series of articles began in the French Communist party newspaper *Humanité* and the *Cahiers du Bolchevisme* that unfolded a full-scale family policy that retained many of the features of the 1933 bill, although significantly dropping others. It amounted to a program very similar to the natalist position of the French Eugenics Society, the program that Brouillard had hoped the Catholic church would support after the 1930 encyclical. The first indications of this policy shift came in a speech by the head of the French Communist party, Maurice Thorez, on 7 October 1935. It was clearly part of a larger change in strategy by the Communists which emphasized cooperation among Left and center groups after the rise of the Nazis in Germany and growth of right-wing movements in France and the rest of Europe. In France the result was the Popular Front coalition, which required the Communists to broaden their appeal to the general French public. The theme of Thorez's speech could not have been more broad in appeal, touching on motherhood, children, and country: "The working class does not want a weak France, with a degenerate people. It wants a hard-working and powerful France. What can be done to achieve it? We want to institute immediately a policy of effective protection for the mother and child" (1935a, p. 196). Thorez also sought to refute one of the common notions about the cause of the declining birthrate in France.

> The sterile and degenerate bourgeois say and have their journalists write that the wives of workers and peasants, the whole of the people of France, do not want children. It is not true. . . . What is true is that they are afraid of the father unemployed, of the mother without work, of not being able to meet the needs of the family. . . . They are afraid of not being able to give birth to children in full health, robust and intelligent instead of being the misfortunates who will only know a life of misery. (1935a, p. 197)

The following month a series of articles appeared in *Humanité* by its editor, former deputy Paul Vaillant-Couturier, entitled "Au secours de la famille." His opening article of 17 November revealed this more moderate position by stating the problem of the family in political terms, with the Right accusing the Left of "destroying, degrading and sterilizing" the family, while the Left accused the Right of being "repopulators" for military or religious reasons. Vaillant-Couturier consciously chose to occupy the middle ground between the two, but, significantly, his front-page article stated the task in eugenic terms: "how to make motherhood a social function of the highest order—by combating misery, low salaries, unemployment, prostitution, slums, clandestine abortions, social diseases, alcoholism, infantile mortality—because upon it depends the continuity and improvement of the species." The next day, Vaillant-Couturier further defined his position by citing Sicard de Plauzoles as an example of those who would resolve these problems by limiting births and Fernand Boverat as an advocate of exactly the opposite course of action. The article ridiculed the latter for representing "the interests of the directing oligarchies . . . which want to produce men above all for purposes of war" and the former for his "scientific preoccupation with the question of human breeding which unfortunately sinks to a Malthusian confusion of the facts."

Before Vaillant-Couturier could elaborate his own position, however, he was flooded with hundreds of letters from readers (if the editor can be believed), which he made the subject of what became virtually a daily column on the front page of the newspaper for the next six weeks. These were exactly the kind of personal stories of ordinary working people that *Humanité* prided itself on reporting—men and women who wanted to marry and have families but could not afford it. Other letters were obviously used to help define the new moderate position of the party. For example, they stressed the value attached to marriage and the family that contradicted many of the radical notions attributed to the Left, such as the portrayal of marriage as a bourgeois institution of slavery for women. In two articles of 22 and 24 November entitled "Lenin Talks on Love" and "Lenin and the Family," Vaillant-Couturier quoted the rather prudish father of the Russian Revolution as follows: "Neither monk nor Don Juan—sport, gymnastics, swimming, exercise, all sorts of physical exercise and varied moral interests . . . are better for youth than endless discussions on sexual questions."

Most of these articles were directed toward the natalists in order to demonstrate that the working classes desired larger families but simply could not afford them. A 30 November article was dedicated "especially to Boverat and the directors of the National Alliance against Depopulation," with quotes from letters and pictures of slums printed on the front page. Another article on 8 December entitled "Law and Money versus Motherhood" complained of inadequate subventions for mothers and families who were evicted from apartments for having too many children. An article on 21 December featured a picture of the recent winner of the Cognacq-Jay Prize, given yearly to exemplary large families in France, but which the article called "an exception without social value which serves to mask the failure of society with regard to the family." The Communists had obviously found a way of beating the bourgeoisie with its own stick of natalism. The clear and unmistakable message of these articles was that the working classes wanted to marry and have families. The problem was the capitalists, who would not hire them or pay them enough money to live decently.

When Vaillant-Couturier finally got around to his long-delayed article of 2 January 1936 on the "Remedies" that the Communist party proposed to these problems, he placed great emphasis on the features of the 1933 legislative proposal calling for support of motherhood and children, with only a few passing references to its call for revoking the statutes against birth control and abortion. But the essentially eugenic viewpoint remained. As Vaillant-Couturier noted in the beginning of the article, "The guiding principle of our proposition resides *in the recognition of motherhood as a social function* [emphasis in original]."

The importance of the revised program to the Communists can be seen in a new bill introduced to the 1936 chamber, sponsored by the now much larger Communist delegation that cooperated with the government of the Popular Front. Gone were the references to legalizing abortion; gone was the call to revoke the 1920 law against contraception. What remained was a bill for "effective protection of maternity and childhood" that hardly anyone could oppose. It retained the call for creation of a national office of the mother and child to coordinate all existing legislation on their behalf, and passage of new legislation that would "protect mothers effectively before, during and after pregnancy," encourage breast-feeding, and protect all children through three years of age (France. *Journal Officiel* 1936).

The Popular Front was not in power long enough to enact this legislation, but the issues remained a part of the Communist party program in the years that followed. Speeches by Thorez and articles in the *Cahiers du Bolchevisme* covered subjects that easily could have been found in earlier issues of *Eugénique,* such as "Depopulation and Childhood Misery," which cited Pinard, Richet, and Fernand Boverat, or "The Battle against Slums" on the effect of poor living conditions on birthrate (Raymond 1936; Chipau 1936). At the end of January 1936, Thorez warned, "The population decreases instead of growing. If this unsettling phenomenon continues or grows, it will be a catastrophe for our country. In a few decades we will be a nation of the elderly, a weak, diminished people on the road to extinction." In other articles Georges Levy, a Communist deputy who had been trained as a physician, reviewed the history of the social hygiene movement in France since 1902, covering in turn the problems of slums, alcoholism, tuberculosis, and syphilis. Although Levy's stated purpose was to show "the human inequality in sickness and death" resulting in death rates in the poorer quarters of Paris that were double those of the rich ones, his language and descriptions were clearly drawn from twenty years of social hygiene and eugenic writing. For example, Levy made much of the concept of "human capital" that Sicard de Plauzole had used in order to justify increased expenditure on health measures:

One forgets too often that the expenses for the protection of public health are excellent investments, because in the future they save the degeneration of the race, sickness and death; and they decrease expenses for hospitalization, insane asylums, welfare, prisons and lost work days." (Levy 1937, p. 55)

Levy also had little doubt that "there is a heredo-alcoholism like heredo-syphilis," which he proved by citing statistics claiming that half the crimes in France resulted from alcoholism (1937, p. 57). In sum, most of the positive, neo-Lamarckian eugenic position is evident in Levy's program, without the negative measures designed to eliminate the undesirable elements.

The Communists' "policy of protection of the family and childhood" was not a program of some right-wing fascist league, but a demonstration of the Communists' "preoccupation with this important problem, both in the parliamentary field as well as the courts." Thorez gave special attention to the question in a 21 November 1938 meeting of the Central Committee of the Communist party by making "protection of the family and childhood" one of the points of the party's Ten-Point Program (1938). Although this program did not support a mandatory premarital exam, it did call for "prenatal consultation." In fact, it is a testimony to how far the Left had moved on these questions that virtually every other point of Pinard's puericulture was included, from "surveillance of pregnant women," including "a longer rest period before and after pregnancy," to the encouragement of breast-feeding (Levy 1939, p. 367). Also included was support of most legislation proposed by natalists for bonuses, subventions, and tax breaks to encourage large families. Thorez even echoed one of their favorite proposals calling for "the advantages accorded to large families to be paid for by taxes on the unmarried and households without children" (1936b, p. 89).

What would cause such a dramatic change in policy by the Left, aside from a genuine response to the sympathetic chord evidently struck by the *Humanité* articles? Natalists were openly skeptical when they first heard of their newfound allies, although Boverat admitted (1936) that the new policy, "whatever its faults denotes serious concern for the population problem." He also noted the fact that the Left had at least put its ideas in the form of a legislative proposal, which was more than could be said of many natalists whose longtime support was "warm but vague in principle."

In the context of the midthirties and the preparations for the creation of the Popular Front, it should not be surprising that such a policy was embraced. It challenged no important part of the programs advocated by the Left, and to those who feared that the party was abandoning its revolutionary mission, Thorez replied simply, "We do not want to take power in a diminished, amputated country. We want to take power with a strong people, a healthy and numerous people" (1935b, p. 82). If legalizing abortion had to be dropped, this was no major revision since the issue had been taken up only briefly in previous years. Moreover, Thorez was very sensitive to the political danger of taking sides on the abortion question.

> We do not wish to repeat on this precise matter the tragic error of our comrades in the German party. For some time they had made abortion one of the essential articles of their program. This article caused them extreme harm. The Nazis went out in the countryside and among the workers and said, "Here are men who wish to weaken our country, to the advantage of foreigners." (1935b, p. 83)

The Left certainly had no qualms about government intervention in the private sphere; more important, the family policy could take advantage of the broad appeal of natalist, social hygiene, and eugenic ideas that had been developing in the first three decades of the twentieth century. In first announcing the change in policy, Thorez complained, "The fascists pretend to be guardians of the family tradition and say, 'The Communists want to destroy the family.'" His family program was consciously designed as a response that, he admitted, "is a veritable turning point in our policy on this question, but it is also the path toward the masses of our country" (1935b, p. 84).

The concluding paragraph of a 1939 article on family policy in the *Cahiers du Bolchevisme* stated:

> What higher goal [is there] to achieve for our party, if its militants set themselves to the task with their habitual ardor; to realize the great work of national renovation, the effective protection of the family and childhood to which are attached at the same time the recovery and future of our country. (Levy 1939, p. 373)

These words just as easily could have introduced the new family policy of the Vichy regime one year later.

Conclusion

The eugenics movement in France demonstrates several features important to an overall understanding of eugenics in modern society. Of course, the most obvious of these is that eugenics was not simply an Anglo-Saxon phenomenon. Any cursory look at the first three international eugenics congresses reveals several participants from other countries of southern and eastern Europe, as well as, later, Latin America and Japan. In France there were organizational, propaganda, and legislative activities that not only supported this international participation, but made eugenics part of the national debate on political and social questions during the first four decades of the twentieth century.

French eugenics also demonstrates that acceptance of Mendelism was not a prerequisite for those whose goal was the biological improvement of the human race. In fact, Mendelian eugenics appeared in France only in the 1930s as part of one of the more extreme proposals for immigration restriction. Although in this case it confirms the link between Mendelism and harsher negative measures, it was exceptional. By and large, France deserves its reputation as the home of a neo-Lamarckian eugenics whose main emphasis was on positive measures. Of prominent importance in the development of this emphasis was the population problem. The decline of the French birthrate in the nineteenth century and the fear of depopulation at the turn of the century worked against proposals for negative measures even though aimed at the "unfit," if they might be a hindrance to general practices of marriage or procreation. Most eugenicists chose to emphasize the positive measures that could increase the quality of all offspring. This had broad support in France because of the widespread belief in the inheritance of acquired characteristics. Accordingly, those wishing to improve the quality of future generations could do so by improving the environment and health of the present generation. The idea had logical, as well as emotional, appeal.

Eugenics in France was not confined to this Lamarckian, positive program. There were many who called for the elimination of dysgenic elements, even though they did not necessarily use Mendelian heredity to justify it. In fact, differing opinions among French eugenicists can be understood better as falling into either an emphasis on positive programs to encourage both propagation by the fit and, thanks to neo-Lamarckism, making the unfit healthier; or elimination of the unfit by harsher negative measures—prohibition of marriage, sterilization, or immigration restriction. The voices calling for the latter existed as early as the 1880s in the writings of Vacher de Lapouge, who did not need Mendelism to justify his proposals. The fact that the early formal structure of a French eugenics society was largely in the hands of neo-Lamarckians and *puericulteurs* should not obscure the existence of Richet's and other French voices that could command a following if conditions changed.

World War I was clearly an important turning point in changing the conditions that surrounded the birth of eugenics in France as well as other European countries, even though its effects were not as dramatic as in Germany and Russia, where revolutions overturned the political and social systems. Ironically, the initial effect of the war was to strengthen French eugenicists in their resolve to follow the same program as before the war. The war losses and added fear of depopulation made social hygiene a popular idea that appeared to tie eugenics to an even broader range of medical and health reform programs that emerged in the postwar years. The problem was that there were too many groups and policies for the eugenicists to control; hence, with many of their issues usurped, French eugenicists were already searching for new approaches before the thirties changed economic and social conditions. The most concrete proof of this was the campaign for a premarital examination law begun by the French Eugenics Society in 1926.

Another reason for the change in French eugenics before the thirties was a turnover in participants. As members of the founding generation of the French Eugenics Society either died or retired from public life, they were replaced by newcomers from different backgrounds whose new approach to eugenics found expression in the meetings and publications of the midtwenties. The very organization of the society also changed in the late twenties when institutional affiliation shifted from the Faculty of Medicine to the Ecole d'anthropologie. These changes did not result in a dramatic change from one interpretation of eugenics to another so much as an opportunity for new and different ideas to be heard.

Developments outside of France had important effects on eugenics in the thirties. The earliest was a change in the policy of the Catholic church that produced the first organized opposition to the movement. Prior to the 1930 papal encyclical, French Catholics, along with their natalist allies, attempted to reach an accommodation with what they perceived as a milder eugenics program of their countrymen. Strains in this alliance would have developed in any case as a result of some of the new ideas being proposed by eugenicists in response to the economic depression of the thirties. For example, increasing unemployment changed perceptions of the population problem and produced calls for legislation tolerating birth control as well as other negative eugenic measures like immigration restriction. Among these, the discussion about the use of sterilization was perhaps most novel. Although the immediate inspiration for discussion was action by eugenicists in the United States and Germany, the fact of serious discussion was most telling of how much the conditions in France had changed. Yet, despite the new debate on negative eugenic measures, the positive program continued to be advocated by supporters in the thirties and even picked up important new support from the French Left after its decision to cooperate with the Popular Front coalition. By 1939 a natalist eugenic family program was an official part of the French Communist party platform.

Eugenics in France did not produce new biological research or statistical studies as did its counterpart movements in England, the United States, Russia, or Germany. The most fruitful new work it can be credited with indirectly inspiring dealt with inherited childhood diseases. Hence, it would be a misnomer to call eugenics a "science" or even a "pseudoscience" in the French context. It was certainly an attempt to apply the new scientific study of human evolution toward a social end, and in this sense eugenics was a sensitive barometer of much broader trends in the twentieth century that transcended narrow political or ideological boundaries. How else can one explain the fact that eugenics was part of the vocab-

ulary of groups ranging from the far Left to the extreme Right in the French political spectrum?

Clearly, the Communists' family program deliberately picked up eugenic ideas
as part of remaking its image of respectability. Right-wing eugenicists may have
been more hardheaded and deliberately provocative in proposing sterilization
and immigration restriction, but their goals could also be encompassed within the
definition of eugenics. The common elements shared by these groups were a concern over decline in modern society, a view of the problem in scientific terms, a
heightened sense of nationalism, and an expectation of the government to play a
role in remedying the situation. Despite the common assumption that the Nazis
and World War II permanently discredited eugenics, these broader trends continue to this day, both in France and in other countries, albeit under different
names. Hence, in judging contemporary developments in genetic engineering or
sociobiology, it would be wise to view them, along with the history of eugenics,
as part of the twentieth-century attempt at biologically based social reform.

Bibliography

NOTE: Reference in the text is made to the following newspapers and periodicals: *Eugénique*
(1913–1926), *Journal des Débats* (1930–1931), and *Humanité* (1935–1936). Reference to parliamentary deliberations as reported in the *Journal Officiel* are listed by year, series, and pages
below.

Abeau, Tilly. 1931. Les ravages de l'Article 317 contre l'avortement [The ravages of Article 317
    against abortion]. *Cahiers du Bolchevisme* 6: 789–92.
Abran, E. 1932. L'avortement doit être légalisé [Abortion must be legalized]. *Cahiers du Bol-
    chévisme* 7: 355–59.
Adam, Frantz. 1929. Que faut-il penser de la stérilisation des aliénés? La doctrine française
    [Thoughts on sterilization of the insane: The French doctrine]. *Siècle Médical* February 1.
Apert, Eugène. 1922. Eugénique et santé nationale [Eugenics and national health]. In *Eugénique
    et Sélection,* ed. Eugène Apert, et al. Paris: Félix Alcan.
_____.1923. Le problème des races [The problem of the races]. *Eugénique* 3: 152–57.
Basch, Victor. 1932. La prophylaxie anticonceptionelle [Birth control]. *Cahiers des droits de
    l'homme* 32: 413–14.
Baudet, Raoul. 1927. L'examen médical avant le mariage [The medical examination before
    marriage]. *Annales Politiques et littéraires* 88: 579–80.
Bonnet, Jean Charles. 1976. *Les pouvoirs publiques français et l'immigration dans l'entre-deux-
    guerres* [French public policy and immigration between the wars]. Lyon: Centre d'histoire
    économique et sociale de la région lyonnaise.
Bouthol, Gaston. 1935. *La population dans le monde* [World population]. Paris: Payot.
Boverat, Fernand. 1931. Qualité et quantité [Quality and quantity]. *Bulletin de l'Alliance Nation-
    ale contre la Dépopulation* 36: 399–403.
_____. 1933. Eugénisme: Dangers des mésures préconisées pour éviter la naissance des enfants
    anormaux [Eugenics: Dangers of the measures recommended to avoid the birth of abnor-
    mal infants]. *Bulletin de l'Alliance Nationale contre la Dépopulation* 38: 69–73.
_____. 1936. Les communistes français et la natalité [French communists and natalism]. *Bul-
    letin de l'Alliance Nationale contre la Dépopulation* 41: 35–42.
Broca, Paul. 1866–1867. Sur la prétendue dégénérescence de la population française [On the sup-
    posed degeneration of the French population]. *Revue des Cours Scientifiques* 4: 305–11,
    321–31.

Brouillard, René. 1930. Eugénique et morale catholique [Eugenics and Catholic morality]. *Inter-diocésaine* (April): 113–19.

———. 1931. Causerie de morale [Moral chat]. *Etudes: Revue Catholique d'Intérêt Général* 207: 441–54, 578–600.

Buisson, Ferdinand. 1921. Les droits de l'enfant [Child rights]. *Cahiers des Droits de l'Homme* 21: 99–104.

Carter, A. E. 1958. *The Idea of Decadence in French Literature.* Toronto: University of Toronto Press.

Chase, Allen. 1977. *The Legacy of Malthus.* New York: Knopf.

Chevalier, Louis. [1958] 1973. *Laboring Classes and Dangerous Classes in Paris during the First Half of the Nineteenth Century.* New York: Howard Fertig.

Chipau, André. 1936. La lutte contre les taudis [The battle against slums]. *Cahiers du Bolchévisme* 13: 1250–57.

Clark, Linda L. 1984. *Social Darwinism in France.* Montgomery, Ala.: University of Alabama Press.

Conry, Yvette. 1974. *Introduction du darwinisme en France au XIXe siècle* [The introduction of Darwinism to France in the nineteenth century]. Paris: Vrin.

Cross, Gary S. 1983. *Immigrant Workers in Industrial France.* Philadelphia: Temple University Press.

Cuénot, Lucien. 1923. Génétique et adaptation [Genetics and adaptation]. In *Eugenics, Genetics, and the Family.* Baltimore: Williams and Wilkins.

———. 1936. Eugénique [Eugenics]. *Revue Lorraine d'Anthropologie* 8: 5–24.

Dennery, Etienne. 1930. *Foules d'Asie* [Asian masses]. Paris: A. Colin.

d'Heucqueville, Georges. 1935. Points de vue sur la stérilisation chirurgicale [Perspectives on surgical sterilization]. *Annales de Médecine Légale* 15: 208–14.

Dubourg, Mgr. 1930. Le véritable eugénisme [True eugenics]. In *L'église et l'eugénisme.* Paris: Editions mariage et famille.

Duclaux, Emile. 1902. *Hygiène sociale* [Social hygiene]. Paris: Félix Alcan.

Dumont, Martial, and Pierre Morel. 1968. *Histoire de l'obstétrique et de la gynécologie* [The history of obstetrics and gynecology]. Lyon: SIMEP Editions.

Fessard, G. 1937. *La main tendue: Le dialogue catholique-communiste: est-il possible?* [The Catholic-Communist dialogue: Is it possible?] Paris: Bernard Grasset.

Fetscher, Rainer. 1924. *Grundzüge der Rassenhygiene* [The essentials of race hygiene]. Dresden: Deutsche Verlag für Volkswohlfahrt.

France. Archives nationales. 1921. (AJ 16, dossier 6273). Procès verbaux de la Faculté de Médecine [Minutes of the Faculty of Medicine]. 17 November and 22 December.

France. Journal officiel. 1933. *Documents parlementaires. Chambre. Annexe* [Parliamentary documents. Chamber. Appendix], 865–74.

———. 1936. *Documents parlementaires. Chambre. Annexe* [Parliamentary documents. Chamber. Appendix], 938–40.

———. 1938. *Documents parlementaires. Chambre.* 530–31.

Godart, Justin. 1933. A l'Association d'études sexologiques [At the Association of Sexological Studies]. *Siècle Médical* (1 March): 10.

Grimbert, Charles. 1930. Les psychopathies ou anomalies mentales et l'eugénisme [Psychopathologies or mental anomalies and eugenics]. *Revue de Philosophie* 25: 129–40.

Hamel, J. 1933. La stérilisation des aliénés et des criminels [The sterilization of the insane and criminals]. *Siècle Médical* (15 March): 6.

Hanson, Eric C. 1982. *Disaffection and Decadence.* Washington, D.C.: University Press of America.

Haury, Paul. 1934. Une natalité suffisante, est-ce la guerre? [An adequate birthrate: Does it mean war?]. *Revue de l'Alliance Nationale contre la Dépopulation* 35: 33–40.

———. 1935. Le destin des races blanches [The destiny of the white races]. *Revue de l'Alliance Nationale contre la Dépopulation* 36: 267–72.

Hayward, J. E. S. 1959. Solidarity: The Social History of an Idea in the Nineteenth Century. *International Review of Social History* 4: 261–84.

———. 1961. The Official Social Philosophy of the French Third Republic: Léon Bourgeois and Solidarism. *International Review of Social History* 6: 19–48.

Heger-Gilbert, Fernand. 1928. La stérilisation des fonctions génitales [The sterilization of genital functions]. *Bulletin de la Société Française Sanitaire et Morale* 28: 65–74.

Heuyer, Louis. 1927. Conditions de santé à envisager au point de vue du mariage dans les maladies mentales et nerveuses et les intoxications [Health conditions to follow in marriage for mental and nervous illnesses and intoxication]. In *Examen médical en vue du mariage*, ed. Georges Schreiber, 123–36.

Jones, Gareth Stedman. 1971. *Outcast London*. Oxford: Clarendon Press.

Jordan, Edouard. 1926. Le certificat médical prénuptial [The premarital medical certificate]. *Pour la Vie* (June): 1.

———. 1931. Eugénisme et morale [Eugenics and morality]. *Cahiers de la Nouvelle Journée* 19: 1–173.

Laffont, A., and J Audit. 1934. Eugénique [Eugenics]. In *Encyclopédie médico-chirurgicale*. Paris: Masson.

Landry, Adolphe. 1913. L'eugénique [Eugenics]. *Revue Bleue* 51: 779–83.

Laumonnier, Jean. 1912. Eugénique [Eugenics]. *Larousse Mensuel* 65: 454–55.

Levrat, Etienne. 1930. Stérilisation et eugénique [Sterilization and eugenics]. *Toulouse Médical* 31: 1–11.

Levy, Georges. 1937. Le problème de la santé publique en France [The problem of public health in France]. *Cahiers du Bolchevisme* 14: 48–61.

———. 1939. Pour une politique de protection de la famille et de l'enfance [For a policy of protection for the family and childhood]. *Cahiers du Bolchevisme* 16: 362–73.

Lewis, Martin Deming. 1962. One Hundred Million Frenchmen: The "Assimilation" Theory in French Colonial Policy. *Comparative Studies in Society and History* 4: 129–53.

Lowenthal. 1935. La lutte contre la dégénérescence: Stérilisation et castration [The fight against degeneration: Sterilization and castration]. *Paris Médical* 95: 250–54.

Ludmerer, Kenneth. 1972. *Eugenics and American Society*. Baltimore: Johns Hopkins University Press.

McLaren, Angus. 1983. *Sexuality and Social Order*. New York: Holmes and Meier.

Manouvrier, Léonce. 1899. Indice céphalique et la pseudo-sociologie [The cephalic index and pseudo-sociology]. *Revue Anthropologique* 9: 233–59, 280–96.

Marchand, Henri Jean. 1933. Evolution de l'idée eugénique [The evolution of the eugenic idea]. Medical dissertation, Bordeaux.

Martial, René. 1933. L'immigration et le pouvoir de résorption de la France [Immigration and the power of absorption of France]. *Revue Anthropologique* 43: 351–69, 449–67.

———. 1934. *La race française* [The French race]. Paris: Mercure de France.

Mauco, Georges. 1932 *Les étrangers en France* [Foreigners in France]. Paris: Armand Colin.

Millet, Raymond. 1938. *Trois millions d'étrangers en France* [Three million foreigners in France]. Paris: Librarie de Médicis.

Murard, Lion, and Patrick Zylberman. 1977. *L'haleine des faubourgs* [The air of the suburbs]. Fontenay-sur-Bois: Recherches.

Nisot, Marie-Thérèse. 1927, 1929. *La question eugénique dans divers pays* [The eugenic question in different countries]. 2 vols. Brussels: G. van Campenhout.

———. 1929. La stérilisation des anormaux [The sterilization of abnormals]. *Mercure de France* 209: 595–608.

Noonan, John T. 1965. *Contraception*. Cambridge, Mass.: Harvard University Press.

Nordau, Max. 1968. *Degeneration*. Trans. and ed. George Mosse. New York: Howard Fertig.

Nye, Robert A. 1984. *Crime, Madness and Politics in Modern France*. Princeton, N.J.: Princeton University Press.

Penel, J. 1930. La stérilisation eugénique en Amérique [Eugenic sterilization in America]. *Hygiène Mentale* 25: 173–88.

Perrier, Edmond. 1881. *Colonies animales et la formation des organismes* [Animal colonies and the formation of organisms]. Paris: Masson.

———. 1922. Eugénique et biologie [Eugenics and biology]. In *Eugénique et sélection*, ed. Eugène Apert et al. Paris: Félix Alcan.

Piéchaud, Ferdinand, and Henri Marchaud. 1934. La loi allemande de stérilisation [The German sterilization law]. *Journal de Médecine de Bordeaux* 11: 103–4.

Pieri, Jean. 1930. La stérilisation [Sterilization]. In *L'église et eugénisme* Paris: Editions mariage et famille.

Pinard, Adolphe. 1899. De la conservation et de l'amélioration de l'espèce [On the preservation and improvement of species]. *Bulletin Médical* 13: 144.

———. 1912. Considérations générales sur la "puericulture avant la procréation" [General considerations concerning "puericulture before procréation"]. In *Problems in Eugenics,* 457–59. London: Eugenics Education Society.

Pope Pius XI. 1930. Encyclical Letter ... On Christian Marriage. *Sixteen Encyclicals of Pope Pius XI.* Washington, D.C.: National Catholic Welfare Conference.

Prestwick, P. E. 1980. French Workers and the Temperance Movement. *International Review of Social History* 25: 44–52.

Quillent, E. 1914. A la santé du prolétariat: L'action antialcoolique [To the health of the proletariat: Anti-alcoholic action]. *Revue Socialiste* 30: 519–26.

Raymond, D. 1936. La dépopulation et la misère de l'enfance [Depopulation and childhood misery]. *Cahiers du Bolchevisme* 13: 101–9.

Richet, Charles. 1919. *La sélection humaine* [Human selection]. Paris: Félix Alcan.

———. 1935. *Au secours* [Help]. Paris: Peyronnet.

Rochard, Jules. 1888. *Traité d'hygiène sociale* [Treatise on social hygiene]. Paris: Hachette.

Ronsin, Francis. 1980. *La grève des ventres: Propagande néomalthusienne et baisse de la natalité française, XIXe–XXe siècles* [The belly strike: Neomalthusian propaganda and the decline in the French birthrate, 19th and 20th centuries]. Paris: Aubier Montaigne.

Rostand, Jean. *Hérédité et racisme* [Heredity and racism]. Paris: Gallimard.

Roubakine, Alexandre. 1935. Sur la prétendue "dépopulation" de la France [On the supposed "depopulation" of France]. *Bulletin de l'Académie de Médecine* 113: 143–47.

Royer, Clémence. 1862. Introduction. In *L'origine des espèces* [The origin of species] by Charles Darwin. Trans. and ed. Clémence Royer. Paris: Guillaume et Masson.

Schiefley, William H. 1917. *Brieux and Contemporary French Society.* New York: G. P. Putnam's Sons.

Schiff, Paul, and Pierre Mareschal. 1931. Hérédité psychopathique et stérilisation eugénique [Psychopathological heredity and sterilization]. *Annales Médico-Légales* 89: 72–78.

Schneider, William H. 1982. Toward the Improvement of the Human Race. *Journal of Modern History* 54: 268–91.

———. 1983. Chance and Social Setting in the Application of the Discovery of Blood Groups. *Bulletin of the History of Medicine* 57: 545–62.

———. 1986. Puericulture and the Style of French Eugenics. *History and Philosophy of the Life Sciences* 8: 235–47.

Schreiber, Georges. 1929. La stérilisation humaine aux Etats-Unis [Human sterilization in the United States]. *Revue Anthropologique* 39: 260–81.

———. 1930. L'hygiène à l'étranger: La stérilisation des indésirables aux Etats-Unis [Hygiene in other countries: The sterilization of undesirables in the United States]. *Revue d'Hygiène* 52: 271–76.

Shaw, George Bernard. 1911. Introduction. *Three plays by Brieux.* New York: Brentanos.

Sicard de Plauzoles, Just. 1922. Le droit à la vie saine [The right to a healthy life]. *Cahiers des Droits de l'Homme* 22: 447–48.

———. 1923. Les droits de l'enfant [The rights of the child]. *Cahiers des Droits de l'Homme* 23: 150–51.

———. 1927a. *Principes d'Hygiène* [Principles of hygiene]. Paris: Editions médicales.

———. 1927b. Le droit à la vie saine [The right to a healthy life]. *Cahiers des Droits de l'Homme* 27: 103.

———. 1928. Le secret médical [Medical confidentiality]. *Cahiers des Droits de l'Homme* 28: 603–10.

———. 1929a. Une thèse interdite [A forbidden thesis]. *Cahiers des Droits de l'Homme* 29: 539.

———. 1929b. Allaitement obligatoire [Mandatory breast-feeding]. *Cahiers des Droits de l'Homme* 29: 229–30.

_____. 1932. L'avenir et la préservation de la race [The future and the preservation of the race]. *Prophylaxie Antivénérienne* 3: 194–219.

_____. 1933. Leçon d'ouverture du cours libre d'hygiène sociale [Opening lesson of public course on social hygiene]. *Prophylaxie Antivénérienne* 5: 137–55.

_____. 1934. Le capital humain: L'eugénique et la question de population [Human capital: Eugenics and the population question]. *Prophylaxie Antivénérienne* 6: 89–127.

_____. 1935. L'avenir de l'espère humain. La surpopulation c'est la guerre [The future of the human race. Overpopulation means war]. *Prophylaxie Antivénérienne* 7: 69–99.

_____. 1936. Declaration des droits de l'enfant [Declaration of the rights of the child]. *Cahiers des Droits de l'Homme* 36: 346–47.

Soloway, Richard A. 1978. Neo-Malthusians, Eugenists and the Declining Birth-Rate in England. *Albion* 10: 264–86.

_____. 1982a. *Birth Control and the Population Question in England, 1877–1930*. Chapel Hill, N.C.: University of North Carolina Press.

_____. 1982b. Counting the Degenerates: The Statistics of Race Degeneration in Edwardian England. *Journal of Contemporary History* 17: 137–64.

_____. 1982c. Feminism, Fertility and Eugenics in Victorian and Edwardian England. In *Political Symbolism in Modern Europe,* ed. Seymour Drescher, et al., 121–45. New Brunswick, N.J.: Transaction Books.

Stebbin, Robert E. 1974. France. In *Comparative Reception of Darwin,* ed. Thomas L. Glick, 117–67. Austin, Tex.: University of Texas Press.

Swarc, G. 1934. La stérilisation sexuelle et l'eugénique [Sexual sterilization and eugenics]. *Hygiène Mentale* 29: 228–34.

Swart, Koenrad. 1965. *The Sense of Decadence in Nineteenth Century France.* The Hague: Martinus Nijhoff.

Talmy, Robert, 1962. *Histoire du mouvement familial en France, 1896–1939* [The history of the family movement in France, 1896–1939]. 2 vols. Paris: Aubenas.

Thorez, Maurice. 1935a. Rapport à l'Assemblé communiste du Paris [Report to the Communist Assembly of Paris]. *Oeuvres* 9: 196–97.

_____. 1935b. Rapport au comité central du parti communiste [Report to the central committee of the Communist party]. *Oeuvres* 10: 82–84.

_____. 1936a. La main tendue aux catholiques et aux Croix de Feu [The open hand to Catholics and the Croix de Feu]. *Oeuvres* 12: 22–23

_____. 1936b. Rapport à la conférence du parti communiste [Report to the Communist party conference]. *Oeuvres* 12: 85–92.

_____. 1938. L'avenir de notre peuple [The future of our people]. *Oeuvres* 16: 84–92.

Tisserand, M. 1939. A propos des mesures appliquées depuis quatre ans en Allemagne pour veiller à la protection de la race [On measures applied in Germany for four years to safeguard the protection of the race]. *Concours Médical* 61: 1087–89, 1161–62.

Toulouse, Henri. 1931a. A l'Associaton d'études sexologiques [At the Association of Sexological Studies]. *Prophylaxie Antivénérienne* 3: 596–607.

_____. 1931b. Le problème humain [The human problem]. *Le Journal,* 2 November.

Turpin, Raymond. 1938. *Premier congrès latin d'eugénique (1–3 août 1937). Rapport* [First Latin Eugenics Congress (1–3 August 1937). Proceedings]. Paris: Masson.

Vacher de Lapouge, Georges. 1896. *Sélections sociales* [Social selections]. Paris: A. Fontemoing.

Vignes, Henri. 1932. Stérilisation des inadaptés sociaux [Sterilization of the socially inadequate]. *Revue Anthropologique* 42: 228–44.

_____. 1934. Les indications de la loi allemande de stérilisation eugénique [Features of the German eugenic sterilization law]. *Presse Médicale* 42: 825–27, 971–73.

Weber, Eugene. 1976. *Peasants into Frenchmen.* Stanford, Calif.: Stanford University Press.

CHAPTER 4

Eugenics in Brazil
1917–1940

Nancy Leys Stepan

As historians of science have shifted their attention away from logical reconstructions of science toward more "naturalistic" views of science as a product of culture and social life, eugenics and genetics have been linked together in scholarship as they once were in reality. As both a social movement and a science, eugenics offers the possibility of testing ideas about the social generation of scientific knowledge, and the last decade has seen considerable debate on the subject (Kevles 1985; McKenzie 1976; Schneider 1982; Searle 1981). In the recent study of eugenics, however, Latin America has been completely ignored by historians. Although this is part of a larger pattern of neglect of Latin American science, it is nonetheless regrettable for two major reasons.

First, leaving out eugenics distorts the modern history of Latin America. Among Latin Americans, because of the historic connection between eugenics and Nazi excesses, and perhaps because of the powerful fiction that Latin America has been relatively free of the racism that has characterized other parts of the world, there is often a tendency to deny that eugenics had any part in modern Latin American history. Yet even a cursory examination of the available sources indicates that eugenics movements were extensive in the region, and shaped science, social thought, and policies in unexpected ways. Between the two world wars, eugenics was associated with a wide array of congresses and conferences, and with social legislation on child welfare, maternal health, family law, the control of infectious diseases, and immigration. Eugenics stimulated some of the first courses on genetics in the region. Medical-legal debates and legislative activities concerning the proper role of the state in the regulation of marriage were permeated with the themes of "eugenical improvement." Eugenics penetrated the meetings of the many Pan American sanitary and scientific conferences held between 1900 and 1940. Eugenics was the subject of two Pan American eugenics congresses, the first held in Cuba in 1927, the second in Argentina in 1934. Latin eugenics movements were also responsible for the creation of the Fédération International Latine des Sociétés d'Eugénique (International Latin Federation of Eugenics Societies), which was founded in 1935 by the initiative of Corrado Gini, president of the Italian Society of Genetics and Eugenics. At the organizing meeting in Mexico City, eugenics societies from Italy, France, Romania, Mexico, Peru, Catalonia, Brazil, and Belgium expressed their intention of participating in the federation, while delegates from twelve other Latin American countries expressed interest and support. In 1937 the Latin federation held its first and only congress in Paris, resulting in the 1938 volume *Congrès Latin d'eugénique: Rapport* (Mac-Lean y Estenos 1952; Marchaud 1933; Nisot 1927).

What kind of eugenics was involved in Latin America? A contrary impulse to denying the reality of eugenic activity in the region is to identify Latin American eugenics with Nazi eugenics of the 1930s. Neither response is historically accurate or interpretively useful. Study of at least one Latin American country, Brazil, reveals that it possessed traits that set it apart, scientifically and ideologically, from Nazi eugenics certainly and, more generally, from the more widely chronicled Anglo-Saxon cases. Further research might lead to a generalization of this conclusion, namely, the existence of a "Latin" type of eugenics, which would include France and Italy as well as various countries of Latin America, distinct from the "Anglo-Saxon" type. The existence of a specifically Latin federation of eugenics points in this direction. Of course, just as we are increasingly aware of very important variations within the Anglo-Saxon eugenics tradition, we could expect there to be important subtypes within Latin eugenics.

A second reason, therefore, for regretting the neglect of Latin America in discussions of eugenics is that it impoverishes not just our understanding of Latin American history, but of eugenics as a putatively world scientistic movement. We can greatly enrich our understanding of the origins, scientific style, and social meanings of eugenics as an international movement by incorporating Latin America into the existing literature.

Structural and Social Origins

This paper examines eugenics in Brazil, the largest country in Latin America and the first to have a significant, organized eugenics movement. Between 1900 and 1940, Brazil was undergoing profound social and political changes caused by late and "dependent" industrialization, urbanization, and massive European immigration. Many of these changes were associated with eugenics in other parts of the world. But in Brazil they were taking place in an "underdeveloped" country whose population was largely Catholic, rural, racially mixed, and illiterate. Because of its tropical climate and "mongrelized" people, Brazil represented all that Europeans regarded as "dysgenic." What would a eugenics movement be like in a country where a tiny elite of mainly European descent ruled a vast and heterogeneous mass of impoverished people? Would the movement become tied to an extreme "race hygiene" movement? In view of the 1988 celebration in Brazil of the one hundredth anniversary of the abolition of slavery, what light does the history of eugenics shed on the racial mythologies and social realities that have shaped Brazil's past?

Interest in eugenics in Brazil predated World War I. The Brazilian term for eugenics, *eugenía* (as distinct from the Spanish term *eugenesia*), was introduced as the title of a medical thesis at the Rio Medical Faculty by Tepedino in 1914. The term, incidentally, had been proposed by the Brazilian philologist João Ribeiro in preference to the other word canvassed in Portuguese, *eugênica*, which some scientists and grammarians wanted (Kehl [1929] 1935, p. 15; Roquette-Pinto 1927, p. 167). The term *eugenía* was further distinguished by carrying the accent over the *i*, in order perhaps to emphasize its similarity to the French *eugenique*, which similarly stressed the end of the word. Be that as it may, *eugenía* was the word routinely used, though often without its accent.

The timing of this eugenic start in Brazil deserves comment. The founding of the first Brazilian eugenics society after the war in 1918, only ten years after the equivalent British society and only six years after the French, suggests how attuned Brazilian scientists were to European scientific developments. Structurally and socially, however, the origins of the eugenics movement related less to European than Brazilian developments. Of these, four were of special importance.

First was Brazil's entry into the world war on the side of the Allies in 1917. The subjects of wartime readiness and discipline, of control and order, of Brazilian capacities and racial capabilities, were much on the minds of the elites (Fausto 1978, pp. 401–26). The European nation-states had long been symbols in Brazil of all that was "civilized" and "advanced," as contrasted with Brazilian "barbarism" and "backwardness." Their collapse generated a new nationalism, based on a desire to project Brazil into world affairs, to define Brazilian realities in Brazilian terms, and to find Brazilian solutions to Brazilian problems. Whereas in Europe the war intensified fears about national degeneration, in Brazil it created a new optimism about the possibility of national regeneration, an optimism balancing more traditional fears of decay. It was a point made by Kehl in calling for a eugenic effort in the country. Throughout the 1920s eugenics was associated with patriotism and the call for a larger role for Brazil in world affairs.

Second, eugenics emerged in Brazil in response to a pressing national issue that Brazilians in the 1920s referred to as "the social question"—the appalling misery and ill health of the working population. This population was largely black and mulatto. Brazil had been the last country in the hemisphere to eliminate slavery; thirty years before, in 1888, the last seven hundred thousand slaves had been emancipated. The former slaves were left to fend for themselves, without education or recompense, in a country undergoing rapid social and economic change. Many of them joined the migration of the poor and unskilled into the cities, to compete on unfavorable terms for jobs with the more than one and a half million white immigrants who entered the country between 1890 and 1920. One result of the wave of migration and immigration was the relatively sudden spurt in industrialism and urbanism in Brazil. The population of São Paulo, for instance, jumped from only 129,409 people in 1893 to 240,000 by 1900—an increase of almost 100 percent in seven years. By 1907 Italians alone outnumbered Brazilians in the city by two to one (Stepan 1976, p. 136). The federal capital to the north-east, Rio de Janeiro, was by this date a city nearing 800,000. Although both cities had undergone extensive remodeling and "civilizing," and both had quite effective public sanitation services capable of dealing with epidemic diseases, endemic diseases were left untreated, morbity rates were high, and the general standard of housing and sanitation of the poor was bad beyond belief.

Poverty, migration, immigration, and unemployment helped usher in a period of radicalized politics, protests, strikes, and work stoppages, climaxing in a huge strike in 1917. The strike demonstrated the political potential of the new industrial working class, but it also demonstrated the power of the municipal and state authorities to use the police and militia ruthlessly, as British and North American visitors remarked, to put down industrial unrest. Traditionally, the educated elites feared violence and danger from blacks and mulattoes, who were portrayed as lazy, sickly, drunk, and in a constant state of vagabondage. Now were added new fears about disorder and violence by foreign-born factory workers (Fausto 1983). The threat of urban violence called into question the adequacy of old-style laissez-faire liberalism to solve "the social question," and suggested new roles for the state in structuring more harmonious relations between employer and employee and by intervening directly in social life. In contrast to British eugenics, which constituted a response to the perception that years of social welfare legislation had apparently failed to improve the mental, physical, and moral conditions of the poor (Stepan 1982, pp. 117–18), Brazilian eugenics was associated with the call for the introduction of such social welfare legislation as a way of improving the Brazilian people, and influenced the form it took.

A third factor in the rise of eugenics was the contemporary state of Brazilian science. Eugenics in Brazil was not associated, as it was in Britain, with controversies concerning the relative merits of biometrics and Mendelian genetics. Even by the 1920s, Darwinian biology and genetics were barely established as fields of scientific research. Brazil had as yet no university departments of science, and biological work was confined to medical schools and agricultural institutes (of which there were very few), and to the Oswaldo Cruz Institute, which had been founded in 1902 as a school of tropical medicine and was perhaps the best-known center of medical research in Latin America at the time (Stepan 1976). But if Brazilians were still largely consumers of science rather than producers, nevertheless the history of eugenics in Brazil must be seen as part of a generalized enthusiasm for science as a "sign" of cultural modernity. The extraordinary success of the sanitation campaigns against smallpox, the plague, and yellow fever, led by Oswaldo Cruz between 1902 and 1917, had given great cachet to the "sanitation sciences" and had stimulated the growth of a scientifically oriented professional and medical class that was increasingly visible and integrated into state and federal policy organizations. "Health" had become very recently a politically acceptable political objective. Eugenics appealed, as it did elsewhere, to an expanding medical profession whose members were eager to promote their role as experts in shaping social life and naively optimistic about their own power to do good. It was a group little given to revolutionary analyses of the economic and racial roots of Brazil's social miseries.

Fourth, the emergence of Brazilian eugenics was conditioned by its racial situation. Brazil was a racially mixed nation, produced out of the fusion of Indian, African, and European peoples. Ever since the transfer of the Portuguese crown from Lisbon to Rio de Janeiro in 1808, race and race relations had been a central aspect of social reality and ideological debates about Brazilian "capacity" and national destiny. And especially since abolition in 1888 and the formation of the First Republic the next year, science had emerged as a tool of growing authority for social, and especially racial, interpretation.

Brazilian doubts about the country's racial identity had been reinforced by racialist interpretations of Brazil from abroad. As a culturally dependent nation, Brazil was strongly influenced by racial ideas from Europe, particularly France. Martins comments that Brazilians had a tendency to "live vicariously their own existence, as though it were a reflection in a mirror" (1978, 5:6). Intellectuals had to contend with the fact that, in text after text of European social and scientific analysis, Brazil was held up as a prime example of the "degenerations" that occurred in a racially mixed, tropical nation (Stepan 1985a). Buckle, Kidd, Le Bon, Gobineau, Lapouge, and various social Darwinists were widely quoted in Brazil for their theories of Negro inferiority, mulatto degeneration, and tropical decay (Martins 1978, 5:84). According to these scientists and others like them, the "promiscuous" crossings that had occurred in Brazil from colonial times until the present had produced a degenerate, unstable people incapable of progressive development.

Many of the Brazilian elite shared this view. The themes of tropical and racial degeneration run through Brazilian medical, bacteriological, and racial writings from the early nineteenth century until well into the "revisionist" period of Freyre in the 1930s and 1940s (Stepan 1976, pp. 57–58). Especially following abolition in 1888, science was increasingly used, as it had been in Europe since the Enlightenment, to define how much "nature" would limit the social and political equality of blacks and mulattoes in the new republic. Raimundo Nina Rodrigues, the founder of "scientific" anthropology in Brazil, was almost as racialist in his outlook as Brazil's severest critics. His anthropological studies revealed to him a complex, multiracial, mixed society that had forged no single, stable, ethnic type and whose foreseeable future was as a black, not white and European, nation ([1894] 1938, pp. 117–44; Correa 1982). Euclides da Cunha's masterpiece of social analysis, *Rebellion in the Backlands* ([1902] 1944), which recounted the story of an armed rebellion by the *sertanejos* of Canudos in the northeastern, poverty-stricken area of Brazil, synthesized the science of his day to argue that the mixture of races "in addition to obliterating the pre-eminent qualities of the higher race, serves to stimulate the revival of the primitive attributes of the lower; so that the mestizo—a hyphen between the races, a brief individual existence into which are compressed age-old forces—is almost always an unbalanced type" ([1902] 1944, p. 85).

Given the above circumstances, eugenics, by definition the science of "racial improvement," was of obvious appeal to an elite convinced of the power of science to create "order and progress" (the motto of the republic) and troubled by the racial makeup of their country. Their interest was, if anything, "overdetermined." Though this interest was never as institutionally consolidated as it was in Europe, nevertheless the language of eugenics and hereditary improvement of "the Brazilian race" had a surprising impact, intellectually and legislatively. In many ways, indeed, eugenics provides an indispensable context for understanding the deepening involvement of the state in the management of "racial health" throughout Latin America between 1920 and 1940.

The Eugenics Movement, 1917–1929

The first eugenics organization in Brazil appeared in the city of São Paulo, which by the First World War had emerged as a major force in national politics and as the capital city of Brazil's most economically powerful state. In 1917 Renato Ferraz Kehl organized a meeting of physicians to discuss Galton's new science of eugenics, prenuptial examinations, and the proposed revisions of the civil marriage code allowing consanguineous marriages (which most physicians opposed, some on eugenic grounds). Following the meeting Kehl sent a circular in December 1917 to city and state physicians proposing the creation of a new scientific society and inviting his colleagues' participation. The Sociedade Eugênica de São Paulo (São Paulo Eugenics Society) held its first meeting on 15 January 1918 (Kehl 1931g).

The society's membership numbered 140. However, as was true of most of the eugenics societies in Europe, the size of the São Paulo Eugenics Society was less important than the character of its membership. Since, according to Love (1980, p. 154), the São Paulo elite remained small in numbers between 1889 and 1937, the society involved many of the medical and professional elite of the city and nearby towns. Of the members, only two were listed without the title of "Dr." (usually signifying in Brazil graduation from either medical or law school), one being a Mr. Rangel and the other the well-known writer Senator Alfredo Ellis. The society had no women members, and only eighteen members from outside the state. In addition, Victor Delfino, the founder of eugenics in Argentina, and Carlos Enrique Paz Soldan, the pioneer of "social medicine" in Peru, were named as corresponding members (*Annaes* 1919, pp. 39–43).

The society sought to project itself beyond the state of São Paulo by asking Rio de Janeiro's Belisário Penna, a well-known sanitation expert, to serve as one of three honorary vice-presidents (the others being professors A. de Sousa Lima and Amancio de Carvalho). The actual president was Arnaldo Vieira de Carvalho, director of São Paulo's new medical school, founded in 1913. Among the society's more important members were Vital Brasil, bacteriologist at the Butantan Institute (later to develop into the best-known snake-serum institute in Latin America); Artur Neiva, a microbiologist who had recently left the Oswaldo Cruz Institute of Rio de Janeiro for São Paulo to take over and remodel the state's sanitation services; Luís Pereira Barreto, well-known physician and positivist; A. Austregesilo, psychiatrist and professor at the Rio Medical School; and the young Fernando de Azevedo, who would go on to a distinguished career in education. Juliano Moreira, director of Brazil's National Mental Asylum, located in Rio de

Janeiro, sent a letter of congratulation to the society and advised it of his own eugenic efforts in the field of mental hygiene (*Annaes* 1919, p. 24). The São Paulo Eugenics Society had an initial success, holding four well-attended meetings between January 1918 and December 1919 in the hall of the Santa Casa de Misericordia, the traditional meeting place of the state's most important scientific society, the Medical and Surgical Society (Sociedade de Medicina e Cirurgia). From the beginning the society defined itself as a learned, scientific organization from which would flow scientific studies, conferences, and propaganda on the physical and moral strengthening of the Brazilian people (*Annaes* 1919, p. 35).

The meetings of the São Paulo Eugenics Society were organized by Kehl, who was to remain the chief propagandist for eugenics in Brazil and whose whole life was to be identified with the movement. Kehl's position as secretary allowed him to orchestrate the meetings. He reminded the society of the advances made in eugenics in Europe and of the need for Brazil to join the advanced world in studying heredity, evolution, and the influence of the environment, economic conditions, legislation, customs, and habits on the Brazilian race. He assured his listeners that eugenics was no longer a utopian fantasy, but a reality of modern scientific nations (Kehl [1929] 1935, pp. 15–27).

In addition to its regular sessions, the society organized several talks that brought eugenics into the public arena, such as Rubião Meira's lecture, "Factors of Degeneration in our Race and the Means of Combating Them," and Kehl's talk to the Young Men's Christian Association. Many of these talks were reprinted in a 1919 volume published by the society, *Annaes de Eugenía*. The small size of the professional and educated class in Brazil and the close contacts between journalism, literature, and medicine gave eugenics a hearing in the daily and weekly press, whose reaction was highly favorable. Eugenics was greeted as a "new" science capable of ushering in a "new social order" via the medical improvement of the human race (*Annaes* 1919, pp. 15–16).

Despite initial enthusiasm, however, the São Paulo Eugenics Society came to an end in 1919, unable to survive Carvalho's death that year and Kehl's departure for Rio de Janeiro (Kehl 1923b, p. xii). With Kehl's departure the locus of eugenics passed north to the federal capital. There Kehl kept the eugenic interest alive by a stream of pamphlets, books, and debates, many of which were reported in the medical press and daily newspapers. By 1947 he had published twenty-six books, the most important being *A cura da fealdade* (1923), *Eugenía e medicina social* (1923), *Lições de eugenía* (1929), and *Aparas eugenicas* (1933). These books were well received and widely reviewed, and many were reissued more than once. In addition to Kehl's writings, Penna's *Saneamento no exército* (1920) was part of the early eugenic effort, as was Monteiro Lobato's *O problema vital,* which was published in 1918 jointly by the São Paulo Eugenics Society and the Pro-Sanitation League of Brazil (Liga Pro-Saneamento do Brasil). In his multivolume study of Brazilian writing, Martins refers to a veritable stream of works on eugenics and related themes in the 1920s and 1930s expressing a nostalgia for hygiene and "purification" (1978, 6:263). According to a bibliography of eugenics prepared by Kehl, seventy-four major publications on eugenics appeared in Brazil between

1897 and 1933 (1933, pp. 261–71). His list left out many books and pamphlets on eugenic themes (e.g., mental hygiene books) as well as much eugenically influenced periodical literature. Kehl's bibliography included twenty-four undergraduate medical theses from the Rio Medical School, for instance, but did not include seven other theses from the São Paulo Medical School appearing between 1919 and 1937 that had "eugenics" specifically in their titles. In fact, Kehl seriously underrepresented the cultural production of eugenics in Brazil by leaving out, for reasons that will be made apparent later, many works that did not fit his own definition of eugenics—for instance, some of the writings of Octavio Domingues.

As the eugenic creed won new converts, the language of eugenics began to infuse scientific discussions of "health." Improvement was now discussed in terms of Galtonian "dysgenic" and "eugenic" factors, fitness and unfitness, and hereditary *taras* (defects). Penna's 1918 book *Saneamento do Brasil* had been devoid of eugenic language; his new book of 1920, based on a series of lectures to the Military Club of Rio and published as *Exercito e Saneamento,* had the very same theme—the disgraceful state of sanitation in Brazil—but now the problem was presented as the hereditary degeneration of the Brazilian people that required a "eugenic" solution.

Though Kehl was unable to organize a new eugenic society, eugenics found a place in the new Liga de Hygiene Mental (League of Mental Hygiene), which was founded in 1922 in the federal capital by Gustavo Reidel, director of the Colônia de Psychopathas do Engenho do Dentro (Engenho do Dentro Psychopathic Asylum) (Freire Costa 1976). The league was organized into twelve permanent sections or committees, each with ten members, as well as twelve regional representatives and twelve Brazilian corresponding members. Many of the members came from the staffs of the state and municipal mental asylums and reformatories. In addition to the work of the committees, the league held regular meetings on a monthly basis. Although its subventions from the municipality of Rio and the federal government, which registered the league as a public utility in 1923 and gave support to its free ambulatory clinics, were not always adequate or secure, from the beginning the league enjoyed considerable success and was a notable addition to the scientifically oriented societies of the federal capital (*Archivos Brasileiros de Hygiene Mental* [hereafter ABHM] 1929, 2: 48–56; 5: 1–3).

Reidel's original goal for the league was to promote the "new" psychiatry, to widen the scope of the psychiatric profession in everyday life, and to realize a program of mental hygiene, particularly for the poor and criminally insane. The league concerned itself with juvenile delinquency, prostitution, alcoholism, venereal diseases, nutrition, and criminality. Like the American psychiatrists whom the mental hygienists in Brazil wished to emulate, the league's members considered themselves "progressive" in the sense of being oriented toward individual psychiatric treatment and, in the case of the criminally insane, toward the criminal rather than the crime (Rothman 1980, pp. 5–6). In fact, their vision of Brazilian society was decidedly conservative (Pereira Cunha 1986). The League of Mental Hygiene presented itself as a professional, scientific, humanitarian organization in keeping with advanced psychiatry in the rest of the world.

The league's purpose, according to its statutes, was to "realize a program of mental hygiene and eugenics in individual, school, professional and social life," and to publicize the pathological conditions caused by syphilis, alcohol, and other factors (ABHM 1929, 2: 39–47; 1941, 13: 91–95). But the league's emphasis on eugenics intensified over the years as a new group of psychiatrists, such as Ernani Lopes (who became president in 1929), took over its leadership. In order to signify the eugenic improvement of mind, a new term, *eufrenia* (euphrenics), was coined (ABHM 1932, 5: 3). This link between psychiatry and eugenics was not surprising, given the hereditarian orientation of psychiatry in Brazil and the psychiatrists' extraordinary concern with the dangers to society caused by the mental illness and social "pathology" of the poor—by crime, delinquency, and prostitution.

Kehl had become active in the league by 1925, and by late 1929 the league's membership included many of the more prominent medical and nonmedical scientists of the city, such as Juliano Moreira himself, the director of the National Insane Asylum; Miguel Couto, president of the National Academy of Medicine and one of the leading clinicians of Rio de Janeiro; Fernando Magalhães, professor of gynecology and obstetrics at the Rio Medical School; Carlos Chagas, protozoologist, discoverer of Chagas' disease *(Trypanosomiasis americana)*, and director of the Oswaldo Cruz Institute; Edgar Roquette-Pinto, eugenicist, physical anthropologist, and director of the National Museum in Rio de Janeiro; the hygienist and pioneer of legal medicine, Afrânio Peixoto; and such specialists in mental illness as Henrique Roxo and A. Austregesilo. The League of Mental Hygiene published its own journal, the *Archivos Brasileiros de Hygiene Mental,* which, after a hiatus between its first volume in 1925 and its second in 1929, appeared more or less continuously throughout the 1930s.

A third strand of eugenics in Brazil was found in medical-legal circles, where the problems of crime and legal responsibility were closely linked to the racial and eugenic issue. Affânio Peixoto wrote widely on eugenic themes, promoting the importance of eugenic medicine for police work and generally advocating cooperation between the legal and medical professions (e.g., 1931, 1942). Meanwhile, Miguel Couto raised eugenic issues concerning immigration in sessions at the National Academy of Medicine (*Boletim da Academia Nacional de Medicina* [hereafter BANM] 1923, 96: 33–34), while Roquette-Pinto's *Seixos rolados* (1927) contained a forty-page chapter entitled "Laws of Eugenics."

These various strands of eugenics came together in the most important public manifestation of Brazilian eugenics of the 1920s, the Primeiro Congresso Brasileiro de Eugenía (First Brazilian Congress of Eugenics) held in Rio de Janeiro in 1929. The centennial celebrations of the founding of the National Academy of Medicine provided the opportunity for Miguel Couto, president of the academy, to call the congress into being. With Roquette-Pinto presiding, it met 1–6 July and was attended by some two hundred professionals, including medical clinicians, officials from the state psychiatric and hygiene institutions and services, medical-legal experts, journalists, and several federal deputies (*Brasil-Medico* 1929, 43: 842–45). Delegates from Argentina, Peru, Chile, and Paraguay were also present, including Paz Soldan, whose 1916 pamphlet *Un programa nacional de política sanitária* had long been regarded by Brazilian eugenicists as a fundamental text.

The conference themes were broad indeed—marriage and eugenics, eugenic education, protection of nationality, racial types, the importance of genealogical archives, Japanese immigration, antivenereal campaigns, intoxicants and eugenics, the treatment of the mentally ill, and the protection of infancy and maternity. The participants passed several resolutions, the most controversial being a call for a national immigration policy to restrict entry into Brazil to those individuals deemed "eugenically" sound on the basis of some kind of medical evaluation.

The success of the congress and the publicity it received in the daily and medical press suggested that eugenics was about to enter a new phase of activity. Already in January 1929, Kehl had begun publishing the monthly journal *Boletim de Eugenía* as a supplement to the medical journal *Medicamenta;* it appeared from July 1929 until December 1931. The League of Mental Hygiene also revived its *Archivos,* moribund since 1925, and intensified its eugenic work, as can be seen by numerous editorials trying to get its antialcohol campaign officially endorsed by the national government. A feeling that the moment was ripe for eugenics was confirmed by the political agitation in the country, agitation that resulted in the "revolution" of 1930.

Before looking at eugenics in the 1930s, however, perhaps it is time to look back, from a moment when eugenics seemed on the threshold of consolidating itself institutionally and ideologically, at the movement of the early and mid-1920s, and ask: What was eugenics in Brazil?

Sanear é Eugenizar: To Sanitize Is to Eugenize

"Apparently the Brazilians interpret the word [eugenics] less strictly than we do, and make it cover a good deal of what we should call hygiene and elementary sexuology [*sic*]; and no very clear distinction is drawn between congenital conditions due to prenatal injury and diseases which are strictly genetic." So wrote the British eugenicist K. E. Trounson in 1931 after studying the eugenic materials sent to him from Brazil by Kehl. Trounson added: "Friction in the family, sex education, and premarital examinations and certificates seem to be the subjects of most interest to Brazilian eugenists, whereas genetics and natural and social selection are rather neglected; the outlook is more sociological than biological" (Trounson 1931, p. 236).

Seen through British eyes, Brazilian eugenics may have seemed an example of misunderstanding or sloppy scientific thinking. Seen from the Brazilian perspective, however, the British missed the underlying logic of their eugenic science, a logic that allowed many Brazilians to claim *sanear é eugenizar*—to sanitize *is* to eugenize (Kehl 1923b, p. 20). Although it may seem to confound the image of eugenics based on the historical experiences of Britain and the United States, Brazilian eugenics exemplifies an important variant of the worldwide movement, one that was pervasive throughout Latin America.

What the British eugenicist apparently failed to notice was that the Brazilian eugenics movement derived scientifically not from Mendelian conceptions of genetics, but from neo-Lamarckian ones. The centrality of Mendelian genetics to modern science and the discrediting of Lysenkoism have made historians over-look the continued vitality of neo-Lamarckian ideas in French and Latin Ameri-can biology and medicine in the 1920s and 1930s. There was nothing new, of course, about the belief in the inheritance of acquired characteristics; on the con-trary, Lamarckism had dominated the science of heredity in Europe and the United States for most of the nineteenth century. What was new to Lamarckism in the early twentieth century was the challenge provided by Mendelian genetics and the association of heredity with the new social goal of human betterment.

A eugenicist's conception of how heredity could be improved depended on his understanding of the nature of that heredity. Despite the eventual success of Men-delian genetics, in the 1910s and 1920s neo-Lamarckians generated a large liter-ature on the inheritance of acquired characteristics as they were forced to come to terms with the findings of the Mendelians. Indeed, some scientists retained their belief in a neo-Lamarckian form of inheritance well into the 1940s, the era of the "new synthesis" of evolutionary biology and Mendelian genetics.

Neo-Lamarckism was particularly prevalent in medical circles. The continued reliance on scientifically refined Lamarckian ideas by physicians in these decades reflected not their stupidity or ignorance, but rather the seeming intractability of certain problems in human pathology. Take, for example, the impact of parental venereal disease on the offspring. Did the child of such parents suffer in "fitness," and was this unfitness transmitted by heredity? Was there not a "hereditary-syphilitic" condition? This was the view of the majority of physicians in France, where Lamarckian views had wide currency and a Lamarckian eugenics move-ment developed (Schneider 1982).

By cultural tradition, Brazilian scientists learned their science from France (see, for example, Meira 1907). Eugenics was no exception, as was made clear at the first meeting of the São Paulo Eugenics Society, where the French Eugenics Society was taken as an organizational model and the French statutes reproduced word for word (*Annaes* 1919). Kehl commented that Brazilians had remained ignorant of eugenics because it was written in German and English (Kehl 1923b, p. vi). Though Kehl himself read German, the names invariably cited by Brazilian eugenicists were French authorities—Pinard, Houssay, Landouzy, Perrier, Morel, Fournier, Richet, Apert, and Moreau.

The neo-Lamarckian basis of the eugenic views of Kehl and many of his Bra-zilian colleagues was often disguised by their constant reference to Galton as the father of eugenics, and to Mendel, and by the absence of direct reference to Lamarck. Kehl often referred to neo-Lamarckian and Mendelian genetics as though they were compatible variations of the same science of heredity (Kehl 1935, pp. 78–99). Indeed, the eclectic style of much of the eugenic writing in Bra-zil, and the uncritical use of European sources—as when Galton's ancestral law was presented without comment in conjunction with Mendel's laws (Kehl 1935, p. 107)—reflected the fact that few physicians in Brazil had learned genetics in medical school or were involved in genetic research.

The reconciliation of Lamarckian-style genetics with the language of modern

Mendelism was not untypical of the times. Bowler notes that the rediscovery of Mendel forced the neo-Lamarckians to sharpen and limit the scope of the inheritance of acquired characteristics (1983, pp. 92–94). Very often, the Lamarckians accepted Mendelian laws of inheritance, leaving a space nonetheless for the idea that somehow an influence from the milieu could permanently alter the germ plasm. The language of the two kinds of inheritance merged, allowing eugenicists to associate themselves with Mendelism, or to use Mendelian genealogical trees for the study of inheritance in families, or the chromosome theory and the idea of the gene, without giving up their deep-seated belief that at least some acquired characteristics were inherited (Monteleone 1929).

We see the French derivation of this neo-Lamarckism most clearly in Kehl's adoption of Forel's theory of "blastophthoria" to explain how intoxicants, venereal diseases, and tuberculosis could cause hereditary decay (Kehl [1929] 1935, p. 141). So unconscious were most Brazilian eugenicists that their movement was based on a scientific misconception that it came as a surprise to them when a new generation of geneticists pointed it out in the late 1920s.

Since Lamarckian eugenicists drew no simple dichotomy between "nature" and "nurture," improvements in nurture could be assumed to improve hereditary fitness over time. This "optimistic" view of the possibility of hereditary improvement could be countered by the correspondingly pessimistic view that the accumulated burdens of past negative environmental influences had created a thoroughly degenerate heredity that was difficult to improve rapidly.

In Brazil in the early 1920s, the optimistic style of Lamarckian eugenics predominated over the pessimistic in public activities. Structurally and scientifically, Brazilian eugenics was broadly congruent with the sanitation sciences and was interpreted by some as simply a new "branch" of hygiene. Hence the Brazilians' insistence that "to sanitize is to eugenize." Olegario de Moura, the vice-president of the São Paulo Eugenics Society, claimed that sanitation was the same thing as that called by some "eugenics," adding it was better to call it sanitation for public understanding, even though eugenics was better "scientifically." Moura equated them as follows: *"Saneamento-eugenía é ordem e progresso"* (sanitation-eugenics is order and progress) (1919b, p. 83). The neo-Lamarckian foundations of eugenics and the broad congruence between eugenics and sanitation were reflected in many of the other Latin American eugenics movements as well. In Argentina, for example, Frias referred to the "methods of positive eugenics for improving the state of public health, combating all sorts of epidemics and endemics—the battle against malaria, tuberculosis, cancer, plagues, venereal infections, alcoholism." At times the public health branch of eugenics was referred to as "indirect eugenic methods" (1941, pp. 149–50).

In addition to its compatibility with sanitation, the neo-Lamarckian style of eugenics was congruent with traditional morality, which gave it further appeal in Brazil. Since the neo-Lamarckian style of eugenics kept open the possibility of regeneration, and a place for moral action, it was an approach that fitted in well with Catholic doctrine and allowed a fusion of moral and scientific language (Nye 1984, pp. 119–22). Poverty, venereal diseases, and alcoholism could be seen as products of both social conditions and moral choice.

Although the causes embraced by Mendelians and neo-Lamarckian eugenicists were sometimes similar, the logic of the two styles was considerably different and often led eugenicists to different, even opposite, conclusions. While Leonard Darwin, president of the Eugenics Society in England, believed eugenicists should aid in the attack on social evils such as alcoholism, he was quite firm in stating that alcoholism was not in itself a eugenic issue since by Mendelian conceptions alcohol did not alter the germ plasm (1926, pp. 83–93). To the neo-Lamarckians, however, alcoholism was a eugenic issue precisely because it was both a symptom and a result of social ills and because the cycle of causes could be interrupted by social action. Instead of a collision between the hereditary movement of eugenics and the environmentally oriented sanitation movement, as occurred in Britain (Searle 1981), in Brazil there was built-in cooperation. In Brazil as in France, therefore, the neo-Lamarckian views of the eugenicists allowed alliances to be forged between them and the more broadly defined sanitation and public hygiene organizations. In Brazil, for instance, Lamarckian eugenics brought in allies from the rural sanitation movement, such as Belisario Penna, whose long journey by horseback in 1913 among the diseased populations of the northeastern states of Brazil had made him a crusader for rural health (Neiva and Penna 1916). As Kehl's eventual father-in-law, Penna was a most useful and strategic addition to eugenics, capable of winning support from like-minded hygienists. Other allies were recruited from the pro-sanitation and nationalistic leagues that had sprouted up in Brazil before and after the war (Moreira 1982). There was a considerable overlap in the membership and style of discourse of the Liga Nacionalista de São Paulo (Nationalist League of São Paulo) and the São Paulo Eugenics Society; indeed, the president of the second, Arnaldo Vieira de Carvalho, was the vice-president of the first.

Since Brazilian eugenicists did not separate nature from nurture, a variety of sanitary reforms could be assumed to improve hereditary "fitness" and therefore to be "eugenic." To the Brazilian medical intelligentsia, already predisposed to promote sanitation as a cure-all for Brazil's woes, eugenics appealed as a scientific extension of the heroic work of figures such as Oswaldo Cruz and Carlos Chagas, as a way of reducing the extraordinarily high infant mortality rates of the poor and the sickly condition of the masses. Even the promotion of sports and physical fitness, which Fernando de Azevedo made his particular cause in the São Paulo Eugenics Society (1919a, 1919b, [1920] 1960) could be seen as "eugenic" because it "improved the race." Eugenics had become a metaphor for health itself.

The reformist, neo-Lamarckian style of eugenics was represented in its purest form, perhaps, in the eugenic antialcohol campaigns of the 1920s. Long seen as a social and moral evil, especially of the poor and black population, alcoholism was reformulated as an "enemy of the race" because the "vice" caused hereditary conditions linked to crime, juvenile delinquency, prostitution, and mental illness in the rural and urban poor. The hygienist and eugenicist Peixoto, for example, said alcohol caused racial degeneration because children of alcoholics were defective and predisposed from infancy to meningitis, convulsions, mental deficiency, madness, and crime ([1933] 1936, pp. 209–11).

Eugenic and psychiatric views of alcoholism came together at the League of

Mental Hygiene. The league tried to educate the public about the evils of intoxicants, which were seen as having a "sterilizing" influence on the masses by causing low reproductive rates, high mortality, and hereditary taint (ABHM 1929, 2: 12–16; 1931, 4: 167–68). In an article in *Brasil-Médico* in 1929, Francisco Prisco commented on the reduction in the size of the working population caused by alcoholism, and its supposedly hereditary consequence (pp. 801–5).

From the league derived the "antialcohol" weeks held in Brazil in 1927, 1928, 1929, and 1931. These were of a quasi-popular character, involving public addresses by such well-known figures as Juliano Moreira, long considered the father of psychiatry in Brazil. In October 1929, following its third "antialcohol" week, the league created a new section devoted specifically to agitating against alcoholism and stimulating public interest and financial support for its work. American-style prohibition, favored by Peixoto (Ribeiro 1950, p. 308), taxes on imported and nationally produced alcohol, and special reformatories for the treatment of the inveterate drunkard were all discussed and promoted by the league between 1925 and 1935 as "eugenic" measures. The league eventually became so identified with antialcoholism that the editors of the *Archivos* later protested that it stood for much more than just that (ABHM 1933, 6: 193–94).

If eugenics tended to merge in the public mind with sanitation, the eugenicists were not without their own special programs that distinguished them from other sanitary reformers. It was the eugenicists, for instance, who offered some of the first public lectures and courses on human heredity and the science of genetics, indicating the ways in which eugenics could serve as a vehicle for the introduction of genetics in countries unfamiliar with the subject. In 1929 a weekly lecture series was offered by Fernando Magalhães at the Academy of Fine Arts (*Boletim de Eugenía* [hereafter BE] 1, 2: 49–50). Somewhat more technical were Octavio Domingues's lectures to medical and agricultural students in 1930 at the agricultural school in the state of São Paulo (Kehl 1931c). Eugenicists were also responsible for organizing popular contests for "eugenic" families, offering monetary awards to children who were found to be hereditarily fit and eugenically "beautiful," thereby best representing the Brazilian "race."

By emphasizing that it was through their effect on the reproductive cells that environmental influences most threatened heredity, eugenicists drew particular attention to human reproduction itself—to sexuality, marriage, and the problem of infectious (and especially venereal) diseases in marriage. As Schneider has pointed out in his study of French eugenics, the Lamarckian eugenicists helped to revive "puericulture" (the cultivation of the child) and to extend its meaning to include "puericulture before birth." The popularization of the word *puericulture* and the new emphasis on child welfare and maternal health in Brazil in the 1920s were closely associated with eugenics (Almeida 1925). The League of Mental Hygiene said that eugenics was intimately tied to puericulture and marriage. We see the same association in the second Pan American Conference of Eugenics and Homiculture, held in Buenos Aires in 1934, where Uruguay's new *Codigo del Nino* (children's code) was held up as a model of eugenic legislation (1934, pp. 137–58).

Through eugenics, too, subjects traditionally outside polite discourse were made respectable. Lectures, some of them quite explicit, were given to educated young men and medical students on sexual hygiene and the prevention of venereal disease (Moura 1919b, pp. 83–89). For young women eugenics meant "dignified" motherhood, with an emphasis on maternal health and prenatal care. Kehl's brochures on how to choose eugenically fit husbands and wives enjoyed a wide circulation (1925). Kehl also used the *Boletim de Eugenía* to solicit answers to a questionnaire on suitable books on sexual and "eugenic education" for young girls in the home and school (1930a, 1930b). The eugenic interest in sex education had little to do with radical views about sexuality or sex roles. On the contrary, Brazilian eugenics was closely tied to conservative, family ideology; many eugenicists were critical of the Brazilian feminists (Hahner 1980) because feminism posed, in the eugenicists' opinion, a threat to the traditional reproductive role of women (Magalhães 1925; Peixoto 1944, p. 236).

Neo-Lamarckian eugenics in Brazil in the 1920s was not solely "optimistic" in style. The São Paulo Eugenics Society had originally divided eugenics into three kinds—"positive," which was concerned with sound procreation; "preventive," which dealt with the conquest of dysgenic factors in the environment (sanitation); and "negative," which aimed to stop the procreation of the unhealthy (Annaes 1919, p. 4). In the 1920s interest in preventive eugenics or sanitation predominated over both positive and negative eugenics. Nevertheless, Brazilian eugenicists did at times discuss abortion, birth control, and even sterilization as eugenic measures for the control of the unfit. Psychiatrists, medical-legal experts, and criminologists were particularly prone to raise the issue of sterilization for the control of reproduction of the "grossly" degenerate individual (Kehl 1923a, 1925b; ABHM 1925, 1: 194; 1931, 4: 245–48; Farani 1931; Cunha Lopes 1934). According to Ernani Lopes, a Dr. Alvaro Ramos had gone so far as to take the advice of Moreira, director of the National Insane Asylum, and undertake "eugenic" sterilizations of women diagnosed as suffering from the sexual derangement known as "perversity syndrome" (ABHM 1931, 4: 246–47).

On the whole, however, the Brazilian medical establishment was deeply conservative on the subject of reproduction and tended to oppose sterilization on any grounds whatsoever. Kehl, himself an advocate of eugenic sterilization for the "grossly degenerate," recalled in 1937 that a decade earlier in Brazil sterilization was considered absurd (pp. 67–73). Even in the more "negative" 1930s, Farani's article in the *Archivos* advocating sterilization was accompanied by a note from the editor explaining that the views expressed represented those of the author alone (ABHM 1931, 4: 169). It is interesting to note that Farani took the neo-Lamarckian line that blastophthoria caused by alcoholism could become hereditary and this justified sterilization in severe cases (p. 172).

Neo-Malthusianism, or "conscientious motherhood," was also discussed by eugenicists, but, once again, the Catholicism of most doctors and their profamily (and class) orientation restrained their enthusiasm (Farani 1931; Kehl 1935, pp. 212–17). The Brazilian doctors shared with their French counterparts a pronatalist ideology, which in the Brazilian case was based on the fear that the vast emptiness of Brazil, the diseased condition of the masses, and their low reproductive rates would prevent Brazil from becoming the powerful, modern nation they dreamed of.

The perceived need for more people in Brazil was an old theme in Brazilian politics and science, and was intimately tied to state policies to encourage immigration (Peixoto 1916, pp. 213–24). Moncorro Filho (1924) typically linked eugenics and puericulture to the problem of reducing high mortality rates and changing the low birthrates of the laboring population. Kehl repeated the adage that Brazil was a country of only 25 million people when it could be one of 500 million, and that what was needed was not birth control, but a restraint on the procreation of the ill (1923b, p. 48). In this context sanitation was seen as a form of "investment," in the traditional, political-economic sense of creating a healthy work force and preventing social revolution caused by misery. Couto spoke for most of his medical colleagues when he argued in 1932 that what Brazil needed was more people, not less—that a *solo fecundado* was a *solo defendido* (a country well peopled was a country well defended). Brazilian pronatalism meant that eugenicists rarely sounded the theme of the putative overfecundity of lower classes and races so crucial to the logic and rhetoric of eugenics movements elsewhere. Instead, they sounded the theme that Brazil was a "vast hospital" filled with diseased people requiring a program of sanitation (Stepan 1976, p. 115).

More appealing to the eugenicists in Brazil were prenuptial medical examinations—a kind of "birth control" without birth control. There was, once again, nothing new about obligatory medical (as opposed to religious) requirements for marriage (they had been written into Danish law as early as 1798, with little effect). Such examinations had many defenders in Europe and the United States in the early twentieth century, from feminists concerned with protecting women from venereal infection in marriage to physicians concerned with protecting children from the effects of parental infections. Eugenicists argued for voluntary or state-imposed medical impediments to marriage by pointing to the supposed hereditary damage that could be eliminated in the population by barring the syphilitic or eugenically unsound individual from marrying. Prenuptial examinations had been part of the eugenicists' goals in Brazil since 1918, when Kehl introduced the subject at the first meeting of the São Paulo Eugenics Society (*Annaes* 1919, pp. 3–7). In the 1920s prenuptial examinations became a subject of wider, if still primarily medical, discussion, as when Fernando Magalhães lectured at the First Brazilian Congress of Hygiene in 1924 on the need for premarital examinations of the dietary, economic, housing, and health conditions of prospective marriage partners.

Many physicians viewed prenuptial examinations as voluntary aids in encouraging large and healthy families. Others hoped to see them introduced as obligatory, state-controlled restraints on "diseased" marriages. Throughout the 1920s Latin American countries debated the possibility of introducing legislation to prohibit the marriage of individuals (usually male) with contagious diseases, or requiring some kind of prenuptial examination on either a voluntary or an obligatory basis (Jimenez de Asúa 1942). Though such proposed legislation in many cases merely extended preexisting health and other legal impediments to marriage, what was new was the vagueness of the diseases named (such as "hereditary epilepsy") and the physicians' confidence in urging state intervention in private life. In 1926 the Brazilian congressman Amaury de Medeiros presented to the commission on public health in the Federal Congress a bill for voluntary prenuptial examinations, which he described as a form of "constructive" (as opposed to negative) eugenics compatible with Brazilian (i.e., Catholic) traditions (Medeiros [1927] 1931). The examinations were directed against people with grave physical defects and transmissible diseases. Though many in the congress opposed the plan, Medeiros had the support of the eugenicists Kehl, Penna, Magalhães, Peixoto, and others on sanitary grounds, though Kehl, for one, hoped to see the examinations made obligatory. Medeiros's death in 1927, however, postponed any legislative action until the 1930s (Porto-Careiro 1933).

Race and the Eugenics Movement in the 1920s

And what about race? In recent years there has been considerable discussion among historians about eugenics, social structure, and race or class ideology. In Britain, for example, class rather than race was at the center of eugenic propaganda, especially the apparent class differentials in "fitness" and fecundity (McKenzie 1976). The eugenicists emphasized the scientific control of the population of the lower classes and the means available for encouraging the growth of the supposedly more eugenic middle classes. Nevertheless, though eugenics was associated with class ideology, no simple relationship existed between the eugenics movement and class. Recent research shows that eugenics in Britain cannot be understood merely as a direct projection or representation of class interests since many of the opponents of eugenics shared the class origins of the eugenicists themselves (Searle 1981).

The Brazilian eugenics movement of the 1920s is a particularly interesting case study of science and social ideology. On the one hand, eugenics was deeply structured by the racial composition and racial anxieties of the country. In a very fundamental sense, eugenics was *about* race and racial improvement, not class. The eugenics movement was "about" race because it focused its attention on the diseases that were seen as specially prevalent among the poor, and therefore mainly black or racially mixed, population. This population was perceived as being ignorant, diseased, and full of vice, with high rates of drunkenness, immorality, mortality, and morbidity. If publicly the word *raça* (race) in the eugenic literature was invariably used in the singular to refer to "the Brazilian people," privately it meant the "black race."

On the other hand, eugenics in the 1920s was not a Nazi-style race hygiene movement, intent on race sterilization or elimination. How could it be when even some members of the elite were uncertain of their own "purity of blood"? Indeed, if anything, in the 1920s eugenics was associated with the effort on the part of many elites to rescue their country from the charge of tropical decay and racial degeneration. By the 1920s a reformulated *ufanismo* (exaggerated pride) in Brazil was characteristic (Skidmore 1974). A more realistic nationalism gripped the nation, based upon the rapid expansion of an export economy built on coffee, industrialization, and the rise of new, professional, middle-class groups who hoped to reform the traditional politics of the republic and launch Brazil as a world power. There developed new concerns about the health of the labor force in the coffee economy and the factories of São Paulo and Rio de Janeiro, about the need to attract white immigrants to work in Brazil, and about the diseased and ignorant condition of the people in the cities and the vast hinterlands.

In this context Brazilians began to reject their traditional dependence on European values and knowledge, and to seek ways to reinterpret their own racial and climatic conditions so as to provide themselves with a more optimistic view of Brazil, in keeping with what they believed to be the country's immense natural resources and special racial makeup. Out of the attempt to reconcile their own limited understanding of social reality with the findings of modern science came a special Brazilian adaptation of the racial science of the day.

In the early part of the twentieth century, many hygienists in Brazil, for example, denied that Brazil's tropical environment was hostile to the white race, or was the cause of tropical disease. A Brazilian thesis of white "acclimation" in the tropics emerged in the work of physicians, running counter to the common European view that, for climatic reasons, the white race was unable to work and thrive in extreme heat, and that the racially mixed population of Brazil was doomed to degeneracy (Stepan 1985a). In his popular book *Minha terra, minha gente* (1916, pp. 207–8) and his technical work *Hygiene* (1917, pp. 68–69), for instance, the professor of public hygiene in the Rio medical school, Afrânio Peixoto, criticized European medical scientists for defaming the Brazilian climate and denied the existence of specifically "tropical" diseases (Skidmore 1974, p. 183).

The new bacteriological and microbiological sciences, represented in the activities of the Oswaldo Cruz Institute in the first two decades of the century, were greeted with enthusiasm by the elites precisely because they addressed directly the issue of tropical "degeneration." For many, the key to a great Brazilian future lay in public hygiene and the sanitation sciences. The identification of eugenics with sanitation was one result of the importance attached to tropical "health" in the 1920s. Peixoto, for example, combined eugenics and sanitation in a characteristic way, calling eugenics a new chapter in hygiene, allowing for health in gestation, physical education, intelligence, and morality. His theme was that prevention was better than cure and that preventive eugenics was the key to a healthy Brazil.

Even more critical to the history of eugenics in Brazil were the Brazilian scientists' efforts to rescue themselves from the charge of mulatto degeneracy. The negative assessment of the mulatto by European and North American scientists was countered by the Brazilian claim that it was through a racial mixing that Brazil would achieve her own "eugenic" future. By the 1920s extreme racialism, though never absent, was becoming the exception rather than the rule. Race relations, it was claimed, were different in Brazil from those in the United States. How different, and why, remain controversial subjects of interpretation, but the absence of legal segregation based on race (because control of social mobility by the white elite was possible via informal, extralegal mechanisms, such as patron-client politics) meant that the Brazilians could claim their national character was based on *o homem cordial* (the warm, privately oriented man), at ease with himself and with others and not given to racial intolerance (Lamounier 1978, p. 144). The middle class was expanding and drawing into itself educated, mixed-race individuals, such as the writer Antonio Machado and the scientist Juliano Moreira, director of the National Mental Asylum and honorary president of the league where eugenic issues were routinely discussed. At any rate, by the 1920s the elites were increasingly ideologically "assimilationist" in public discourse, even if privately and socially racialist and discriminatory.

By the 1920s, then, the overt expression of extreme racialism ran against the grain of social and ideological developments. Against a background of deep worry that Brazil had failed to achieve a homogeneous national type, and that the country was in fact menaced by racial degeneration, the thesis of *"branqueamento"* ("whitening") began to acquire more positive meanings and to shape the eugenic movement in interesting ways (Skidmore 1974, pp. 64–77). According to this thesis the past history of miscegenation between the three "races" that peopled the country—the Indian, the Negro, and the European—had prevented the development of racial conflict and patterns of segregation that characterized race relations in the United States. Moreover, racial mixing was seen as a cause not of degeneration, but of regeneration because it brought about a steady whitening of the population by natural means (Monteleone 1929, pp. 113–14). The remaining "pure" Negro and the indigenous Indian populations were disappearing, it was argued, owing to natural and social selection against them, high mortality and low reproduction rates, and social "disintegration" following emancipation. Meanwhile, white immigration was seen as a vehicle for rapidly increasing the proportion of whites, while crosses between mulattoes and whites favored whitening because of the whites' biological superiority and because mulattoes preferred partners whiter than themselves.

A scientific defense of the whitening thesis was provided in 1911 by the director of the National Museum, João Batista Lacerda, in a paper he prepared for the First Universal Races Congress in London. The following year, Lacerda calculated from Brazilian census data that by the year 2012 the Negro population would be reduced to zero and the mulatto to only 3 percent of the population (Skidmore 1974 p. 67)! A later statement of the whitening thesis was given in 1920 by the racially minded and very popular writer Oliveira Vianna when he argued in *Populações meridionais do Brasil* (1920) that via the "regressive influence of ethnic atavisms" and crossing with whites there would over time be a filtering out of mulattoes and the development of a clear biological predominance of whites over Negroes and mestizos.

The growing intellectual and political popularity of the whitening "myth" in the 1920s and 1930s is more important than its sociological accuracy, though the kinds of social and class relations that gave it support are worth further study. The large scale of white immigration in the southern, least black parts of Brazil in the last decade of the nineteenth century and the first two decades of the twentieth century played its part, as did the extraordinarily high infant mortality rates among the poor black and mulatto populations. The intelligentsia's faith in the power of "whiteness" to dominate over "blackness" was reinforced by the continued "success" of informal mechanisms of social control of black mobility, as well as more institutionalized forms of repression, such as the use of the police to keep the social and racial "order," until well into the 1930s when social and power relations were reorganized within the new authoritarian state.

In short, doubts about the racial situation in Brazil were giving way to a cautiously optimistic racial interpretation of the "social problem" that influenced the ways in which the new science of eugenics entered into scientific discourse and social debate. The whitening myth clearly rested on the idealization of whiteness; it represented a kind of wishful thinking by an elite in charge of a multiracial society in an age dominated by racism, a yearning for a real sentiment of *brazilidade* (Brazilianhood) in a country torn by racial and class cleavages. It was a reassurance that "aryanization" (to use a word popularized by Vianna) could be a reality in Brazil. If faith in whitening seriously limited Brazil's vaunted racial liberalism, nevertheless, in the context of the times, the thesis did allow a more positive evaluation of the contribution of the mulatto, if not the Negro, to Brazilian cultural and social life. Vianna's negative assessment of the mulatto was countered by Lacerda's view that the mulatto in Brazil possessed an intelligence above that of the Negro, and that Brazil's racial history was no impediment to a sound future.

As the whitening thesis gained ground in the 1920s and 1930s as the unofficial ideology of the Brazilian elite, many Brazilians turned their attention away from racial pessimism toward education, social reform, and sanitation as answers to the "national problem." The result was a eugenics movement that, while grounded in racialist ideology, was subtly directed away from overt racialism.

For instance, Belisario Penna, one of the leaders of the eugenics movement, was a conservative and a critic of what he saw as the corrupt politics of the republic and its misguided faith in democracy and egalitarianism. As a student at the Oswaldo Cruz Institute in 1913, he and Artur Neiva had taken a long journey on horseback through the backlands of Brazil, recording the devastation caused by hookworm, Chagas' disease, malaria, and malnutrition in the racially mixed and poverty-stricken people of the northeast (Neiva and Penna 1916). The journey turned Penna into a propagandist for the strategic significance of sanitation in the economic and social regeneration of the country (Skidmore 1974, pp. 182–83). His *Saneamento do Brasil* (Sanitation of Brazil), a vitriolic condemnation of the inability of the federal system of government to marshal the resources or administer effectively a national program of action against disease and malnutrition in the rural areas, made him a well-known figure in medical and eugenic circles. To Penna, race was not what made the *sertanejos* (people of the backlands) and *caboclos* (people of mixed Indian and Negro race) incapable; he saw epidemic and endemic disease as the cause. To sanitize was, for him, to eugenize. This emphasis on sanitation and public health was maintained by Penna into the late 1920s. In "Eugenía e eugenismo" (Eugenics and eugenism) (1929), he reiterated that social conditions were much more important for health than were race or the climate of the region.

Perhaps even more emblematic of eugenics in Brazil was "Jeca Tatu" (literally "backwoods hog" or armadillo, meaning a typical country hick), a fictitious literary figure introduced by the writer Monteiro Lobato to signify the backward condition of the Brazilian race. Jeca Tatu was a poor, ignorant, racially mixed individual. By 1918, however, Monteiro Lobato had changed his mind about the meaning of Jeca Tatu. His book *O problema vital (The vital problem)* was written expressly to popularize sanitation as the salvation of Brazil in an effort to focus attention away from racial explanations of social disintegration (Skidmore 1974, pp. 183–84). In this work he revised his essay on the decadence of Jeca Tatu, which he had first analyzed in terms of race and later in terms of epidemic disease. "Jeca Tatu is made, not born," he wrote (Skidmore 1974, p. 271). If you gave Jeca Tatu food and eliminated his parasites, wrote Kehl, he would become a "Jeca Bravo" (Kehl 1923a, p. 203).

Azevedo concurred. He argued that the racial makeup of the people of Brazil was no impediment to the success of eugenics, and claimed that Jeca Tatu was racially one and the same as the successful *bandeirante* (colonists) who had cleared the territory of São Paulo and made it great. Their differences were not racial, but social and hygienic. Eugenics, he claimed, called for the elimination of poisons, not people (Azevedo, 1919b, pp. 132–33).

Eugenics in the 1920s, in short, sought to identify itself with sanitation. The language of eugenics was a language less of selection and genetics than of reform of public health. The British eugenicist Trounson was right—the Brazilians interpreted the word less strictly than did the British, and made it cover "a good deal of what we call hygiene."

Lamarck versus Mendel: A Scientific Divide

By the late 1920s eugenics seemed poised for expansion. But beneath the surface, divisions were beginning to appear in the movement—divisions that, despite the public responsiveness to eugenics, would prevent it from achieving wider consensus and institutional security. These divisions were of two sorts—scientific and ideological. Scientifically, the rift concerned neo-Lamarckian versus Mendelian genetics. Ideologically, it concerned the issue of race, confirming that in Brazil race, indeed, lay at the heart of the eugenic matter. How these scientific and ideological issues intertwined in Brazil was surprising and tends to challenge our traditional expectations about the social policies that flow from science.

Taking first the scientific division, we have seen that despite its veneer of Galtonianism and Mendelism, Brazilian eugenics was French in intellectual origins and neo-Larmarckian in outlook. The majority of the eugenicists were physicians, not practicing research scientists, which was to be expected in a country where the professional career of research science was only just becoming institutionalized and where medicine was a standard route to professional status. As physicians, most of the Brazilian eugenicists were in clinical practice; few had firsthand knowledge of modern genetics, or read German or English fluently. In many ways they were very similar to their French counterparts and, like them, were unconsciously rather than consciously neo-Lamarckian in their genetic assumptions.

The neo-Lamarckism of the Brazilian eugenicists did not, however, remain unchallenged. By the second half of the 1920s, a new generation of biological scientists had emerged, most of them employed in the country's new scientific institutions and beginning to acquire considerable sophistication concerning the scientific divide between Anglo-Saxon Mendelism and Latin neo-Lamarckism.

Some of them were unwilling to let the Lamarckian opinions of the eugenicists be passed over in silence.

One of the first, if not the first, medical thesis on Mendelism was written in 1918 (Viana). Of the Mendelian critics involved in eugenics, however, of most importance was the anthropologist Edgar Roquette-Pinto, director of the National Museum of Anthropology from 1926 to 1936 and president of the First Brazilian Congress of Eugenics in 1929, whose scientific opinions thus carried considerable weight. Better remembered for his generally "scientific" (mainly anthropometric) approach to the study of Brazilian racial types than for specific results, Roquette-Pinto's endorsement of Mendelian genetics resulted in a specific attack on the neo-Lamarckian assumptions of his fellow eugenicists.

In his 1927 book of essays, Roquette-Pinto defined eugenics as an artifical selection of human heredity based on three components of modern genetics—cytology, biometry, and experimental biology. By 1928 he was citing the American geneticist Charles Davenport on the chromosome theory to argue that "every educated person knows that, actually, the celebrated 'influence of the environment' has been reduced to very restricted limits." "The majority of biologists," he commented, "do not believe that the environment is capable of influencing hereditary characters, all of which are dependent on the germ plasm. The environment—it is currently believed—only modifies the somatoplasm, the part of living things that does not become part of inheritance" (1933, p. 35). Roquette-Pinto believed that the confusion between eugenics and sanitation, so evident at the 1929 First Brazilian Congress of Eugenics, would be cleared up by the time of the next eugenics conference, when the real subject of eugenics—Mendelian inheritance—would be made central.

A second locus of Mendelian eugenics was the Agricultural School of Piracicaba, which had been founded outside São Paulo in 1901 to improve the production of animals and plants of commercial value to the state. Carlos Teixeira Mendes was the school's professor of agriculture; his interest in plant breeding and selection may have explained his early adoption of Mendelism and the new science of hybridization (Teixeira Mendes 1917). In 1918 he gave the first lectures in Brazil on Mendelian genetics in his own Department of Agriculture and in the zootechnical department headed by Octavio Domingues. Domingues was a major disseminator of Mendelian genetics in Brazil and, if not an original research scientist, an important figure within the eugenics movement.

As a member of the American Genetical Association and the Eugenics Society of London, Domingues held to a strictly Mendelian genetics. His eugenic texts, notably *A hereditariedade em face de educação* (Heredity in the face of education) (1929), *a Hereditariedade e eugeniá* (Heredity and eugenics) (1936), and *Eugenía: Seus propósitos, suas bases, seus meios* (Eugenics: Its propositions, its bases, its methods) (1942), were among the first to review systematically and in an up-to-date fashion North American, British, and European genetics for scientists and the general reading public. Domingues cited (among others) Galton, Pearson, Punnett, Morgan, Davenport, Castle, Conklin, and Jennings, as well as the French biologists Cuénot and Guyénot.

In his analysis of contemporary genetic theory, Domingues made an extensive criticism of the neo-Lamarckism of his fellow eugenicists, calling it a deformation of science caused by too great a dependency on France, though he also noted that even within French genetics one could find critics of Lamarck, citing Cuénot and Guyénot as examples (1936, pp. 145–50). In 1930 he complained in the *Boletim de Eugenía* that, with few exceptions, Brazilians were ignorant of genetics. In 1936 Domingues could still claim that few in Brazil had ever heard of Thomas Hunt Morgan (p. 139).

Of his fellow eugenicists, Domingues commented in 1929:

> Our cultivators of eugenics are following an erroneous path, confusing eugenics with individual and social hygiene, with gymnastics, individual physical development, with sports—subjects which ally them with the science of Galton, but which is not really eugenics. (p. 139)

He reiterated his warning seven years later:

> Among us, when our hygienists proudly recommend cleanliness, good hygienic habits, abstinence from alcohol, smoking, drugs of any kind, or rational gymnastics, they praise these recommendations thinking that what is acquired in a lifetime is transmitted to offspring. Therefore one way to improve the race genetically is to adopt these measures, so that in a few years our people will be transformed into pure Hellenes: beautiful bodies and Greek physiognomies! (1936, p. 147)

Domingues countered this neo-Lamarckian view of eugenics with a Mendelian one. Like Roquette-Pinto he emphasized the significance of heredity in human life, distinguished between biological and social inheritance, and called for a program of eugenics *and* "eutechnics," or general sanitation, in Brazil that would create the healthy environment in which the genetically fit could thrive.

A third example of the new Mendelians was André Dreyfus, considered one of the pioneers of Mendelian genetics in Brazil. In the 1930s Dreyfus transferred from the São Paulo Medical School to the first "modern" university in Brazil, the University of São Paulo, and helped transform it into the country's leading center of genetic research by the 1940s. In his paper presented at the First Brazilian Congress of Eugenics, "O Estado Actual do Problema da Hereditariedade" (The current state of the problem of heredity), Dreyfus reviewed Mendel's laws of inheritance and their recent experimental confirmation, pointing out that they had given genetics a wholly new orientation. Dreyfus noted that such alternatives as Galton's law of ancestral inheritance, which Kehl had repeatedly cited, were now taken seriously only by researchers "distant from the positive results of genetics" (1929, p. 60). All efforts to confirm neo-Lamarckian notions experimentally had failed, and as a result the belief of "various eugenists that a favorable environment, good food, instruction, will be able to influence the hereditary patrimony" had "sadly to be abandoned" (Dreyfus 1929, p. 91). In 1943 Dreyfus invited the Russian-born geneticist Theodosius Dobzhansky to São Paulo to train what became the first group of drosophila researchers. The fact that Dreyfus felt it important to repeat his censure of neo-Lamarckian inheritance in his lectures on genetics two years later, in 1945, indicates the extraordinary persistence in Brazil of the belief in the transmission of acquired characteristics.

While the scientific critique of neo-Lamarckism came as no surprise to some Brazilian biologists (Andrade Filho 1925), it took many of the delegates to the eugenics congress of 1929 aback. Levi Carneiro, presiding over the section on education and legislation, remarked in his address, "Educação e eugenía" (Education and eugenics), that Roquette-Pinto's denial of neo-Lamarckian hereditary transmission negated the importance of alcohol or venereal disease for racial (i.e., hereditary) improvement, as well as prenuptial exams, which Roquette-Pinto had said lay outside eugenics because they would prevent infectious diseases unrelated to heredity. Carneiro spoke for the majority of eugenicists in defending both prenuptial exams (which the congress had already endorsed) and neo-Lamarckian inheritance (citing Richet and Houssay), but he was uncertain enough of his grounds to admit that the influence of the environment was not entirely established (1929, p. 112). As Carneiro suggested, the denial of the transmission of acquired characteristics called into question the rationale of the antialcohol campaigns with which the League of Mental Hygiene was so closely identified. Perhaps not surprisingly, the founder of the league, Gustavo Reidel, continued to doubt that Mendel's laws applied to the human species, maintaining that, as far as he was concerned, mental discord and mental illness had a direct hereditary effect on offspring, so that a eugenic program of mental hygiene was fully justified (Carneiro, 1929, p. 112).

Perhaps the most interesting example of the effect of the Mendelian critique on neo-Lamarckian eugenics was that of the tireless promoter and leader of Brazilian eugenics, Renato Kehl. By the late 1920s and early 1930s, Kehl himself had become frustrated by the confusion in the public mind between eugenics and sanitation. As he explained later, allies from the sanitation movement had been useful at the beginning of the eugenics campaign, when public knowledge of eugenics and heredity was slight (Kehl 1933, p. 22) and when he himself was not very clear about the distinction between sanitation and eugenics (1937, p. 45). But when the elite embraced personal hygiene, physical excercise, and even organized sports as "eugenic," Kehl began to protest that no amount of hygienic reform could improve the hereditary stock of Brazil. His neo-Lamarckism narrowed, and he began to emphasize negative eugenic measures and to cite the German and Scandinavian race hygienists with approval. Yet as he moved toward a more negative and racialist eugenics, Kehl found it hard to abandon the neo-Lamarckism that had dominated his thinking for so long. By 1929 Kehl conceded that syphilis and tuberculosis did not cause hereditary conditions as he had previously believed, but only congenital damage limited to a single generation ([1929] 1935, p. 147). He agreed, that is, that only rarely did these "racial poisons" (to use the British terminology) actually modify the reproductive cells and therefore the hereditary character of offspring.

But Kehl's retreat from neo-Lamarckism was less than at first appeared. In *Lições de eugenia* (Lessons in eugenics), which appeared in 1929 in time to be circulated at the First Brazilian Congress of Eugenics, Kehl's review of theories of heredity continued to be eclectic, reminding us of the state of flux in which hereditary theory was perceived in Latin circles. The neo-Lamarckism of Cope and Giard, the neo-Darwinism of Weismann, the preadaptationism of Cuénot, the mutationism of de Vries, and the chromosome theory of Morgan were all presented to Kehl's Brazilian readers, with little selection between them ([1929] 1935, pp. 78–99). Moreover, Kehl's use of the term *eugenismo* to describe *all* the activities that aided eugenics, including education, sanitation, sports, legislation, and hygiene, blurred the very distinction he sought to draw *between* eugenics and sanitation, or eutechnics (1937, pp. 46–47). His continued reliance on neo-Lamarckian concepts was also revealed by his insistence that antialcoholism was central to eugenics because alcohol could affect not only the physiology of the reproductive cells, but heredity itself ([1929] 1935, p. 171; 1930e). As late as 1937, in Kehl's manifesto *Por que sou eugenista* (Why I am a eugenist), we find similar references to the effect of chronic diseases and toxins on the germ plasm—that is, to what he termed "blastophthoric disorders"—though he now excluded yellow fever at least as blastophthoric in its effects (pp. 59–61).

Lamarck, Mendel, and Race: An Ideological Divide

The scientific divide between the Lamarckian eugenicists and the Mendelians in Brazil would have mattered less to the eugenics movement (and would be less interesting to the historian) had it not been closely associated with an ideological division over the direction in which eugenics should move. This division centered around race and whether eugenics would move toward a more negative, German-style eugenics. It pitted the neo-Lamarckian eugenicists on the racialist side against the more modern "Mendelians" on the antiracialist side.

The roots of the more negative eugenics that surfaced in the 1930s lay in the past. Kehl's eugenics had always been more negative and racialist than that of the majority of his eugenicist colleagues. But this negativism had been camouflaged by the need to bring into the eugenics movement allies from sanitation and from clinical medicine, few of whom were initially knowledgeable about either genetics or eugenics. The existence of a tradition of polite, nonracist discourse also checked the public expression of naked racialism. By the late 1920s and early 1930s, however, a more extreme, Anglo-Saxon-style eugenics became more widely expressed.

There were many causes of the new appeal of negative eugenics. Greater familiarity with German and American eugenics played a part. The passage of the eugenically inspired immigration law in 1924 in the United States generated considerable eugenic discussion in Latin America. At the first Pan American Conference on Eugenics and Homiculture, held in Havana in 1927, the Latin American delegates voted to give each state the right to control immigration as it saw fit and in ways harmonious with each country's perceived ethnic composition (p. 323). American eugenic sterilization laws were also discussed. Kehl began to provide short abstracts in German of the articles published in the *Boletim de Eugenía,* suggesting the existence of a German readership for eugenics in Brazil such as could be found in the German-speaking colonies of the south and southwest. By 1929 Kehl openly praised the eugenicists in Germany for their "courage" in eugenic matters, predicting that one day the state would control all reproduction (a footnote to the 1935 edition of *Lessons in Eugenics* noted that his prediction had been borne out) (pp. 25, 32). He also claimed that the Brazilian Commission of Eugenics was modeled on the German Society for Race Hygiene established on 18 September 1931.

Another possible factor in the emergence of a more negative eugenics in Brazil was the development in the late 1920s and early 1930s of antidemocratic, organic-statist ideologies, culminating in the founding of the Ação Integralista (Integralist party) in 1935 (Trindade 1974, 1975). Though a connection between eugenics and the integralists has yet to be established, the emphasis on natural hierarchy, the family, and the role of the state in structuring social relations suggests certain similarities between a conservative eugenic movement and corporatist ideology. The Catholic orientation of the integralists, however, presented a real barrier to the penetration of extreme eugenics, especially following the papal encyclical "Casti Conubii" in 1930 condemning sterilization and eugenics.

Finally, the late 1920s saw the slowing of white immigration to Brazil, raising concern in the minds of some people about Brazil's racial destiny. Without a continuous influx of white blood, they asked, what would be the result of Brazil's vaunted racial miscegenation? By the time Kehl wrote *Aparas eugenicas: Sexo e civilização* (Eugenical fragments: Sex and civilization) (1933), a "semiological book of genital-social ills" (p. 7), his mood was pessimistic. Brazil, he believed, was a "demoralized" republic, in search of *homems validos* (sound people). He was determined to draw the line between sanitation and eugenics—to deny that "to sanitize is to eugenize, to educate is to eugenize." The need in Brazil, Kehl argued, was less for exercise, education, and even general hygiene, none of which could affect the germ plasm, than for a true eugenics based on, among other things, the sterilization of degenerates and criminals, the imposition of obligatory prenuptial examinations, and the legalization of birth control (1933, pp. 49–50). In the pages of the *Boletim de Eugenía,* the language of selection, virtually absent from the eugenic literature of the 1920s, was now much more in evidence, as was that of class (e.g., Decroly 1929). Concern was expressed about the class differentials in fertility, the social costs of philanthropy, and the burden on the state of mediocrity and unfitness (Kehl 1929b, 1931b, 1931e; Cunha Lopes 1931). Eugenics, it was argued, should be concerned with the rational and state management of the population, the encouragement of the reproduction of the eugenic upper and middle classes, and the prevention of the reproduction of the less eugenic lower classes. The whole tone of eugenics, as presented by Kehl and his allies, was changing, bringing it much closer to the North American movement.

Nowhere was the shift toward a more pessimistic and negative eugenics more noticeable than on the subject of race. References to "our race" *(a nossa raça)*, or "the Brazilian race" were replaced with those to "the white and black races." The number of articles on the dangers of racial mixing increased in the *Boletim de Eugenía* and dominated the later editions of Kehl's books (e.g., Kehl 1929c; Silva 1931). The Scandinavian and German race hygienists Mjöen and Lundborg were not only quoted with approval, but selections from their writings translated into Portuguese and reproduced (Mjöen 1931; Lundborg 1930, 1931). Even the term *race hygiene* began to be used (e.g., Kehl's footnote to Mjöen 1931). Mulattoes were now described as heterogeneous, unstable elements disturbing the national order. That Brazil was achieving a whitening via racial miscegenation was, to Kehl, a cause not for celebration but for sadness. Kehl advised against racial and class crossings, while protesting his lack of racial and class prejudice ([1929] 1935, pp. 136, 240–41).

Yet in espousing a negative, neo-Lamarckian, racialist eugenics so appealing to the private and sometimes public worries of the Brazilian elite, Kehl was writing against powerful scientific and ideological currents that were pulling Brazilian eugenics in a different direction and would prevent it from becoming the race hygiene movement Kehl now envisaged (Castiglione 1942, pp. 7–8). Many of the Brazilian Mendelians opposed the association of eugenics not only with Lamarckism, but with its racialism as well. While Kehl called for a negative and racialist eugenics based on the transmission of acquired characteristics, several of the Mendelians called for a more voluntaristic, less racialistically oriented eugenics in which eugenics and sanitation worked together for the improvement of "the race."

One of Kehl's Mendelian opponents was Domingues. In 1929 he called the Brazilian mulatto a product of normal and healthy Mendelian hybridization and Brazil a "special and precious" example of racial mixing. If the mestizo was at times inferior, he wrote, it was no more so than the supposed pure races of Europe. Domingues's continued commitment to the whitening ideology was revealed by his use of Mendelian laws to argue that, on the basis of the Mendelian inheritance of skin color and Brazil's racial ratios (he believed whites dominated over blacks), through continued racial mixing Brazil would, over time, naturally whiten. A mulatto people, that is, could produce white offspring because in Mendelian inheritance factors controlling color were not merged or blended, but were preserved and recombined (Domingues 1929, pp. 89–91, 132, 136). Though ready to defend the eugenic value of birth control and even sterilization on an individual and not racial basis (1936, pp. 25–30), Domingues preferred a positive eugenics based on fostering a eugenic conscience in individuals through education, whereby individuals with hereditary defects would refrain from reproduction; he opposed any control of reproduction by the state (1929, p. 147).

Domingues, then, was both a Mendelian and less racially inclined than Kehl. His views on race and racial mixture are particularly revealing of the ways in which the whitening ideology interacted with eugenic ideology in the late 1920s. Domingues interpreted racial mixture not as a cause of racial degeneration, but as a biologically adaptive process that would allow a true "civilization" to develop in the tropics. We see here a curious foreshadowing of Giberto Freyre's thesis of "racial democracy," with its reliance on racial biology and its positive view of racial mixture as, in itself, a form of eugenização (eugenic improvement).

The Mendelian anthropologist Roquette-Pinto played an even more public role than Domingues in keeping eugenics out of the hands of the strident racialists. His contact with Franz Boas in New York in 1926 possibly was a factor in turning Roquette-Pinto into an ardent defender of the value of the mulatto to Brazilian culture. He challenged the views of Kehl, Mjöen, and others on mulatto degeneracy as not scientifically established, and at the First Brazilian Congress of

Eugenics also criticized Kehl's *Lessons in Eugenics* (which Kehl had circulated to the conference members) as not representing, in its more extreme views, the outlook of the congress. In his essays on Brazilian anthropology, he invoked Jennings' *Prometheus* (1925) to warn against hasty eugenicists who lacked scientific data (Roquette-Pinto [1933] 1978, p. 52).

In a subtle inversion of the use made by Davenport of Mendelian genetics to warn against the dangers of racial crossing (Kevles 1985, p. 47), Roquette-Pinto argued that Mendelian crosses between whites and blacks were a healthy process of whitening. But even without further crosses with whites, he maintained, mulattoes already contained white genes and could produce offspring so white that even a trained anthropologist like himself could not distinguish them from Europeans (Roquette-Pinto 1927, pp. 61–62, 174, 202). Having measured over two thousand male Brazilians, he had, he said, a good basis for such a judgment. He added, however, that the goal of eugenics was not to whiten, but rather to educate all people, white and black, to the importance of heredity, so that the eugenically minded individual, aided by state-run programs of sanitation, would participate voluntarily in the "purification" of race. Eugenics itself, he stated, was an area "where the state does not penetrate" (Roquette-Pinto 1927, p. 205). The result of the various scientific and social constraints at work in Roquette-Pinto's writings was a eugenics in which suggestions for a positive and reformist approach to heredity were linked with eutechnics, or reform of the environment.

The political and scientific disputes within Brazilian eugenics surfaced at the eugenic congress of 1929. The debate on race sparked by the paper "O problema eugênico da immigração" (The eugenic problem of immigration) by the racially minded (and Mendelian) congressman Azevedo Amaral dominated the proceedings. Discussion spilled over into the second and third days. So heated was the debate that Amaral's proposals had to be reformulated and voted on as two separate proposals: the first to restrict entry of non-Europeans in general, the second specifically to restrict the entry of blacks. The issue, of course, turned on the subject of the value of race crossing. Amaral was joined by the mental hygienist Fontanelle, the clinician Xavier de Oliveira, and the president of the National Academy of Medicine, Miguel Couto, in claiming that race mixture led to degeneration. They were opposed by Roquette-Pinto, by the anthropologist Froés de Andrade, by Penna and Magalhães, and by the physiologist Miguel de Osorio, all of whom either defended race crossing or opposed immigration restriction based on ethnic or racial criteria (Primeiro Congresso Brasileiro de Eugenía [herafter PCBE] 1929, pp. 16–42).

As president of the congress, Roquette-Pinto played a strong part in forcing the issue—which was, he said, not one of race at all, but of hygiene. Penna seconded him. Magalhães reminded the conference members that Brazil's past was based on the mestizo and added: "We are all mestizos and would therefore exclude ourselves" (PCBE 1929, p. 20). "We do not believe," said Froés da Fonseca, "that eugenizing the Brazilian people is a racial problem" (1929, p. 79).

Roquette-Pinto (1933, pp. 109–13) especially defended the eugenic worth of the Japanese against attacks by such eugenicists as Couto, who had long called for restriction of Asian immigrants on eugenic grounds (BANM 1925, 96: 33–34).

Roquette-Pinto and his supporters were quite willing to concede the need for some kind of individual selection of immigrants, as had been proposed in 1925 by Moreira, Pacheco e Silva, and others. This selection would be on health grounds, and to ensure the entry of people willing to learn Portuguese and adapt to Brazilian ways so that Brazil could achieve national unity. What was opposed was a *racial* immigration selection, which was seen as being based on nothing but out-of-date and unscientific prejudice.

At the 1929 congress racial etiquette triumphed over private belief. In a conference full of controversial subjects, Azevedo Amaral's proposals were among the few not to be endorsed in their original form. His proposal for a national policy of immigration exclusion based on race was rejected by the participants attending the session by twenty-five votes to seventeen (PCBE 1929, pp. 20–21).

Eugenics in the Estado Nôvo

For many of the countries of Latin America, the 1930s was as extreme a period politically as it was for Europe. The decade opened with the breakdown of the First Republic in the so-called revolution of 1930. The revolution was a product of new social forces, some radical in orientation, some conservative, all critical of the control of politics by the traditional and mainly landed oligarchy. It brought to prominence Getúlio Vargas, a politician from Rio Grande do Sul, who took over the presidency and prepared the way for a Constituent Assembly in 1933 and the drafting of a new constitution for Brazil. These events at first seemed to offer the promise of new political and social opportunities and the space for institutional experimentation. The period saw the creation of new federal departments, notably the first Ministry of Labor. At a time when Brazil was suffering the effects of the world depression and the rapid decline in world coffee prices, the collapse of established ways of doing things and the search for new ones seemed to offer a new prospect of consolidating eugenics at the national level. This took the form of renewed efforts to create legislation against alcoholism and requiring prenuptial examinations (Porto-Carreiro 1933). The League of Mental Hygiene extended its services by creating the first "infants euphrenic clinic" (ABHM 1934, 7: 65). In 1931 the ever-energetic Kehl created the Commissão Central Brasileira de Eugenía, a commission whose task was to promote eugenics and to lobby the members of the Constituent Assembly on eugenic legislation. Its ten permanent members were Kehl, its chairman; Penna; Ernani Lopes, president of the League of Mental Hygiene; Gustavo Lopes, an assistant in the national Department of Public Health; Porto-Carreiro, professor of public medicine at the University of Rio de Janeiro; Cunha Lopes, from the National Psychopathic Assistance; Toledo Pizo, Jr., professor of zoology at the School of Agriculture at Piracicaba; Octavio Domingues, the school's professor of zootechnics; Achiles Lisboa, a hygienist and eugenicist; and Caeta Coutinho, inspector of pharmacies in the Department of Public Health.

The commission gained political visibility the year it was founded when Penna became director of the Department of Public Health within the new Ministry of

Education and Public Health. Penna's appointment gave hope to the eugenicists that antialcohol legislation would finally be made the law of the land (ABHM 1931, 4: 167–68). In addition, Roquette-Pinto and Kehl were invited to serve on a special commission organized within the Ministry of Labor to advise on eugenics and the problems of immigration.

By 1937, however, the period of political experimentation and limited parliamentary democracy had come to an end. Seven years after the revolution of 1930, and only three years after the 1934 constitution was created by elected representatives, Getulio Vargas was able to consolidate his power in a new corporatist state, the Estado Nôvo. It lasted through the late 1930s and the Second World War, ending with a military coup in 1945. The Vargas era has continued to evade easy ideological and political definition (Putnam 1941, 1942; Lowenstein 1942; Levine 1970; Chaui and Franco 1978). Though it ended the First Republic, continuities with the past were marked. Originally interpreted as a Brazilian version of European fascism, the authoritarian and corporatist Estado Nôvo combined a baffling mixture of regressive and progressive elements.

On the one hand, after a period of political experimentation that saw the founding of the first mass parties on the democratic Left (the Aliança Nacional Libertadora [National Liberation Alliance]) and the Right (the Integralists), political repression increased, especially after 1935 and the stiffening of Vargas's control over the political system and the state. This control involved the policing of "dissidents" on the Left and the Right, and the eventual suppression of political parties and imprisonment of many of the party leaders. By 1938 all political parties had been eliminated. Whatever the Estado Nôvo was, the fascist parties as such had no political role in it.

Socially, the Vargas regime also saw the extension of the power of the national state to manage and control "social-problem" groups such as the mentally ill, prostitutes, and juvenile delinquents. On the one hand, it was in this period that a state system of identification was discussed by the medical-legal expert Leonidio Ribeiro, who opened a new Instituto de Identificação (Institute of Identification) in the federal capital in 1933 and worked closely with the right-wing police chief of the city, Felinto Müller, to bring "up-to-date," "scientific" techniques of identification and treatment of the "pathologically" criminal to Brazil (Ribeiro 1934). On the other hand, under Vargas Brazil undertook a policy of incorporation into the state of new social groups, notably the urban-based, industrial working class, who were rewarded with new social-welfare and labor legislation and the creation of new Ministry of Labor in return for corporatist controls and social acquiescence (Flynn 1979, pp. 100–103).

It was in this context that eugenics survived into the 1930s in Brazil. The complexity of the Vargas regime was matched by that of the eugenics movement—in its scientific orientation (neo-Lamarckian and Mendelian), racial ideology (ranging from segregationist to assimilationist), and proposed social policies (public hygiene, maternal protection, labor legislation, immigration control). The formation of a new national department of health in 1934, the emphasis on child welfare (however ineffective in practice), the restriction on female and child labor (also illusory), and the attention paid to the health of mothers cannot be understood independently of the history of eugenics in Brazil. Although these developments were obviously the result of far more than merely eugenic pressures, the eugenicists actively pressed their views during the debates of the Constituent Assembly of 1933, and were effective in translating some of their eugenic concerns into legislation and new cultural and social institutions. Many of these legislative and cultural innovations persisted into the Estado Nôvo. The areas of success and failure throw considerable light on the ideological character of the Vargas years.

First, the eugenicists' proposal to the Constituent Assembly to make the "promotion of eugenical education" the responsibility of the national state won acceptance in the constitution of 1934 (Kehl [1929] 1935, pp. 235–43). Given the identification of eugenics with "health," this outcome was perhaps unexceptional and certainly meant little in a country where probably 90 percent of the population was illiterate and primary schooling was woefully inadequate. The "eugenical education" clause is more significant for the symbolic importance attached to eugenics than for its practical results.

Second, as the Catholic church grew closer to the Brazilian state in the 1930s (Della Cava 1976; Todaro 1974), winning important constitutional concessions such as the legality of church marriages and the prohibition on divorce, extreme eugenicists such as Kehl found the ideological environment unpropitious for programs of state-sanctioned sterilization of the "unfit" or "grossly degenerate" supported by Peixoto, Fontanelle, Pacheco e Silva, and Leitão da Cunha (Kehl [1929] 1935, p. 225). Efforts by eugenicists and by radical workers and doctors to legalize abortion in exceptional cases, or birth control for eugenic (Kehl [1929] 1935, pp. 212–17; Ribeiro 1942, p. 323) or other reasons, were equally thwarted. Despite the efforts of the eugenicists to assure politicians that eugenics was neither a substitute religion nor contrary to Catholic faith, sterilization, birth control, and abortion were perceived as anti-Catholic measures and remained illegal in Brazil until very recently.

More successful, and more acceptable to Catholic sentiment, was the *nubente* clause, which required prospective marital couples to present proof of their mental and physical "health" before marriage, a requirement written into the constitution in 1934 eight years before its equivalent in France. The Brazilian law was qualified by the statement that its application would take into consideration the regional conditions of the country (probably a reference to the absence of any administrative apparatus that would oversee its application or indeed the absence of adequate numbers of health officials anywhere but in the larger cities). Since a very large number of unions in Brazil were extralegal, the effectiveness of any such legislation was doubtful. Whether such a law should be called "eugenic" was also a moot point; several of the eugenicists recognized that it was not since it did not screen for supposed "hereditary" defects, but rather infectious ones like venereal disease. The law was not, at any rate, put into effect, and disappeared from the constitution of 1937 of Vargas. This explains the continued call throughout the 1930s and 1940s for adequate prenuptial examinations (Roxo 1939–1940).

That strand of eugenics that emphasized "sanitation," whether public hygiene or reproductive hygiene of a neo-Lamarckian sort, also found a place in the new state. From the point of view of many eugenicists, the passing of new social security measures, the creation of unemployment benefits and pensions, the extension of aid to pregnant women, the introduction of maternal benefits for large families, and protective labor legislation (such as restricting women's working hours) were all welcome parts of "eugenic" improvement (Kehl [1929] 1935, pp. 233–34). Many of the eugenicists admired the extensive programs of "family protection" carried out by Mussolini in Fascist Italy, a program whose profamily, pro-Catholic, antiabortion, and antifeminist orientation was highly congruent with Brazilian eugenic ideology (Ribeiro 1937).

The admiration expressed for the social and worker legislation introduced in 1934 and later by eugenicist extremists like Kehl was partly *faute de mieux* since the legislation involved no eugenic selection based on class or race, and no discrimination between the supposedly eugenically "worthy" and "unworthy" when it came to benefits. The president of the Eugenics Society of London, Leonard Darwin, after all, had asserted that prenatal care did not fall within the scope of eugenics, and that nonselective public assistance promoted racial decay (1926, pp. 416–19).

Eugenics also found a home of sorts in the Estado Nôvo when Vargas made puericulture a tool for the incorporation of the masses into the state and for the generation of nonpartisan, nonpolitical, patriotic sentiment (Lowenstein 1942, p. 305). Physical education and team sports were encouraged in the schools as an instrument of "levelling ethnic disparities" (Lowenstein 1942, p. 193). The language of "eugenetics," "eufrenics," "eugenics," and "dysgenics" was widely used in Brazil to describe child and maternal welfare activities in the 1930s. In 1937 Vargas founded the Instituto Nacional de Puericultura (National Institute of Puericulture) under the new Ministry of Education and Health. Functioning within the Arthur Bernades Hospital, in collaboration with a laboratory for the study of infant biology, medical doctors carried out eufrenic (eugenic mental hygiene) examinations of children sent to them for study by juvenile detention centers, as well as prenatal consultations. The institute survived as an independent organization until 1946, when it was incorporated into the University of Brazil. In the issues of its bulletin, one can trace the institute's shift from puericulture, eufrenics, and "eutrophics" in the 1930s to "child welfare" by the 1940s.

Lastly, some eugenicists made their way into the new Ministry of Labor and the state-run clinics for infants, children, and "delinquents," where they studied the hereditary "pathologies" of the Brazilian race.

The most interesting example of the way in which eugenics intertwined with the new state in the 1930s concerned race and nationality. According to Lowenstein the Vargas state was marked by the "desire to create a homogeneous consciousness of nationhood as the basis of social and political life" (1942, p. 188). New state apparatuses were developed to help create such consciousness, to mobilize patriotism, and to create a sense of national unity. Given this ideological orientation, the deliberate public usage of the language of racism, the evocation of antagonism or difference, or the recognition of the reality of racial discrimination was avoided, especially after Brazil entered the war against Germany. Even before then, however, the notion that racial and cultural fusion was the solution to Brazil's racial and social makeup had become the unofficial ideology of the state, maintained in the teeth, as it were, of actual deep racial and class divisions. National identity and homogeneity were to be forged at home by the incorporation into the state of the strategic industrial workers, who rewarded Vargas with their support, and by an exclusive nationalism that resulted in a series of laws restricting the number of foreigners who could hold jobs in Brazilian firms and making Portuguese the sole language of instruction in schools. Vargas's destruction of the integralist movement, which in other ways was ideologically congruent with the Estado Nôvo, was in part a result of his fear that the party threatened "Brazilianization" by its identification with German fascism. Eventually, Vargas also suppressed the use of foreign-language newspapers, foreign flags, and, insofar as was possible, the foreign identification of the German colonies.

In these circumstances the racialist eugenicists' rejection of racial amalgamation within Brazil as the solution to Brazil's racial identity found relatively few adherents. Fusion, through racial and cultural means, so that blackness would disappear and whitening occur, was in itself taken to be "eugenic."

On the other hand, immigration restriction, long a goal of some eugenicists, was popular with the politicians in the 1930s because of the growing public endorsement of a eugenically aided process of racial fusion and whitening within Brazil. The decline of European immigration in the late 1920s and the rise of Japanese immigration were critical to the eugenicists' claims that Brazilianization and the forging of national unity at home needed to be protected from outside threats, especially threats from ethnic or national groups whose physical or cultural characteristics they claimed would disturb the natural process of unification and homogenization (Poés de Andrade 1925). It was a point made by Penna at the eugenics congress in 1929, when he worried out loud about colonists who settled in Brazil in large numbers and refused to adapt themselves, linguistically and culturally, to Brazilian ways (PCBE 1929, p. 18). We see an expression of the same worry in Fernando de Azevedo's classic work on Brazilian culture, published in 1943, in which the author referred to Japanese and German colonies as "cysts in the national organization" (p. 37).

Penna agreed in 1929 with Roquette-Pinto and other eugenicists involved in the debate on racial restriction that the problem was not really racial and eugenic but political. But the worries about national unity—about *how* Brazil was to create a single nation and ethnicity—served to unite eugenicists and politicians who otherwise had rather different outlooks on the racial issue. It gave the eugenicists their greatest legislative success. Eugenicists' concern about the fitness of immigrants, whether cultural or racial, found a receptive climate in the Constituent Assembly of 1934, where several eugenicists played an active role in drafting the new constitutional provisions. Miguel Couto and Xavier de Oliveira rehearsed for the assembly the eugenic arguments they had made in 1929 for a racial selection of immigrants, attacking especially the Japanese as contributing to a "racial mosaic" in the country (Navarro 1950, pp. 137–38; Brasil, Annaes da Assembleia Constituente 1935, 4: 490–93, 546–48). Antonio Pacheco y Silva argued that restriction was both a eugenic and a public health measure, presenting data to show that Japanese and Italian immigrants brought new diseases into the country (Castiglione 1942, p. 14). To other members of the Constituent Assembly, immigration restriction was necessitated by the problem of high unemployment at home (Mitchell 1983). From welcoming white immigration as a source of eugenization, Brazil was now about to close its doors to immigrants in the name of protecting the process of eugenization at home. The result of the various arguments was a "eugenic" immigration law, setting racial quotas (including one for blacks) as well as economic and other tests of fitness for entry, for the first time in Brazil. The immigration clauses were retained in the 1937 constitution of the Estado Nôvo, ratifying the commitment to whitening in the national state and reinforcing the myth of national unity.

Within Brazil, however, the racial ideology that achieved national consensus by the late 1930s was not that of Kehl but of the Brazilian sociologist Gilberto Freyre. His writings provided the key ideas that dominated domestic interpretations of Brazilian history and nationality for the next thirty years. Freyre had been trained at Columbia University, where he had come under the influence of Franz Boas (Stein 1961) and from him learned an antiracialist and "cultural anthropological" orientation. Freyre also referred to Brazilian sources for his views, such as Roquette-Pinto's statement in 1929 that the Brazilian type was not racially inferior but sickly (Freyre 1963, p. xxvii). Freyre's intention was to oppose the exaggerated biological racism of writers like Oliveira Vianna, and to introduce more sociological analyses of the Brazilian "problem." In a series of classic works of Brazilian history and sociology, beginning with *Casa grande e senzala* [The masters and the slaves] in 1933, Freyre emphasized the reality of Brazil's racial and cultural diversity, defended Brazil's racial "harmony," contrasted it with the racial conflict and patterns of segregation of the United States, and argued that Brazil was unique in creating out of racial mixture a "luso-tropical" civilization in the New World.

Though Freyre's work represented a subtle subversion of the racial thought in his country and a critique of traditional racial pessimism, it did not constitute a fundamental break with the past (Medeiros 1980). Freyre maintained, in effect, that far from being eugenically unfit, as Vianna and others had claimed, "eugenically" superior Africans had merged freely in a racial democracy with a Portuguese people culturally suited to the tropics, and with the Indian, to produce a racially mixed people of increasing ethnic and "eugenic" soundness. Although the racial valuations had changed, the structure of the argument—with its emphasis on race rather than class or economic factors—had not.

Freyre's failure to uncover the deep racial prejudices and social structures that marginalized blacks and mulattoes in Brazil's social system (a failure for which he was roundly criticized by a new generation of Brazilian social scientists in the 1960s) is not the issue here (Viotta da Costa 1970). The point is that the racial and social friction in Brazil in the late 1920s and 1930s provided the context in which eugenics could survive. The variant of eugenics identified with public hygiene and compatible with racial whitening and the myth of racial democracy gained support. The side of eugenics identified with the negative and pessimistic race hygiene movements of Europe and the United States did not.

Eugenics thus found itself strangely placed in Brazil, its scientific and social complexion rendering any simple conclusions about the relation between science and social life impossible. Scientifically, neo-Lamarckian-style genetics dominated in medical circles until the 1940s (Couto 1935; Bandeiro de Mello 1940). Gilberto Freyre, in the 1934 edition of *Casa grande e senzala,* defended neo-Lamarckism not in the form of the theory of blastophthoria, but in the work of Kammerer. In his visits to Brazil in the 1940s to study drosophila in tropical climates, Dobzhansky remarked on how many Brazilians still believed in the inheritance of acquired characteristics (1980, pp. 113, 194, 226). Not until the late 1940s did Mendelism finally replace neo-Lamarckian ideas.

Ideologically, as Kehl and some of his associates turned in admiration to Nazi eugenics in the 1930s (without giving up their Lamarckism), other Brazilian intellectuals began to "discover" blacks, to study their contributions to Brazilian culture, and to move away from biological racism toward a more culturally oriented sociological "racism" in which eugenics still found a place (Levine 1973–1974). The Manifesto of Brazilian Intellectuals against Racism of 1935, signed by, among others, Roquette-Pinto, Artur Ramos, and Gilberto Freyre, represented the most public identification of Brazilian racial traditions with the antiracism of leading British scientists (Ramos 1935, pp. 177–80). Ironically, faith in whitening, itself based on the racialist assumption of the superiority of the European race, rendered an extreme eugenics unnecessary in Brazil.

Conclusion

The history of eugenics in Brazil is of analytic and comparative interest for several reasons. First, it shows that scientific discourse was a constituent element of Brazil's modern history. Eugenics as a theme, as a language of analysis, and as a set of social policies was not an exclusively Anglo-Saxon phenomenon, but was instrumental in structuring debates and actions in Brazil, a country then remote from active genetic research but finely attuned to science as a symbol of modernity.

Second, the history of eugenics in Brazil shows that eugenics cannot be understood merely in terms of the Anglo-Saxon variant—the evidence shows that it constituted a different variant, scientifically and ideologically. The world eugenics movement helped shape Brazilian debate, but eugenics was also reshaped in Brazil and adapted to suit its intellectual topography and social agenda, becoming a major element in the ideological reformulation of what race meant for the Brazilian future.

Third, the Brazilian case is important for the light it sheds on the relationship between science and social ideology. Historians have tended to associate the belief in the inheritance of acquired characteristics with a reformist style of social ideology. Neo-Lamarckian concepts allowed for the possibility that through changes in the social environment, such as the elimination of toxins, permanent hereditary change could be effected. And in fact, in the 1920s Lamarckian eugenicists in Brazil and France tended to be of the "soft," optimistic variety. Recent research, however, shows that the relationship between Lamarckism and social thought is more complex than it first appears (Graham 1978). Examples exist of conservative ideology based on Lamarckian genetics. Late-nineteenth-century American scientific racism, for instance, was founded upon Lamarckian ideas of inheritance (Stocking 1968); similarly, within the field of sexual science, Maudlsey's conservative arguments about women's nature and women's role in society were based on Lamarckian views (Sayers 1982, pp. 17–18).

In the case of Brazilian eugenics, the existence of both a "soft" and a "hard" eugenics within the neo-Lamarckian tradition, and the opposition that developed between the neo-Lamarckian racialists and the Mendelian antiracialists (the latter admittedly based on a covert racialism) suggest that the inherent logic of science does not determine its social meanings and outcomes. Rather, science and social ideologies become linked in culturally and historically specific ways that need to be examined in context.

Finally, the history of eugenics in Brazil suggests it would be very interesting to extend the study of eugenics to other countries of Latin America. Latin America was far from monolithic—politically, socially, or ideologically. On the basis of Brazilian findings, one would expect that eugenics in each country would be shaped by local social, economic, and racial, as well as scientific, circumstances.

In the 1920s and 1930s, for example, postrevolutionary Mexico combined a semiofficial embrace of Vasconcellos's vision of a superior mestizo or "cosmic" race, born out of the fusion of Caucasian, Indian, and African peoples, with an unofficial, real marginalization of the Indian and nonacculturated mestizo. In Argentina, a country of large-scale European immigration in the early twentieth century and one in which the indigenous Indian had been virtually exterminated, debate in the 1920s and 1930s revolved around which fraction of the European "race" best represented Argentinian nationality. In Cuba eugenics appears to have been somewhat untypical for the region in its close ties to North American eugenics and its "harsher" ideologies and policies. The connection between Cuban eugenics and North American influence in the island is clearly worth exploring. In short, the eugenic issue was embedded in nationally specific political and social debates concerning nationalism, national identity, class, race, child welfare, and immigration.

Although these Latin American eugenics movements diverged individually from one another, they seem at the same time to have shared a number of common features. Whether they collectively represent a peculiarly Latin American form of eugenics or instead manifest a broader Latin style of eugenics shared with France and Italy, only further research can determine.

Bibliography

Albuquerque, José de. 1930. Doenças familiaes e exame prénupcial [Family diseases and the prenuptial exam]. *Boletim de Eugenía* 2, 20: 51–52.

Almeida, Lino de. 1925. *Doenças veneraes e a puericultura pré-natal* [Venereal diseases and pre-natal puericulture]. Rio de Janeiro: Typ. Coelho. Thesis, Faculdade de Medicina.

Amaral, A. J. de Azevedo. 1929. O problema eugênico da immigração [The problem of eugenic immigration]. In Primeiro Congresso Brasileiro de Eugenia, *Actos e Trabalhos,* 327–40.

Andrade, Geraldo do. 1929. Concepcionismo inconsciente e mortalidade infantil [Involuntary conception and infantile mortality]. *Boletim de Eugenía* 1, 12: 55–56.

Andrade Filho, Joaquim José de. 1924. *Da genohygia no Brasil* [Geohygiene in Brazil]. Rio de Janeiro: Typ. do Commercio. Thesis, Faculdade de Medicina.

Annaes de Eugenía [Annals of eugenics]. 1919. Sociedade Eugênica de São Paulo. São Paulo: Revista do Brasil.

Armstrong, C. Wicksted. 1933–1934. A Eugenic Colony: A Proposal for South America. *Eugenics Review* n.s., 6: 91–97.

Arrom, Domingo Durán. 1935. Curso de medicina social [A course in social medicine]. *Boletim do Ministério do Trabalho* 7: 279–86; 9: 256–74.

Austregesilo, A. 1937. A mestiçagem no Brasil como factor eugénico [Race mixture in Brazil as a eugenical factor]. In *Novos Estudos Afro-Brasileiros*, ed. Gilberto Freyre. Rio de Janeiro.

Azevedo, Fernando de. 1919a. Meninas feias e meninas bonitas: Eugenía e plastica [Ugly girls and pretty girls: Eugenics and plastic surgery]. In *Annaes de Eugenía,* 149–53.

———. 1919b. O segredo de marathona [The secret of the marathon]. In *Annaes de Eugenía,* 115–35.

———. [1920] 1960. *Da educação física: O que e, o que tem sido e o que devería ser* [Physical education: What it is, what it has been, and what it should be]. 3d ed. São Paulo: Edições Melhoramentos.

———. [1943] 1950. *Brazilian Culture: An Introduction to the Study of Culture in Brazil.* New York: Macmillan.

———. 1955. *As ciências no Brasil* [The sciences in Brazil]. 2 vols. Rio de Janeireo: Edições Melhoramentos.

Bandeiro de Mello, Nelson. 1939–1940. Alcoolismo e hereditariedade [Alcholism and heredity] *Archivos Braileiros de Hygiene Mental* 12: 84–91.

Barreto, Luís Pereira. 1919. Meninas feias e meninas bonitas: Eugenía e esthetica [Ugly girls and pretty girls: eugenics and aesthetics]. In *Annaes de Eugenía,* 139–43.

Basile, Renato. 1980. A história da biologia no estado de São Paulo [The history of biology in the state of São Paulo]. *Ciência e Cultura* 32:1303–9.

Beiguelman, Bernardo. 1979. A genética humana no Brasil [Human genetics in Brazil]. *Ciência e Cultura* 31: 1198–1217.

Bowler, Peter J. 1983. *The Eclipse of Darwin: Anti-Darwinism in the Decades around 1900.* Baltimore: Johns Hopkins University Press.

Braga, Edgard. 1931. Fundamentos do exame médico pré-nupcial [The fundamentals of the medical pre-nuptial examination]. *Boletim de Eugenía* 3, 28: 38.

Brasil. 1935. *Annaes da Assembleia Nacional Constituente* [The annals of the National Constituent Assembly] 4: 490–93, 546–48.

Caldas, Mirandolino. 1929. As nossas campahnas [Our campaigns]. *Archivos Brasileiros de Hygiene Mental* 2: 57–60.

———. 1930. A hygiene mental no Brasil [Mental hygiene in Brazil]. *Boletim de Eugenía* 3, 3: 69–77.

Campos, Francisco do. 1931. A eugenía e a reforma do ensino [Eugenics and the reform of education]. *Boletim de Eugenía* 3, 28: 34–35.

Carneiro, Levi. 1929. Educação e eugenía [Eugenics and education]. In Primeiro Congresso Brasileiro de Eugenía, *Actos e Trabalhos,* 107–16.

Castiglione, Teodolindo. 1942. *A eugenía no direito da familia* [Eugenics in family law]. São Paulo: Saraiva e Cia.

Castro Faria, L. de. 1959. *A contribuição de E. Roquette Pinto para a antropologia Brasileira* [The contribution of E. Roquette Pinto to Brazilian anthropology]. Rio de Janeiro: Museu Nacional.

Catálogo-diccionário das theses inauguraes [Dictionary-Catalogue of Inaugural Theses]. 1935. São Paulo: Bibliotheca da Faculdade de Medicina da Universidade de São Paulo.

Chaui, Marilena, and Maria Sylvia Carvalho Franco. 1978. *Ideologia e mobilização popular* [Ideology and popular mobilization]. Rio de Janeiro: Paz e Terra.

Congrès Latin d'Eugénique. 1938. *Rapport* [Report]. Paris: Masson.

Correa, Mariza. 1982. *As illusões da libertade: A Escola Nina Rodrigues e a antropologia no Brasil* [The illusions of liberty: the Nina Rodrigues School and anthropology in Brazil]. Ph.D. dissertation, Universidade de São Paulo.

Couto, Miguel. 1925. Problemas médico-sociaes do Brasil [Medical-social problems of Brazil]. *Medicamenta* 4: 14–15.

_____. 1935. *Clinica médica* [Clinical medicine]. 3d ed. Rio de Janeiro: Flores e Mario.

Cunha, Euclides da. [1902] 1944. *Rebellion in the Backlands.* Tr. Samuel Putnam. Chicago: University of Chicago Press.

Cunha, Ovidio da. 1935. Introducção à Africanologia brasileira [Introduction to Afro-Brazilian studies]. *Boletim do Ministério do Trabalho,* no. 9: 286–91.

Cunha Lopes. 1931. Pesquisas genealógicos [Genealogical research]. *Boletim de Eugenía* 3: 53–54.

_____. 1934. *Da esterilização em psichiatria* [Sterilization in psychiatry]. Rio de Janeiro: Separata dos Archivos Brasileiros de Neuratria e Psichiatria.

Darwin, Leonard. 1926. *The Need for Eugenic Reform.* New York: Appleton.

Decroly, O. 1929. A selecção dos bem-dotados [The selection of the well-endowed]. *Boletim de Eugenía* 1: 49–50.

Delfino, Victor. 1919. Por la raza y por la patria [For race and for country]. In *Annaes de Eugenía,* 187–89.

Della Cava, Ralph. 1976. Catholicism and Society in Twentieth Century Brazil. *Latin American Research Review* 11: 7–50.

Dobzhansky, Theodosius. 1980. *The Roving Naturalist: Travel Letters of Theodosius Dobzhansky.* Ed. Bentley Glass. Philadelphia: American Philosophical Society.

Domingues, Octavio. 1929. *A hereditariedade em face de educação* [Heredity in the face of education]. São Paulo: Companhia Melhoramentos.

_____. 1930a. Os programmas de ensino e a genética [Programs of instruction and genetics]. *Boletim de Eugenía* 2, 13: 50–51.

_____. 1930b. Saúde, hygiene e eugenía [Health, hygiene and eugenics]. *Boletim de Eugenía* 2, 18: 48–51.

_____. 1931. 'Birth control,' esterilização e pena de morte ['Birth control,' sterilization and the death penalty]. *Boletim de Eugenía* 3, 30: 36.

_____. 1936. *Hereditariedade e eugenía, suas bases, theorias, suas applicações práticas* [Heredity and eugenics, its bases, theories, applications]. Rio de Janeiro: Civilização Brasileira.

_____. 1940. *A margem da zootécnia: Estudos e Ensaios* [At the margin of zootechnology]. Rio de Janeiro: Alba.

_____. 1942a. *Eugenía: Seus propósitos, suas bases, seus meios (em cinco lições)* [Eugenics: Its Propositions, its bases, its methods (in five lessons)]. São Paulo: Editora Nacional.

_____. 1942b. *Vamos a criar galinhos* [Let us create chickens]. Rio de Janeiro: Ministerio da Agricultura.

Dreyfus, André. 1929. O estado actual do problema da hereditariedade [The current state of the problem of heredity]. In Primeiro Congresso Brasileiro de Eugenía, *Actos e Trabalhos,* 87–97.

_____. 1945. Curso de genética, com aplicação à orquidologia [A course in genetics, with application to the study of orchids]. *Boletim do Círculo Paulista de Orquidófilos* 2: 51–58, 69–78, 89–102, 109–17, 125–32, 141–46, 157–64.

Falcão, Waldemar. 1941. *O ministério do trabalho no Estado Nôvo* [The ministry of labor in the Estado Nôvo]. Rio de Janeiro: Imprensa Nacional.

Farani, Alberto. 1931. Como evitar as proles degeneradas [How to prevent degenerate offspring]. *Archivos Brasileiros de Hygiene Mental* 4: 169–79.

Fausto, Boris. 1983. Controle social e criminalidade em São Paulo: Um apanhado geral (1890–1924) [Social control and criminality in São Paulo: A general summary (1890–1924)]. In *Crime, violência e poder.* Ed. Boris Fausto, 193–210. São Paulo: Brasiliense.

Flynn, Peter. 1979. *Brazil, A Political Analysis.* Boulder, Colo.: Westview Press.

Freire Costa, Jurandir. 1976. *História de psichiatria no Brasil: Um corte ideológico* [History of psychiatry in Brazil: An ideological analysis]. Rio de Janeiro: Editora Documentário.

Freyre, Gilberto. [1933] 1934. *Casa grande e senzala* [The masters and the slaves]. 2d ed. São Paulo: Imprensa Paulista.

———. 1963. *The Masters and the Slaves: A Study in the Development of Brazilian Civilization.* 2d ed. New York: Knopf.

Frias, Jorge A. 1941. *El matrimónio, sus impedimentos y nulidades: Derecho comparado* [Matrimony, its impediments and invalidity: Comparative law]. Cordoba, Argentina: "El Ateneo."

Froés da Fonseca, A. 1929. Os grandes problemas de anthropologia [The great anthropological problems]. In Primeiro Congresso Brasileiro de Eugenía, *Actos e Trabalhos,* 63–86.

Godoy, Paulo de. 1927. *Eugenía e selecção* [Eugenics and selection]. São Paulo: Editorial Helios Ltda. Thesis, Faculdade de Medicina.

Graham, Loren R. 1977. Science and Values: The Eugenics Movement in Germany and Russia in the 1920s. *American Historical Review* 82: 1133–64.

Hahner, June E. 1980. Feminism, Women's Rights, and the Suffrage Movement in Brazil. *Hispanic American Historical Review* 15: 65–111.

Jimenez de Asúa, Luis. 1942. *Libertad de amar y derecho de morir: Ensayos de un criminalísta sobre eugenesia e eutanasia* [The freedom to love and the right to die: Essays of a criminal lawyer on eugenics and euthanasia]. Buenos Aires: Losado.

Kehl, Renato Ferraz. 1919a. Conferencia de propaganda eugenica [Address on eugenical propaganda]. In *Annaes de Eugenía,* 67–79.

———. 1919b. Darwinismo social e eugenía [Social Darwinism and eugenics]. In *Annaes de Eugenía,* 177–83.

———. 1919c. Que é eugenia? [What is eugenics?]. In Sociedade Eugenica de São Paulo. *Annaes de Eugenía,* 219–23. São Paulo: Revista do Brasil.

———. 1923a. *A Cura da fealdade: Eugenía e medicina social* [The cure of ugliness: Eugenics and social medicine]. São Paulo: Monteiro Lobato.

———. 1923b. *Eugenía e medicina social: Problemas da vida* [Eugenics and social medicine: Problems of life]. 2d ed. Rio de Janeiro: Livraria Francisco Alves.

———. 1923c. *Melhoramos e prolonguemos a vida: Valorisação eugenico do homen* [Improvement and prolonging life: The eugenic valorization of man]. Rio de Janeiro: Livraria Francisco Alves.

———. 1925. *Como escolher uma boa esposa: Ensaios de Eugenía* [How to choose a good wife: Essays in eugenics]. Rio de Janeiro: Livraria Francisco Alves.

———. 1929a. Eugenía no Brasil: Esboço histórico e bibliográphico [Eugenics in Brazil: Historical and bibliographic outline]. In Primeiro Congresso Brasileiro de Eugenía, *Actos e Trabalhos,* 45–61.

———. 1929b. Limitação da natalidade [Limitation of birth]. *Boletim de Eugenía* 1, 12: 89–90.

———. 1929c. Questões de raça [Questions of race]. *Boletim de Eugen´a* 1, 6/7: 51–52.

———. 1930a. Aos nossas leitores: Inquerito sobre a educação sexual da infancia e da mocidade [To our readers: An inquiry into sexual education in infancy and youth]. *Boletim de Eugenía* 2, 19: 52.

———. 1930b. Inquerito sobre educação sexual [Inquiry into sexual education]. *Boletim de Eugenía* 2, 24: 45–47.

———. 1930c. Linhagems: Paes e avos [Pedigrees: Fathers and grandfathers]. *Boletim de Eugenía* 2, 14: 49.

———. 1930d. Nova theoria sobre a hereditariedade [New theory of heredity]. *Boletim de Eugenía* 2, 23: 51–53.

———. 1930e. Qual o mechanismo da hereditariedade normal e morbida? [What is the mechanism of normal and morbid heredity?]. *Boletim de Eugenía* 2, 16: 50–51.

———. 1931a. A campanha da eugenica no Brasil. *Archivos Brasileiros de Hygiene Mental* 4: 94–96.

———. 1931b. Campanha da eugenía no Brasil: Um interessante inquérito [The eugenics campaign in Brasil: An interesting inquiry]. *Boletim de Eugenía* 3, 28: 32.

———. 1931c. Casamentos e natalidade nas classes media e inferior [Marriage and birth rates in the middle and lower classes]. *Boletim de Eugenía* 3, 35: 165.

———. 1931d. Eugenics Abroad. *Eugenics Review* 23: 234–36.

———. 1931e. Eugenics in Brazil. *Boletim de Eugenía* 3, 28: 36.

———. 1931f. Os erros de filantropia [The errors of philanthropy]. *Boletim de Eugenía* 3, 32: 49.

———. 1931g. The first eugenics movement in Brazil. *Boletim de Eugenía* 3, 28: 35–36.

———. 1933. *Aparas eugenicas: sexo e civilização (Novos diretrizes)* [Eugenical fragments: Sex and civilization (new policies)]. Rio de Janeiro: Francisco Alves.

———. [1929] 1935. *Lições de eugenia: Refundida e aumentada* [Lessons in eugenics: Reformulated and augmented]. 2d ed. Rio de Janeiro: Editor Brasil.

———. 1936. Eugenics in Brazil. *Eugenics Review* 27: 231–32.

———. 1937. *Pour que sou eugenista: 20 anos de campagna eugenica, 1917–1937* [Why I am a eugenist: Twenty years of the eugenics campaign, 1917–1937]. Rio de Janeiro: Francisco Alves.

———. 1942. *Catecismo para adultos* [Catechism for adults]. Rio de Janeiro: Francisco Alves.

Kevles, Daniel J. 1985. *In the Name of Eugenics: Genetics and the Uses of Human Heredity.* New York: Knopf.

Lacerda, João Batista de. 1911. The *Métis*, or half-breeds, of Brazil. In *Papers on Inter-racial Problems Communicated to the First Universal Races Congress held at the University of London, July 26–29, 1911,* ed. G. Spiller, 377–382. London.

Lamounier, Bolivar. 1978. Raizes do Brazil [Origins of Brazil]. *Revista Senhor Vogue* (April): 141–45.

Levine, Robert E. 1970. *The Vargas Regime: The Critical Years, 1934–1938.* New York: Columbia University Press.

———. 1973–1974. The First Afro-Brazilian Congress: Opportunities for the Study of Race in the Brazilian Northeast. *Race* 15: 185–95.

Lippi Oliveira, Lucia, et al. 1982. *Estado Nôvo: Ideológia e poder* [The Estado Nôvo: Ideology and power]. Rio de Janeiro: Zahar.

Lopes, Ernani. 1929. Relatório [Report]. *Archivos Brasileiros de Hygiene Mental* 6: 139–57.

Lopes, Juana M. de. 1933. Em torno de exame pré-nupcial [On the subject of the prenuptial examination]. *Archivos Brasileiros de Hygiene Mental* 6: 139–57.

Love, Joseph. 1980. *São Paulo in the Brazilian Federation (1889–1937).* Stanford, Calif.: Stanford University Press.

Lowenstein, Karl. 1942. *Brazil under Vargas.* New York: Macmillan.

Lundborg, Hermann. 1930. Biologia racial: Perspectivas e pontos de vistas eugénicas [Racial biology: Perspectives and eugenical points of view]. *Boletim de Eugenía* 2, 14: 50–51.

———. 1931. Cruzamento de raças [Crossing of races]. *Boletim de Eugenía* 3, 34: 125–27.

Lurie, Edward. 1960. *Louis Agassiz: A Life in Science.* Chicago: Chicago University Press.

Machado, N. Moreira. 1919. A syphilis e o casamento [Syphilis and marriage]. In *Annaes de Eugenía,* 231–36.

McKenzie, Donald. 1976. Eugenics in Britain. *Social Studies of Science* 6: 499–532.

Mac-Lean y Estenos, Roberto. 1952. *La eugenesa en America* [Eugenics in America]. Mexico: Universidad Nacional.

Magalhães, Bernardo de. 1919. Eugenía: Seus fins, factores dysgenicos a combater [Eugenics: Its goals, and the combat of dysgenic factors]. In *Annaes de Eugenía,* 157–73.

Magalhães, Fernando. 1925. Os inimigos da raça: Molestias evitáveis e intoxicações euphorísticas [Enemies of race: Avoidable diseases and euphoristic intoxicants]. *Medicamenta* 4: 13–15.

Magarinos, José. 1934. Immigração [Immigration]. *Boletim do Ministério do Trabalho,* no. 5: 265–70.

———. 1935. O homem [Man]. *Boletim do Ministério do Trabalho,* no. 12: 287–97.

Marchaud, Henri-Jean. 1933. *L'Évolution de l'idée eugénique* [The evolution of the eugenical idea]. Bordeaux: Imprimérie-Librairie de l'Université.

Martins, Wilson. [1915–1933] 1978. *Histórica da inteligência Brasileira* [History of the Brazilian mind]. 6 vols. Rio de Janeiro: Editora da Universidade de São Paulo.

Mayr, Ernst, and William B. Provine, eds. 1980. *The Evolutionary Synthesis: Perspectives on the Unification of Biology.* Cambridge, Mass.: Harvard University Press.

Medeiros, Amaury de. [1927] 1931. O exame prénupcial [The prenuptial examination]. *Archivos do Instituto Medico-Legal* 2: 71–86.

Medeiros, Maria Alice de Aguir. 1980. Casa grande e senzala: Uma interpretação [Masters and slaves: an interpretation]. *Dados* 23: 215–36.

Meira, João Florentino. 1907. *Neo-Lamarckismo* [Neo-Lamarckism]. Rio de Janeiro: Typ. Carvalhoes. Thesis, Faculdade de Medicina.

Meira, Rubião. 1919. Factores de degeneração de nossa raça: Meios de combatel-os [Factors in the degeneration of our race: Methods of combating them]. In *Annaes de Eugenía* 49–64.

Miceli, Sergio. 1979. *Intelectuais e classe dirigente no Brasil, 1920–1945* [Intellectuals and the ruling class in Brazil, 1920–1945]. Rio de Janeiro: Difel.

Mitchell, Michael. 1983. Race, Legitimacy and the State in Brazil. Paper presented at Latin American Studies Association Meetings, 29 September–1 October, at Mexico City.

Mjöen, Jan Alfred. 1931. Cruzamento da raça [Race crossing]. *Boletim de Eugenía* 3, 32: 49–54.

Monteleone, Pedro. 1929. *Os cinco problemas da eugenía Brasileira* [The five problems of Brazilian eugenics]. Rio de Janeiro: Thesis, Faculdade de Medicina.

Moreira, Juliano. 1925. A seleção individual de immigrantes no programma da hygiene mental [Individual selection of immigrants in a program of mental hygiene]. *Archivos Brasileiros de Hygiene Mental* 1: 109–15.

Moreira, Silvia Levi. 1982. *A Liga Nacionalista de São Paulo: Ideologia e atuação* [The Nationalist League of São Paulo: Ideology and action]. M.A. thesis, Universidade de São Paulo.

Moura, Olegario de. 1919a. Saneamento do Brasil: Eugenisação no Brasil [Sanitation of Brazil: Eugenic improvement of Brazil]. In *Annaes de Eugenia* 239–40.

———. 1919b. Saneamento-eugenía-civilização [Sanitation-eugenics-civilization]. In *Annaes de Eugenia* 83–90.

Navarro, Moacyr. 1950. *Miguel Couto Vivo* [Miguel Couto alive]. Rio de Janeiro: Editora A. Noite.

Neiva, Artur, and Belisario Penna. 1916. Viagem científica pelo norte da Bahia, sudoeste de Pernambuco, sul de Piauhí, e norte á sul de Goiáz [Journey in the north of Bahia, southwest of Pernambuco, south of Piauhi, and north to the south of Goias]. *Memorias do Instituto Oswaldo Cruz* 8: 74–224.

Nina Rodrigues, Raimundo. [1894] 1938. *As raças humanas e a responsibilidade penal no Brasil* [Human races and penal responsibility in Brazil]. 3d ed. Ed. Afrânio Peixoto. Rio de Janeiro: Editora Nacional.

Nisot, M. T. 1927. *La question eugénique dans divers pays* [The eugenic question in various countries]. Brussels: Librairie Faile.

Nye, Robert A. 1984. *Crime, Madness and Politics in Modern France: The Medical Concept of National Decline*. Princeton, N.J.: Princeton University Press.

Odalia, Nilo. 1977. O ideal de branqueamento da raça na historiografia brasileira [The ideal of whitening the race in Brazilian historiography]. *Contexto* 3: 127–36.

Pacheco e Silva, A. C. 1925. Immigração e hygiene mental [Immigration and mental hygiene]. *Archivos Brasileiros de Hygiene Mental* 1: 27–35.

Pan American Conference on Eugenics and Homiculture of the American Republics. 1927. *Actas de la Primera Conferencia Panamericana de Eugenesia y Homicultura de las Republicas Americanas*. Havana: Republica de Cuba.

Pan American Conference on Eugenics and Homiculture of the Americas. 1934. *Actas de la Segunda Conferencia Panamericana de Eugenesia e Homicultura de las Republicas Americanas*. Buenos Aires: Frascoli y Bindi.

Paz Soldan, Carlos Enrique. 1916. *Un programa nacional de política sánitaria* [A national program of sanitary policies]. Lima: Estado-Nunez.

Peixoto, Afrânio. 1916. *Minha terra, minha gente* [My country, my people]. São Paulo: Francisco Alves.

———. 1917. *Hygiene*. Rio de Janeiro: Francisco Alves.

———. 1924. As doenças evitáveis [Avoidable diseases]. *Medicamenta* 3: 21–24.

———. [1933] 1936. *Criminologia* [Criminology]. São Paulo: Editoria Nacional.

———. 1944. *Eunice ou a educação da mulher* [Eunice or woman's education]. Rio de Janeiro: W. J. Jackson.

Penna, Belisario. 1918. *Saneamento do Brasil* [The sanitation of Brazil]. Rio de Janeiro: Revista do Tribunaes.

———. 1920. *Exército e saneamento* [The army and sanitation]. Rio de Janeiro: Revista de Tribunaes.

———. 1929. Eugenía e eugenismo [Eugenics and eugenism]. *Boletim de Eugenía.*

Pereira Cunha, Maria Clementina. 1986. *O espelho do mundo: Juquery, a história de um asilo* [Mirror of the world: Juquery, a history of an asylum]. São Paulo: Paz e Terra.

Pinheiro Guimarães, Francisco. 1935. *A hereditariedade normal e patologica* [Normal and pathological heredity]. Rio de Janeiro: Livraria Francisco Alves.

Poés de Andrade, Geraldo de Souza. 1925. *O Japonez a luz de biologia: Considerações anthropologicas e ensaios eugénicos* [The Japanese in the light of biology: Anthropological considerations and eugenical essays]. Rio de Janeiro: Thesis, Faculdade de Medicina.

Porto-Careiro, J. P. 1933. O exame pre-nupcial como factor eugénico [The prenuptial examination as a eugenical factor]. *Archivos Brasileiros de Hygiene Mental* 6: 87–94.

Portugal, Oswaldo. 1919. Prophilaxia das molestias venereas: Conselhos aos moços [The prophylaxis of venereal diseases: Advice to young men]. In *Annaes de Eugenia* 93–111.

Primeiro Congresso Brasileiro de Eugenía (First Brazilian Congress of Eugenics). 1929. *Actos e trabalhos.* Rio de Janeiro.

Prisco, Francisco. 1929. Alcohoolismo [Alcoholism]. *Brasil-Médico* 43: 801–5.

Putnam, Samuel. 1941. Vargas Dictatorship in Brazil. *Science and Society* 5: 97–116.

———. 1942. Brazilian Culture under Vargas. *Science and Society* 6: 34–57.

Ramos, Arthur. 1935. *Guerra e relação de raça* [War and race relations]. Departmento União Nacional dos Estudos.

Ribeiro, Leonidio. 1931. Os Problemas médico-legais em face de reforma da polícia [Medical-legal problems in relation to the reform of the police]. *Archivos do Instituto Medico-Legal e Cabinete de Identificação,* 1: 11–26.

———. 1937. Jubileu do Professor Afrânio Peixoto, 1906–1936: Homenagem dos "Archivos" [The jubilee of professor Afrânio Peixoto, 1906–1936: Homage of the "Archives"]. *Archivos de Medicina Legal* 7: 293–303.

———. 1942. *O novo codigo penal e a medicina legal* [The new penal code and legal medicine]. Rio de Janeiro: Jacintho Editora.

———. 1950. *Afrânio Peixoto.* Rio de Janeiro: Edições Conde.

Roquette-Pinto, Edgar. 1927. *Seixos rolados: Estudos Brasileiros* [Tumbling pebbles: Brazilian studies]. Rio de Janeiro.

——— [1933] 1978. *Ensaios de antropologia Brasiliana* [Essays in Brazilian anthropology]. Rio de Janeiro: Editora Nacional.

Rossi, Luciano. 1926. *Questões de Hygiene Social* [Questions of social hygiene]. Rio de Janeiro: Thesis, Faculdade de Medicina.

Rothman, David J. 1980. *Conscience and Convenience: The Asylum and Its Alternatives in Progressive America.* Boston: Little Brown.

Roxo, Henrique. 1939–1940. Problemas de hygiene mental [Problems of mental hygiene]. *Archivos Brasileiros de Hygiene Mental* 12: 49–51.

Sayers, Janet. 1982. *Biological Politics.* London: Tavistock Publications.

Schneider, William. 1982. Toward the Improvement of the Human Race: The History of Eugenics in France. *Journal of Modern History* 54: 268–91.

Searle, G. R. 1981. Eugenics and Class. In *Biology, Medicine and Society, 1840–1940,* ed. Charles Webster, 217–42. London: Cambridge University Press.

Silva, Luis L. 1931. Cruzamento do branco com o preto [Crossing between white and black]. *Boletim de Eugenía* 3, 30: 35–36.

Skidmore, Thomas E. 1974. *Black into White: Race and Nationality in Brazilian Thought.* New York: Oxford University Press.

Stein, Stanley J. 1961. Freyre's Brazil Revisited: A Review of New World in the Tropics: The Culture of Modern Brazil. *Hispanic American Historical Review* 41: 111–13.

Stepan, Nancy Leys. 1976. *Beginnings of Brazilian Science: Oswaldo Cruz, Medical Research and Policy, 1880–1920.* New York: Science History Publications.

———. 1982. *The Idea of Race in Science: Great Britain, 1800–1960.* London: Macmillan.

———. 1985a. Biological Degeneration: Races and Proper Places. In *Degeneration: The Dark Side of Progress,* ed. J. Edward Chamberlin and Sander L. Gilman, 97–120. New York: Columbia University Press.

———. 1985b. Eugenesia, genética y salud pública: El movimiento eugenésico Brasileño y mundial [Eugenics, genetics, and public health: The eugenics movement in Brazil and the world]. *Quipu: Revista Latinoamericana de História de las Ciencias y la Técnologia* 2: 351–84.

Stocking, George W., Jr., 1968. *Race, Culture, and Evolution: Essays in the History of Anthropology.* London: Collier-Macmillan.

Teixeira Mendes, Carlos. 1917. *Melhoramento de variedades agricolas* [Improvement of agricultural varieties]. Piracicaba: Typ. de Livraria Americana.

Tepedino, Alexandre. 1914. *Eugenía (esboço)* [Eugenics (outline)]. Thesis, Faculdade de Medicina, Rio de Janeiro.

Todaro, Margaret. 1974. Integralism and the Brazilian Catholic Church. *Hispanic American Historical Review* 54: 431–52.

Trindade, Helgio Henrique. 1974. Plinio Salgado e a revolução de 30: Antecedentes [Plinio Salgado and the revolution of '30: Antecedents]. *Revista Brasileira de Estudos Politicos* 39: 9–56.

———. 1975. A Ação Integralista Brasileira: Aspectos históricos e ideológicos [The Brazilian Integralist party: Historical and ideological aspects]. *Dados* 10: 25–60.

Trounson, K. E. 1931. The Literature Reviewed. *Eugenics Review* 13: 236.

Varigny, H. de. 1931. Da eugenía [On eugenics]. *Boletim de Eugenía* 3, 25: 49–53.

Viana, Luiz. 1918. *Em torno de Mendelismo* [On the subject of Mendelism]. Niteroi: Medical thesis.

Vianna, Oliveira. 1920. *Populações meridionais do Brasil* [The southern populations of Brazil]. Rio de Janeiro.

———. [1922] 1938. *Evolução do povo Brasileira* [The evolution of the Brazilian people]. 2d ed. São Paulo: Editora Nacional.

———. [1932] 1934. *Raça e assimilação* [Race and assimilation]. 2d ed. São Paulo: Editora Nacional.

Viotta da Costa, Emilia. 1970. *Da monarchia à república: Momentos decisivos* [From monarchy to republic: Decisive moments]. São Paulo: Editora Grijalbo.

CHAPTER 5

Eugenics in Russia
1900–1940

Mark B. Adams

In recent years the history of eugenics has proved a useful way to study the subtle relationships between science and society. However, despite the explosion of literature on eugenics movements in many countries, the history of eugenics in Russia has remained virtually unexplored. In an important article published more than a decade ago comparing the German and Soviet movements in the 1920s, Loren Graham showed in a dozen pages what interesting comparative issues are raised by the Russian case (Graham 1978), but no detailed study of Russian eugenics has yet been published that begins to approach the available literature on the British, American, and other national cases in analytic sophistication or detail.

In the Soviet Union, too, despite the profusion of excellent works on the history of Soviet biology that have appeared recently, eugenics remains a delicate subject. One of the few articles on the subject presents its history in the context of work on human genetics (Efroimson 1967). Even book-length biographies of the scientific founders of eugenics treat the matter gingerly, seeking, in carefully sculpted language, to acquit their subjects of any wrongdoing (Astaurov and Rokitskii 1975; Medvedev 1978). The more usual practice in Soviet histories of Russian biology is to ignore eugenics altogether. More than a decade ago, the catalog of Nauka Publishers announced the forthcoming publication of a two-volume history of human genetics in the USSR by I. I. Kanaev; however, it has never appeared. For reasons that will later become clear, the history of Soviet eugenics remains a sore subject.

Versions of sections of this chapter are currently in press (Adams 1989a, 1989g, 1990). A fuller account of this history from the 1890s through the present day will be available in a book I am completing on hereditarian thought in the Soviet Union, tentatively entitled *Nature and Nurture in the USSR: The Science and Politics of Human Heredity*.

However understandable, the dearth of studies on the Russian case is none-theless lamentable: the very reasons that make the history of eugenics in other countries interesting make the history of Soviet eugenics especially so. First, in studies of Western eugenics movements, the relation of eugenics to politics, class structure, and social change has often been highlighted; some have seen it as a conservative response by empowered social classes faced with perceived threats and social disruption. Russia can provide an ideal test case for some of these hypotheses. In the period 1900–1917, tsarist Russia represented one of the most conservative political elites in Europe; further, from 1900 through 1940, the coun-try faced social turmoil and instability on a previously unprecedented scale—from the 1905 revolution, through the First World War, two 1917 revolutions, a bloody civil war, the New Economic Policy (NEP) of the early and mid-1920s, the "Great Break" of 1929–1932, and the imposition of Stalinism and the purges in the 1930s. Against this background the case of Russian eugenics affords us an excellent opportunity to study the relation of eugenics to politics and to explore the influence of the social setting on scientific development.

Second, the conservative ideological and political dimensions of eugenics have been emphasized by many historians, and some have paid special attention to the so-called "reform" or "Bolshevik" eugenics that was developed by leftist biolo-gists in the 1930s as an alternative to the conservative politics of the British and American movements (e.g. Paul 1984). But a decade earlier, a eugenics movement was being created in the first Marxist-Leninist state, and there were concerted attempts to create a Bolshevik eugenics by real Bolsheviks. For purposes of under-standing the complex relation between eugenics and political ideology, then, the Russian case assumes special importance.

Finally, the relation of eugenics to developments in genetics in America, Brit-ain, and elsewhere has been given much attention. From this perspective the Rus-sian case is especially interesting on three counts. First, in the 1920s and 1930s, the Soviet Union developed a thriving genetics community and became a world leader in population genetics, agricultural genetics, and mutation research, and so the relationship between genetics and eugenics there is of some interest. Second, the Soviet Union was the first country in the world to "ban" eugenics (1931) and was also the first in the world to institutionalize the discipline of "medical genet-ics" (1934–1935); in the USSR, then, the historical relationship between eugenics and medical genetics can be studied in sharp relief. Finally, the rise of Lysenkoism in the Soviet Union in the 1930s—with its assertion of Lamarckian inheritance, its denial of the validity of genetics, and its involvement in the repression of Soviet geneticists—is one of the most troubling phenomena in the history of twentieth-century science. Because Russian eugenics also lay at the boundary of genetics and society, its history may throw a new light on the Lysenko phenomenon.

Accounts of Russian science face a dual challenge posed by Western stereo-types of the Soviet Union. Frequently, Western readers either admire the Soviet Union or strongly dislike it, and they tend to assume that Russian science is either essentially the same as Western science or totally different. In any case, Soviet developments may seem to be irrelevant to the history of science proper—either because they are basically the same as Western developments, in which case we already know about them, or because they are utterly peculiar and anomalous, and hence can contribute little to our understanding of the science we know. Social histories of science in single countries—the United States, Britain, France, Germany—sometimes assume great amounts of knowledge about the country in question not often shared by the general reader; in the Soviet case the problem is especially acute.

The challenge, then, is to see whether Soviet science, with all its peculiarities and special features, can be understood using the same terms and analytic frameworks that have proved useful in studying Western science. The history of eugenics lends itself to this task. As a field imported into Russia from abroad, eugenics illustrates the diffusion and adaptation of foreign ideas and models to different national conditions. As a new field in the Russian context, eugenics allows us to study the process of self-definition and discipline building, the role of scientific entrepreneurs and professional networks, and the relationship between science and its social patrons. Finally, as a field surviving in a social context wracked by political, ideological, and administrative turmoil, it can illustrate the ways science is institutionalized, and how it manages to maintain its integrity in the face of social change. The history of eugenics in Russia encompasses many striking and unique features, to be sure. Yet, I believe, it can be understood in terms of the patterns of institutional and disciplinary history familiar to Western historians of science.

The Origins of Russian Eugenics, 1900–1920

The Russian eugenics movement officially began in 1920—that is, under the Soviet regime, shortly after the revolution of 1917, just as the bloody Russian civil war was drawing to a close. However, those origins reflected the opportunistic realization of certain prerevolutionary agendas by individuals whose careers reflected the trends of late tsarist science. The origins of eugenics in Russia, then, can only be understood in the context of science and society in prerevolutionary Russia.

The Russian setting

In 1725 Peter the Great created the Academy of Sciences in his new Baltic capital of St. Petersburg.[1] For more than a century thereafter, however, its operating languages were Latin, French, and German, and its activities were dominated by visiting or resident scholars from Europe. For a time, until she found them politically subversive, Catherine the Great cultivated Enlightenment ideas and turned her court into a kind of Parisian salon. Following the Napoleonic Wars, which brought Russian troops and Tsar Alexander into the heart of Europe, there was new interest in European science and technology. In the late eighteenth and early

1. The city was renamed Petrograd in 1914, Leningrad in 1924.

nineteenth centuries, native agendas focused on the exploration of the vast Russian frontier: the military conquests that created the Russian Empire were increasingly accompanied and followed by state-sponsored surveys of the cartography, geography, geology, natural history, and ethnography of its new territories. Tsarist cultural politics often vacillated between fascination with European ideas (often accompanied by perceptions of Russian "backwardness") and autocratic nationalistic isolation (accompanied by perceptions of Russia's unique character and status). Throughout this time Russia was ruled by the tsar, his advisors, and his state bureaucracy. Its nobility and gentry had been originally granted their holdings through state service, and vast numbers of Russian peasants were tied to the land as serfs.

Russia's defeat in the Crimean War in the 1850s triggered vast changes. Nicholas I had run a regime founded on nationalism, autocracy, religious orthodoxy, and a widespread belief in the invincibility of the Russian army; its defeat raised fundamental questions about Russia's competitiveness in a modern world. Following the accession of Alexander II, the early 1860s saw staggering reforms, among them the emancipation of the serfs, the creation of an independent judicial system, and major changes in the administration of local and provincial governments.

These events unleashed forces that would dominate Russian life for the remainder of tsarist rule and beyond. Culturally, the period 1865–1910 produced the great literature, music, and art for which Russia is famous—by Borodin, Tchaikovsky, Dostoyevski, Tolstoy, Chekhov, and many others. Economically, tsarist governments launched a series of attempts to solve the "rural problem," industrialize Russia, develop steel and railroads, and encourage foreign investment and native entrepreneurship; with these efforts came the growing importance of merchants and industrialists in the civic life of Russia. Politically, tsarist regimes sought to control the pace of change and to ensure stability. At the same time, there was increasing political instability, and the period saw a series of radical parties and revolutionary movements, hundreds of assassinations of tsarist officials (including the "Great Reformer" himself, Alexander II), and a growing disaffection by the intelligentsia, many of whom looked to Western political and parliamentary models.

Eugenics, degeneration, and late tsarist science

In many respects the development of Russian science and technology following the Great Reforms exhibits striking parallels with the development of American science after the Civil War. In Soviet literature, scientists of the period are often characterized as political "progressives" or even radicals, but this judgment is anachronistic. True, in the 1860s some sciences—Darwinism, physiology, psychology—were occasionally embraced by political radicals, but the most substantial growth in Russian science occurred in the late 1870s and 1880s, at the height of political reaction. Indeed, around 1880, the number of scientific and technical periodicals published in Russia began to grow exponentially, and by the turn of the century had far outstripped other kinds (Adams 1965).

This is not difficult to understand. While conservative tsarist censors were suspicious of the political and ideological "baggage" that often seemed to accompany contemporary English, German, and French science, they were not hostile to science or technology itself; indeed, the development of Russian science and technology had become vital to the country's economic development and military strength (Todes 1984). In the late nineteenth century, the Academy of Sciences for the first time became dominated not by Germans, but by native Russians. Universities opened new chairs, and science professors who had previously trained only a few graduate students were now creating whole schools. Moscow University and (St.) Petersburg University remained the centers of higher education, but there was a great expansion in provincial science, and important departments of mathematics, chemistry, and bacteriology arose in such centers as Kazan, Odessa, Kiev, and Kharkov. Indeed, these and other provincial cities account for much of the increase in scientific publication in the late nineteenth century. Postgraduate training abroad in leading scientific centers became common practice, and Paris and German science departments and medical schools were flooded with Russian students, many of them women (Koblitz 1988). New technical schools were opened. Varieties of positivism developed that encouraged a view of science as objective and apolitical (Utkina 1975).

In the period 1890–1910, members of a new generation of younger Russian scientists and scholars were beginning their careers. They were products of the excellent Russian university system who often completed advanced training in Europe, where they absorbed the new trends in European thought generally, including the music of Wagner, the philosophies of Bergson and Spengler, and the new work on radioactivity, economics, sociology, anthropology, experimental biology, and rejuvenation, as well as other enthusiasms of the day. As it happens, they came to scientific maturity at a time of great political uncertainty. Following Russia's humiliation in the Russo-Japanese War of 1904–1905, the social upheavals (known inaccurately as "the 1905 revolution") led the tsarist government to grant limited reforms, including the election of a parliament, or duma. Many political parties were formed to cultivate the new democratic spirit.

Many members of this scientific generation became politically active, seeking to resist tsarist control and crackdowns at the universities, and looked increasingly to the West for the latest in cultural, scientific, institutional, and political ideas. The more enterprising used their contacts to press for reform in Russia and for the development of new scientific, technological, and academic institutions; most had liberal democratic sympathies, and some were active in antitsarist politics. In Moscow they clustered around Shaniavsky University, the Beztuzhev courses, the Lebedev institute, and other privately or municipally funded institutions. To gain support for their enterprises, they appealed to Russian industrialists and philanthropists, and resurrected the journal *Priroda* as a mouthpiece, where they published accounts of the latest Western and Russian research and reported on scientific societies, institutions, and funding patterns in Germany, France, Britain, and America.

For many young Russian scientists, both the scientific and social conditions fostered great interest in what science could contribute to the solution of pressing social issues. Of course, scientific discussion of the relative contributions of the "biological" and the "social" in shaping human mental and physical traits arose in many countries around the turn of the century in connection with attempts to establish the scientific basis and legitimacy of the "social sciences" and as part of an efflorescence of interest in interdisciplinary research generally. But in Russia the period 1900–1930 witnessed an almost unparalleled profusion of new interdisciplinary theories and fields. The period saw the enunciation and development of V. I. Vernadsky's "biogeochemistry"; P. P. Lazarev's "biological physics"; N. N. Semenov's "chemical physics"; N. I. Vavilov's "law of homologous series," which created a periodic table of variations modeled on Mendeleev's chemical periodic table; and A. G. Gurvich's theory of "mitogenetic rays," which held that dividing cells give off a form of radiation that triggers other cells to divide, like radioactive decay. As we might expect, the interface of the biological and the social proved an especially active area. N. Semashko and others adapted German "social hygiene" into a distinctly Soviet variant (Solomon 1989). The plant breeder N. I. Vavilov argued for the creation of a "science of selection" that would synthesize botany, agronomy, anthropology, and archaeology (Adams 1978). The botanist V. N. Sukachev wrote a text for the new field of "plant sociology" (Adams 1990).

These trends can be seen in the behavior of contemporary science publishers. In the period 1905–1917 translations of many Western scientific works were published in Russia, a number of which dealt with experimental biology, genetics, and evolution. An early book by T. H. Morgan was published in Russian translation (Morgan 1909), for example, as was Punnett's classic book *Mendelism* (1912); a giant compendium of foreign Mendelian research also was published (Bogdanov 1914). In Saint Petersburg the firm "Obrazovanie" commissioned many dozens of anthologies for its series on "new ideas" in biology, psychology, sociology, physics, philosophy, and so forth. In Moscow a comparable role was played by the firm of M. and S. Sabashnikov. Many of the collections they tried to assemble dealt with the relationship of the biological and the social, including volumes on the evolutionary origins of human beings, the biological basis of behavior, and sociological works based on biology. One area they found especially promising was eugenics.

Galton's *Hereditary Genius* had appeared in Russian translation in 1875 (Galton 1875). Over the next three decades, European ideas filtered into Russia, some through translations, others brought back by the many Russian students for whom Continental experience was a standard part of postgraduate training. In writings addressed to both professional audiences and the general public, various Russian psychiatrists, physicians, and intellectuals reported on European developments and expressed their own views on fertility *(rozhdaemost')*, on congenital *(vrozhdennyi)* defects and diseases, and on the general problem of human degeneration *(vyrozhdenie)*. Many of these works were published by psychiatrists or neurologists, and concentrated on the problem of "degeneration" as reflected in madness, crime, psychopathology (e.g., Iudin 1907; Bekhterev 1908; Liublinsky 1912; Sholomovich 1913; Ukshe 1915), and occasionally alcoholism (e.g., Alekseev 1914).

By 1915 the word *evgenika* was in use in occasional books and essays devoted to the subject (e.g., Blium 1909), although one could occasionally find the alternate Russian rendering, *evgenetika* (Kravets 1914a). In addition to psychiatry, sanitationists and public health physicians also showed some interest (e.g., Karaffa-Korbut 1910; Gamaleia 1912). Of special interest was the Western literature on eugenics that treated it as one of the new areas of experimental biology as, for example, books by Charles Davenport, one of which appeared in Russian translation by a scientific publisher in 1913 (Davenport 1913). For the Russian eugenics movement was to be launched in Russia not as part of psychiatry or hygiene, but as a socially responsible, socially relevant branch of the new experimental biology.

The founders: Kol'tsov and Filipchenko

For the new generation of Russian scientists beginning careers in the prerevolutionary decades, new interdisciplinary research fields had a special appeal. Intellectually, of course, such fields provided some of the most exciting science of the day. But they also served the professional career interests of young Russian scientists: by importing new scientific trends into Russia from abroad and presiding over their development, young scientists could legitimate the new and independent places they were seeking for themselves in the structure of Russian science. The central roles in creating Russian eugenics were played by two entrepreneurial and prolific members of this scientific generation, two zoologists who had been converted to the new experimental biology in Europe and had returned home to cultivate it in their respective cities.

In Moscow the key figure in creating Russian eugenics was Nikolai Konstantinovich Kol'tsov (1892–1940) (Adams 1980a, 1980b). He had traveled to Europe, and specifically the Naples Station, to study invertebrate morphology in the 1890s—before the rediscovery of Mendel's laws in 1900—so the new "experimental biology" to which he became converted did not yet include Mendelism, although he became aware of it on subsequent trips. A docent at Moscow University, he was active in liberal politics and resigned in 1909 in protest over the actions of Minister of Education Kasso. He began teaching full-time at the Beztuzhev Courses for Women and at Shaniavsky University, private institutions partially underwritten by the Moscow city duma. At both places he developed laboratories where he trained students who specialized in one of the new experimental disciplines: limnology, experimental psychology, biometrics, blood chemistry, hormone research, organ transplantation, physicochemical biology, developmental mechanics, cytology, and genetics.

Beginning in 1914, he joined L. A. Tarasevich, a leading bacteriologist, as coeditor of the new popular-science journal *Priroda* based in Moscow. As editor, Kol'tsov reviewed Western developments in experimental biology, but he also reported on the funding and organization of science in other countries, detailing the emergence of the Kaiser Wilhelm institutes in Germany and the lavish efforts of the Carnegie and Rockefeller foundations in the United States. Kol'tsov had close family ties to the Moscow merchantry, which had been so active in charity, education, civic reform, and democratic politics (Rieber 1982). After lobbying the Moscow merchantry for funds to support his various research enterprises, in 1916 he managed to create an Institute of Experimental Biology under the auspices of the privately funded Moscow Scientific Research Institute Society. His institute was endowed with a large grant from the will of Russian railway magnate G. M. Mark and was located in a large house in the city's merchant quarter. At that time genetics was a relatively minor component of its overall program.

As the Russian capital, St. Petersburg, with its ready access to the Baltic, was Russia's "window to the West." Not surprisingly, it had the lion's share of Russia's scientific institutions. There eugenics developed under the auspices of Iurii Aleksandrovich Filipchenko (1882–1930) (Adams 1989d). Filipchenko spent 1911–1912 working with Richard Hertwig in Munich and at the Naples Station, where he became acquainted with the latest biological trends, especially Mendelism (Zavarzin 1930). There he collected material on crustacean embryology. He returned to St. Petersburg in 1912 and defended his master's thesis on the development of Apterygota and the genealogical relation between insects and millipedes and centipedes. In 1913 he was awarded a master's degree in zoology and comparative anatomy from St. Petersburg University, became preparator of its zootomical cabinet, and was appointed to its faculty as a *privat-dozent.*

Upon his return to Russia, Filipchenko gave up his work on invertebrate embryology and converted to the new experimental biology and especially genetics, to which he devoted himself for the next decade and a half. On 18 September 1913 he opened Russia's first genetics course at the university, entitled "The Study of Evolution and Heredity." In 1914 he put together anthologies on hybridization and sex determination for the series "New Ideas in Biology," which included his own translations of works by De Vries, Plate, Lotsy, Hertwig, Correns, and others, supplemented with his own essay reviews. In the period 1913–1917 he published a number of popular articles on the new biology in leading contemporary journals and two books based on his lectures (Filipchenko 1913, 1915, 1916, 1917).

Filipchenko took up the study of heredity in mammals in 1913, when he was appointed to the physiological division of the Veterinary Laboratory of the Ministry of Internal Affairs as assistant to I. I. Ivanov (1870–1932); he worked there through 1916. A student of Pavlov and a pioneer researcher in artificial insemination, Ivanov had been appointed to head the division when it was founded on 19 May 1908. In July 1910, the division organized a zootechnical station at Askania-Nova, the large estate of F. E. Faltz-Fein in the southern Russian steppes that had been donated in 1904 as a wildlife park. There Ivanov conducted important hybridization work on various domesticated and wild varieties of bison, cattle, and other ungulates. At the time Ivanov was interested in Richard Hertwig's theories of sex determination and probably chose Filipchenko as his assistant for that reason.

Together with Ivanov, Filipchenko taught a course at the Veterinary Laboratory beginning in 1913, where he lectured on Mendelian genetics, biometrics, the mutation theory, cytogenetics, and sex determination. Under Ivanov's direction Filipchenko investigated the effects of mammalian sperm on the sex determination of offspring and studied hybrids of rats and mice produced by artificial insemination. He mastered the various measures and indices for studying skull characteristics and sought to apply them to the study of cattle. He had expected that the inheritance of such quantitative characteristics would not follow Mendelian laws, since he believed they were determined by the cytoplasm. However, after three years of studying the skulls of parental and hybrid forms, he concluded that different breeds varied in mean indices, and that, in crossing, these mean indices are transmitted following Mendelian laws. In 1916 Filipchenko published a popular volume on the origin of domesticated animals. His doctoral dissertation of 1917, "Skull Inheritance and Variation in Mammals," led to joint publications with Ivanov in German and Russian and was awarded the Von Baer Prize of the Russian Academy of Sciences for 1919 (Filipchenko and Ivanov 1916; Filipchenko 1916–1917).

Filipchenko's expertise in the new biology and his study of mammalian crania led to connections with Bekhterev's Psychoneurological Institute. He taught a course there beginning in 1914, was elected its professor of vertebrate anatomy the next year, and served as its academic secretary through 1920. Filipchenko's readings on genetics, craniometry, the inheritance of quantitative characteristics, and neurology brought him into contact with the eugenics work being developed in the United States and Europe. He began giving popular lectures on eugenics in 1917 and, in 1918, published his first popular article on the subject (Filipchenko 1918). Over the next seven years, he would write prolifically for popular audiences, producing no less than four books on eugenics between 1921 and 1925, including a comparative biography of Galton and Mendel (Filipchenko 1921a, 1921b, 1924a, 1925a).

Revolutionary opportunities

Because of the effects of world war, revolution, and civil war, 1917–1922 was a time of both hardship and opportunity for young Russian scientists. On the one hand, scientific work and its financial support were disrupted, there were shortages in food, paper, and equipment, and much scientific talent died or emigrated. On the other hand, for those who remained, the system had opened up and there were new opportunities for professional advancement.

We can observe these effects in the careers of Filipchenko and Kol'tsov. Filipchenko was awarded a doctoral degree in zoology and comparative anatomy from Petrograd University in 1917; on 18 December of that year, he was appointed salaried docent in zoology, and within a year had been promoted to professor. In 1918 he organized the university's Laboratory of Genetics and Experimental Zoology and became its director; in 1919 the laboratory became a university department and Filipchenko became its chairman. Like other scholars of the period, Filipchenko took on extra jobs, both to earn extra income during times of privation and to fill in for those who had emigrated or fallen victim to war, famine, or disease. He served as *privat-dozent* in zoology at the Advanced Courses for Women in Petrograd until its merger with the university (1918–1919); as senior zoologist at the Zoological Laboratory of the Russian Academy of Sciences (1918–1921); and as professor of zoology in the Chemical Pharmaceutical Institute (1919–1922). In 1920 he helped found the Natural Scientific Research Institute at

Peterhof (which became the biological research base for Leningrad University) and served as its academic secretary and head of its Laboratory of Genetics and Experimental Zoology for the next decade. During this period he also wrote drafts of many of his books. In times of hardship and famine, he also worked closely with Maxim Gorky on the Commission to Improve the Living Conditions of Scientists (KUBU).

Kol'tsov was a decade older and better established than Filipchenko, so for him the Revolution was a mixed blessing. He became a professor at both the first and second Moscow universities in 1918. On the other hand, the institute he had worked so hard to establish and fund now lost its endowment, and the expensive equipment needed for physicochemical laboratory science became difficult to obtain. Then, too, Kol'tsov had been politically active before the Revolution; in August 1920 he was briefly under arrest as a counterrevolutionary, but his network of prerevolutionary contacts, including Semashko and Maxim Gorky, was soon able to secure his release. For Kol'tsov as for Filipchenko, the special opportunities and constraints of the postrevolutionary years made eugenics an opportune field. Eugenics fit ideally the new emphasis on science as a way of undermining religion and improving the human condition; it entailed a scientistic, materialist, biosocial concept of human beings; it sought to apply the results of genetics to benefit society; and it emphasized the human power to shape the future.

For its founders eugenics was not only scientifically intriguing and professionally useful; it also had strong visionary appeal. In characterizing eugenics both Filipchenko and Kol'tsov initially took their lead from Galton and his concept of a "civic religion"—one that, at a trying time, provided hope for a better future. In his first programmatic statement on eugenics, Kol'tsov echoed Galton's call, making some minor adjustments to the new revolutionary order. Contrasting the ideals of Islam and socialism, Kol'tsov concluded:

> The ideals of socialism are bound up with our earthly life: but the dream of creating a perfect order in the relations between people is also a religious idea, for which people will go to their deaths. Eugenics has before it a high ideal which also gives meaning to life and is worthy of sacrifices: the creation, through conscious work by many generations, of a human being of a higher type, a powerful ruler of nature and creator of life. Eugenics is the religion of the future and it awaits its prophets. (1922a, p. 27)

Although equally scientistic, Filipchenko never indulged in such effulgent rhetoric. Nonetheless, Kol'tsov's poetics captured something of the early spirit and appeal of eugenics throughout Russia. For readers who had just lived through epidemics, civil war, and famine, it held forth the prospect of a better future guaranteed by the authority of science. For biologists it defined a central role in helping their society. For potential patrons it offered a reason to fund research at a time of severe shortages. For isolated scientific workers in enclaves throughout the war-torn regions of the former empire, it served to inspire, rally, and recruit. In all, it was a stirring call for the creation of something new. Of course, its creation would entail an important role for the author as well as substantial financial support for his own scientific institute, its programs, and its students. There was nothing contradictory in a religious credo, much less a worldly one, that called for sacrifice from the faithful and at the same time managed to serve the interests of its prophets.

Getting organized

In addition to its ideological utility and visionary appeal, however, there were also important practical reasons that made eugenics an opportune field. A mainstay of eugenic research was the collection and analysis of genealogical and anthropometric data from questionnaires and archives. Such work fit the times: all it required was "paper and initiative" (Medvedev 1978, p. 43). More practically, at a time of famine, war, and inflation, it paid and fed students and supported research collectives that were in danger of dissolving. It also helped to support other, more theoretical scientific work in laboratory biology and genetics. It is instructive to follow the ways Kol'tsov and Filipchenko used eugenics to obtain social support and funding from public authorities for their broader scientific agendas.

When he became fully aware of the importance of new Western work on genetics in 1918 and 1919, Kol'tsov quickly moved the field to the forefront in his plans, since breeding experiments were relatively cheap and easy to perform. In addition, genetics promised immediate practical benefit in animal breeding at a time of famine. Kol'tsov thus was able to support his student Alexander Serebrovsky through the Commissariat of Agriculture and to obtain funds for a poultry-breeding station linked to his institute. More importantly, with the loss of his institute's endowment, Kol'tsov managed to gain support from the Commissariat of Public Health [Narkomzdrav], headed by his friend Semashko, and his entire institute became part of its system of research institutions (GINZ).

Since it demonstrated the utility of genetics to human health, eugenics was especially useful in solidifying the links between Kol'tsov's institute and its new patron. In the summer of 1920, then, Kol'tsov created within his institute its new Eugenics Section (Evgenicheskii otdel). In September he wrote to Filipchenko proposing that they combine efforts in developing the field. In a meeting on 1 November, however, they agreed that Filipchenko would organize something in Petrograd completely independent of Moscow (Medvedev 1978, p. 44). Given his location, his contacts, and his research orientation, Filipchenko approached the Russian Academy of Sciences, located in Petrograd. During the prerevolutionary war years, the academy had created a Commission on the Study of Natural Productive Forces (KEPS). On 14 February 1921 the Bureau of Eugenics was established under the auspices of KEPS, to be headed by Filipchenko. He appointed to the bureau's staff Jan Janovich Lusis, Denis Karl Lepin, and A. I. Zuitin, three Latvians who were his senior students at the time (Filipchenko 1922a).

Just how useful eugenics was in supporting the more general research agendas of Filipchenko and Kol'tsov may be seen in the way eugenics functioned in their respective institutions. The Bureau of Eugenics supported Filipchenko's cohort of students in genetics—D'iakonov, Zuitin, Lepin, Lusis, and, after 1924, Dobzhansky, Kerkis, and Medvedev. No matter what kind of genetics they studied, they were all partially funded through the bureau. The utility of the imprimatur "eugenics" for supporting "genetics" may be gauged from the following fact: when the research unit was named the Bureau of Eugenics (1921–1925), its bulletin included both eugenic and genetic research; when it was named the Bureau of Genetics and Eugenics (1925–1927), it published no eugenic research, only animal and plant genetics. Filipchenko made a point of emphasizing that the bulletin of his bureau published only scientific research carried out by its staff. For Filipchenko, then, eugenics was a "civic religion" and a research interest, but it was also a way of obtaining patronage from the academy for a "socially relevant" field in order to support his research, his students, and his institutional agenda.

For Kol'tsov eugenics served parallel functions. Work within the Eugenics Section of Kol'tsov's Institute of Experimental Biology had a highly interdisciplinary character, encompassing not only genealogical research, but also studies of blood chemistry and blood groups, as well as studies of behavioral genetics in mice by Kol'tsov's wife and former student, Maria Sadovnikova (Adams 1980a, 1980b). Students trained in animal genetics often worked in eugenics as a matter of course. For example, V. V. Sakharov won a degree at the institute in 1924 for two concurrent pieces of research—one on mutations in fruit flies, the other on pedigrees of Russian musicians (Sakharov 1924; Sakharov and Serebrovsky 1925). In the late 1920s he served as secretary of the Russian Eugenics Society while continuing his laboratory studies. Thus, although his research agenda was broader and more "physicochemical" than Filipchenko's, Kol'tsov also used eugenics to legitimate his growing research enterprise and the support it was receiving from its principal patron, Narkomzdrav.

The Structure of Russian Eugenics, 1920–1930

There is often a great temptation to simplify the historical treatment of eugenics by reducing its complexities to some "essential" characteristics. However, this approach runs the considerable risk of ignoring the remarkable diversity of the people, institutions, and interests that the field encompassed. Even well-established disciplines consist of clusters defined by professional and institutional affiliation, theoretical orientation, and problematics. In the case of Russian eugenics, these clusters had more reality than the putative discipline they constituted, for in the early 1920s, Russian eugenics was a new imported field that its organizers were seeking to create in Russia under postrevolutionary conditions. To do so, they sought to tie together appropriate professional, institutional, and regional groups into a unified network and to provide it with a viable and legitimate Russian identity.

The Russian Eugenics Society

In the fall of 1920, weeks after he had created the Eugenics Section of his institute and roughly at the time he was in communication with Filipchenko, Kol'tsov brought up the idea of forming a Russian eugenics society at a meeting of a Narkomzdrav commission on which he sat. The commission, which was working on the creation of a division of race hygiene in its Social Hygiene Museum, included a number of physicians active in social hygiene and several officials of the public health commissariat—Bogoiavlensky, Viktorov, Dauge, Zakharov, Martsynovsky, Mol'kov, Prokhorov, Sysin, Shifman, and Iudin (on their backgrounds, see Solomon and Hutchinson 1989). A preliminary planning meeting was held on 15 October 1920 at the House of Sanitary Education (Bunak 1922b).

At the founding meeting of the Russian Eugenics Society a month later on 19 November 1920, it was clear that the society's governance would be dominated by Kol'tsov and his students. At the gathering, held at the Institute of Experimental Biology, Kol'tsov was elected president. The society's bureau was established, consisting of Kol'tsov, psychiatrist T. I. Iudin and anthropologist V. V. Bunak as regular members, and N. V. Bogoiavlensky and A. S. Serebrovsky as temporary members (both were promoted to full members the following year). In 1923 M. V. Volotskoi assumed the post of secretary of the society. Bunak, Serebrovsky, and Volotskoi were all young protégés of Kol'tsov who worked in his institute. Its statutes established the Russian Eugenics Society under the auspices of Narkomzdrav and its State Scientific Institute of Public Health (GINZ); subsequently, the society was also granted recognition by the education commissariat, Narkompros. During its first year the society held nineteen meetings at which twenty-six papers were presented, fourteen by workers affiliated with the institute (Bunak 1924a). During its second year it held thirteen sessions and heard twenty-four papers. In subsequent years, except for occasional joint meetings with other professional groups and a few public sessions held at the House of Scientists (Dom uchenykh), the society customarily met one or two Fridays a month during the academic year at Kol'tsov's institute and was administered from the institute's Eugenics Section.

At the time of its founding in late 1920, the Russian Eugenics Society had thirty members; by the end of 1921, the number had risen to eighty-three; by the end of 1922, to ninety-five. These numbers are significant when we consider that membership in the Russian Eugenics Society was not open to the general public and consisted only of active members, most of whom were biologists, physicians, psychiatrists, or health officials. Beginning in 1922 the society began publishing a journal, *Russkii Evgenicheskii Zhurnal,* which issued an average of three numbers annually through early 1930. Especially in the journal's early years, a substantial number of its articles were based on papers presented at society meetings. Topics of discussion included the genealogy of Russian notables, blood chemistry, demography, criminality, Mendelism, sterilization, and various organizational questions such as the mounting of expeditions and the establishment of anthropometric stations.

Russia had been largely cut off from Western developments since 1916, so there was great interest among society members in genetic and eugenic developments in Germany, France, Scandinavia, Britain, and America. On behalf of the society, Kol'tsov wrote to Davenport and other prominent Western eugenicists to report on Soviet developments and to ask for publications. Kol'tsov was anxious to bring the Russian society into the international eugenics organizations. On 2 December 1921 Kol'tsov was elected the official representative of the Russian Eugenics Society to the International Commission of Eugenics; this was confirmed at the commission's 1922 meeting in Brussels, where Russia became one of twenty-two cooperating countries and one of only fifteen fulfilling the commission's requirements ("Membership" 1924).

Center and periphery: Regional clusters

With the difficulties of travel and communication during and immediately after the civil war and famine, it was to be expected that a new professional society based in Moscow would reflect the work of the Moscow community. Understandably, given the role of Kol'tsov and his institute in the eugenics society during its early years, investigators from other cities occasionally alluded to the Russian Eugenics Society as an organizational mouthpiece for Kol'tsov's group (e.g. Filipchenko 1924a, p. 180). Perhaps as a courtesy, Kol'tsov invited Filipchenko to serve as coeditor of the *Russkii Evgenicheskii Zhurnal,* but aside from book reviews, reports, and an occasional article, Filipchenko had little to do with the journal's contents.

However true of the society, Russian eugenics was not an exclusively Moscow enterprise. In other major cities, local eugenics societies were created in the early 1920s through the entrepreneurship of influential professional figures who had had an earlier interest in eugenics. Generally, they became affiliated with the central society in Moscow only after several years. The earliest to do so, in late 1924, was the Leningrad branch of the Russian Eugenics Society, made up of Filipchenko and the staff of his bureau. Although located in the capital and more active than those in other cities, the Leningrad group is exemplary of the general pattern.

For example, the Saratov Eugenics Society was formed on 29 December 1923 under the presidency of professor and psychiatrist M. P. Kutanin and was officially approved by the provincial administration on 28 March 1924. By the end of that year, it had held four meetings and had forty-four members, principally local psychiatrists, gynecologists, and obstetricians, but also including A. A. Bogomolets (future academician in microbiology) and G. K. Meister (plant breeder and president of the Lenin All-Union Academy of Agricultural Sciences in the mid 1930s until he was repressed). In 1925 the group held six meetings. Only at the end of that year did the Saratov Eugenics Society become the Saratov branch of the Russian Eugenics Society, and in 1926 it became affiliated with Saratov State University (Kutanin 1927).

Conditions in the Ukraine's two principal cities differed. The Odessa branch of the Russian Eugenics Society was formed in the mid-1920s under the presidency of Nikolai Kostiamin, professor of hygiene at the Odessa Institute of Hygiene and a physiological chemist specializing in blood chemistry. Its researches apparently focused on genealogical notes and pedigrees of "highly gifted and talented families" (Kostiamin 1925). Kiev was anomalous. It had a number of prominent workers in eugenics, notably A. A. Krontovsky. In addition, genetics and eugenics had been taught at the city's Veterinary Zootechnical Institute by professor I. Klodnitzky, a student of August Weismann and specialist in the alternation of generations in the plant louse. However, in 1925 he complained that "there is no eugenics office in Kiev," noting that "conditions for these sciences here are very hard and much worse than in Petrograd and in Moscow"; that his courses in eugenics and genetics had been eliminated; and that "there is no sense in organizing a society" since there is "no money as neither the government nor any institutions are interested in the question" (Klodnitzky 1925).

By contrast, in Sverdlovsk, G. V. Segalin had a successful enterprise that might well have formed a eugenics society but did not. As head of the Psychotechnical Laboratory of the Ural Polytechnique Institute, Segalin published a periodical with "Practical Medicine" entitled *Klinicheskii Arkhiv Genial'nosti i Odarennosti (Evropatologii)* that was, as explained on its title page, dedicated to "questions of the pathology of personalities gifted with genius, and also questions of creative endowments associated with psychopathological deviations." Although Segalin seemed the primary theoretician of the journal, his editorial board of some twenty physicians included the young neurologist S. N. Davidenkov. The first volume was published in 1925, the last a few years later. The journal's pages included articles dealing variously with the suicidal, schizophrenic, or otherwise disturbed minds of Pushkin, Lermontov, Esenin, Skriabin, and Maxim Gorky. Although both Segalin's group and the Russian Eugenics Society shared a preoccupation with studies of the hereditary basis of both genius and pathology, so far as I can tell they never became affiliated (Segalin 1925a, 1925b, 1926).

Professional networks

We have already noted the central involvement of experimental biologists, animal geneticists, and physicians in Russian eugenics. In addition, the eugenic involvement of psychologists and psychiatrists was especially marked. The most renowned was V. M. Bekhterev, founder (1908) and director of the Psychoneurological Institute in St. Petersburg, whose staff helped distribute hundreds of questionnaires for the Eugenics Bureau's study of the Petrograd intelligentsia. In addition, psychiatrist P. I. Liublinsky coedited the journal of the Russian Eugenics Society beginning in the mid-1920s, and psychiatrist T. I. Iudin was one of the society's most active and prolific members, publishing books and numerous articles on eugenics, psychopathology, and constitution. Another link between psychiatry and eugenics was institutionalized by the A. Ia. Kozhevnikov Society of Neurologists and Psychiatrists, an important professional society with many distinguished members, based in Moscow. On 16 December 1927 the Kozhevnikov society created its own Genetics Bureau under the direction of the prominent young neurologist and eugenicist S. N. Davidenkov. The group was based at the Psychiatric Clinic of the (First) Moscow State University, and held four meetings in 1928, hearing papers on hereditary syphilis, gigantism, and narcolepsy (Davidenkov 1928).

Eventually, Kol'tsov even managed to get the Nobel Prize–winning physiologist Ivan Pavlov into the act. In 1926 Pavlov published an account indicating that experiments conducted in his laboratory confirmed the Lamarckian inheritance of acquired characteristics. Kol'tsov's wife, Maria Sadovnikova, had done much research in the genetics of behavior in mice, and when the Kol'tsovs went over the data with Pavlov, they convinced him that those data were perfectly compatible with genetics. Pavlov was apparently converted. He withdrew his earlier claim in a public letter to *Pravda* and later organized an experimental station at Koltushi, outside of Leningrad, devoted to "the genetics of higher nervous activity." After Pavlov's death in 1936, the station was headed by Pavlov's leading student, L. A. Orbeli (Adams 1980a).

Anthropologists were also involved in the 1920s eugenics movement. Following the European tradition, in Russia "ethnology" (the study of *ethnos,* or culture) was distinguished from "anthropology," a more biological discipline modeled after Broca's school, which roughly corresponds to what is known in the United States as physical anthropology. The anthropologist V. V. Bunak became involved in eugenics through his interest in craniometry and his attempts to develop new biometric techniques for measuring skull characteristics. He met Kol'tsov through Moscow University and worked for a time in the Eugenics Section of Kol'tsov's institute. During the 1920s Bunak edited the leading Russian journal of anthropology and chaired the department of anthropology at Moscow University until 1930 (Bunak 1940).

Research agendas

The research problems pursued by Russian eugenicists also tended to form thematic clusters, some reflecting underlying institutional and professional patterns. An analysis of the contents of eugenics publications during the 1920s reveals four principal themes.

First, a dominant theme, especially in the early 1920s, was the genealogy of talented individuals and families. Papers were published exploring the pedigrees of such Russian notables as Chaadaev, Samarin, Herzen, Kropotkin, Trubetskoi, Lermontov, Witte, Shafirov, Pushkin, Bakunin, von Baer, the Aksakovs, and even the Romanovs, the former imperial family. In addition, collective genealogies were done of the Decembrists; professors in Odessa, Leningrad, Moscow, and Saratov; artists; musicians; literary figures; and the entire membership of the Russian Academy of Sciences from the date of its founding. This was a primary preoccupation of the Filipchenko group, but was also important in Moscow, Sverdlovsk, Odessa, and elsewhere. The work involved the formulation of genealogical and anthropometric questionnaires and library and archival research, so it had the advantage of being inexpensive and independent of technical equipment.

Second, a large body of published work concerned the contribution of heredity to various nervous, mental, or behavioral diseases, dysfunctions, or conditions. These included schizophrenia, manic depression, muscular dystrophy, epilepsy, stuttering, alcoholism, criminality, and syphilis. An important organizing concept was the idea of a human hereditary "constitution," a subject developed most fully in a book by Iudin (1926). Here much of the work was done by physicians, whose methods were genealogical and clinical. A principal outlet for this work was Segalin's journal, which published articles on the mental illnesses of Gorky, Esenin, Skriabin, Lermontov, and others. This work was conducted by psychiatrists or neurologists; their approach was psychiatric, their method, the explication of diaries and creative works of music and literature. Others, such as Kol'tsov's wife, studied behavioral genetics in mice and monkeys, following the methods of experimental psychology.

Third, a sizable body of work concerned the inheritance of various physical diseases and conditions. Many were pathological, including gigantism, syndactyly, marbling bone disease, split foot, clubfoot, ulcers, hernia, diseases of the urogential system, and endemic goiter; such studies relied principally on the methods of pathology and the clinic. A distinct subset of this group dealt with blood chemistry, blood groups and agglutination, and such diseases as leukemia, hemophilia, and pernicious anemia; here the work of the physician was supplemented by that of the physiological chemist. Eye color, the shape and color of the iris, and hair color were also studied, using the methods of chemistry and genealogy and occasionally the techniques of ophthalmology.

Finally, papers were published on what might be termed "population policy." These included papers by demographers, statisticians, and other social scientists concerning the effects of war, marriage, law, education, custom, and race on fertility and mortality. Some of these papers summarized Western literature and policy, others presented new data and their implications, and still others sought to analyze the geographical and racial distribution of various traits (e.g., eye color, blood type), diseases (e.g., endemic goiter), or behaviors (e.g., criminality).

As is evident, Russian eugenics in the 1920s was not a single field with some essential defining characteristics. Rather, the term *evgenika* served more as an umbrella under which a diverse group of people found legitimacy and support as they struggled for a common disciplinary identity. Professionally, eugenics involved a mix of experimental biologists, animal geneticists, hygienists, physicians, psychiatrists, anthropologists, and demographers. Institutionally, it comprised a loose federation of groups in Moscow associated the Kol'tsov institute, the university, the medical school clinic, public health, and anthropology; groups in Leningrad associated with the Academy of Sciences, the university, and the Bekhterev institute; and sundry other enclaves in Saratov, Odessa, and elsewhere.

In light of the diversity of intellectual, social, and thematic clusters of which Russian eugenics was made up, and the sheer variety of people and concerns that passed "in the name of eugenics," we may well wonder what they all had in common, aside from the rhetoric of human biological improvement. How could this splay of workers and interests be harmonized into a coherent Russian field worthy of public support?

Russifying eugenics

In introducing eugenics into Russia, both Filipchenko and Kol'tsov had taken their lead from Galton. As Mandrillon has pointed out, a preoccupation with hereditary talent reflected the central values of the Russian intelligentsia (1987). Yet, at times, Russian eugenics seemed a distinctly foreign import. When the journal of the Russian Eugenics Society first appeared, for example, on its title page was the society's emblem: a genealogical chart of the linked pedigrees of Charles Darwin and Francis Galton. A larger and more detailed version, complete with black-and-white circles and squares, hung prominently on the wall behind Kol'tsov's desk in the director's office of his institute throughout his lifetime.

In adapting an imported eugenics to fit postrevolutionary Russian circumstances, the Russian eugenicists, like their counterparts elsewhere, sought a native forerunner. They found one in the person of a certain Florinsky, who had published a book in St. Petersburg in 1866 entitled *The Improvement and Degeneration of the Human Race.* The work did not depart notably from contemporary European literature on degeneration, and had been largely ignored in Russia until it was unearthed by Russian eugenicists in the early 1920s. In terms of legitimating the new field's native roots, the identity of the author posed something of a problem, since the title page of the original listed the author as "F. Florinsky," well known in his day as a religious idealist philosopher. However, a lengthy footnote to a 1924 article on the subject concluded that this was undoubtedly a mistake, since the clinical examples cited in the work surely made it the product of Vasilii Markovich Florinsky (1833–1899), a professor at the Medico-Surgical Academy in St. Petersburg (Volotskoi 1924a). The 1866 book was republished in 1926 with its authorship corrected (Florinskii 1926).

However, adapting eugenics to Soviet conditions in the early 1920s required more than a legitimate nineteenth-century Russian precursor—it needed a legitimate Soviet identity that assured it a place in the ecology of knowledge in postrevolutionary Russia. If Kol'tsov's characterization of eugenics (and socialism) as an earthly "civic religion" had seemed strategically apt in 1921, it soon became unhelpful. Another way of characterizing eugenics was as an international "movement," and various accounts of the history and state of the eugenics movement *(evgenicheskoe dvizhenie)* appeared in the Soviet Union in the 1920s (Liublinskii 1926a). However, this term, too, was not without its drawbacks. In particular, why should an international movement be funded by Narkomzdrav's severely strained budget?

Given the character of its Soviet practitioners, their institutions, and their funding, many eugenicists felt that the legitimacy of eugenics ultimately depended on its recognition as a new scientific discipline in its own right. Among them was professor T. I. Iudin, who began *Evgenika,* his 240-page textbook for the field, with an opening chapter entitled "Eugenics as an Independent Scientific Discipline: Its Limits, Methods, and Tasks." There he addressed the relation of eugenics to other established disciplines: "Biologists call eugenics 'human genetics,' anthropologists call it 'social anthropology,' sociologists call it 'political demography,' physicians call it 'constitutional pathology,' Moll and Forel call it 'sexology,' hygienists call it 'part of social hygiene' or 'racial hygiene,' Haeckel calls it 'gonionomy'." Clearly, argued Iudin, a field related to so many existing disciplines but distinct from all of them merited disciplinary status in its own right. He concluded that although analogous to medicine, eugenics was a union of "genetics" and "sociology" and "will become the biology of social types" (Iudin, 1925, p. 6).

Iudin's characterization of eugenics as a combination of genetics and sociology went to the heart of its identity problem. In its early years, at least, the field may have allowed geneticists to talk sociology, but very few sociologists were talking genetics. A report on the activities of the Russian Eugenics Society during its first year applauded the interest of academics, biologists, and medics in the field, but lamented the dearth of sociologists and other social scientists and activists (Bunak 1922b). However, the call for the involvement of social activists was answered soon enough.

Bolshevik Eugenics

The term *Bolshevik eugenics* has occasionally been used in Western literature to characterize the views expressed in the 1930s by some British and American leftist biologists who were critical of the British and American movements, enamored of Soviet Russia, and devoted to the creation of socialist eugenics—such men as J.B.S. Haldane and especially H. J. Muller (e.g., Paul 1984). However, throughout the entire life span of the Russian eugenics movement (1920–1930), such figures viewed the Soviet experiment, however admiringly, from afar. In the 1920s there was indeed a "Bolshevik eugenics." However, its home was not Texas or London, but Moscow.

Although the Russian Eugenics Society existed in a Bolshevik state, not all Russian eugenics was Bolshevik; indeed, for most of the decade, the movement was dominated by liberal intelligentsia. Within the Russian movement, then, Bolshevik eugenics was but one strain, elaborated by a small group of younger Marxist eugenicists centered in Moscow. But Bolshevik eugenicists proved disproportionately important to the movement: occupying key posts within the society, they were both visible and vocal and played an important role in legitimating eugenics as a new field in Russia in the 1920s. As we might expect, there were real tensions between liberal and Bolshevik eugenics over matters of both theory and social policy. Nonetheless, in many respects the two variants were symbiotic. The bourgeois founders provided the movement with scientific respectability; the Bolsheviks, with ideological legitimacy and social relevance. Their role is best understood by considering a dilemma faced by the movement.

Whatever its self-image and ideology, Russian eugenics was a new field that was patronized by the public health commissariat because of its actual or potential usefulness. Kol'tsov may have found a sympathetic friend in Semashko, but the commissar had to balance his own agenda against competing interests within his enterprise and justify it against the claims of other commissariats to scarce resources. The Commissariat of Public Health was responsible, after all, for public health; why spend severely limited resources on a new field when they might better go to well-established medical specialties or to fields of obvious practical importance—nutrition research, for example, or epidemiology? Eugenicists had sought to legitimate their new field by alluding to its vital practical importance; yet, although they might diagnose human genetic diseases, they could not show how to cure or even treat them. Russian eugenic literature was replete with the rhetoric of human biological improvement and visions of a better, if distant, future. But what of the more immediate future? Funded as public health, eugenics could consolidate its relationship with its singular patron by demonstrating its immediate usefulness.

Like discipline builders elsewhere, however, the founders of Soviet eugenics felt a tension between reformist zeal and academic and professional caution. For those eugenicists like Filipchenko who regarded their research mission as primary, the agenda was clear: learn more. This meant conducting research in genetics, biometrics, human biology, physical anthropology, blood chemistry, psychiatry, and the human constitution; establishing anthropometric stations to collect data; correlating and integrating information from clinics and hospitals; and staying in touch with the latest foreign developments. As for educating the public, most eugenicists endorsed the publication of popular works and the establishment of new courses in universities and medical schools, although teaching eugenics to women in secondary schools provoked considerable resistance. But, for those who sought a guide to immediate practical action, it was less clear what eugenics had to offer.

Initially, the founders of Soviet eugenics followed Western models closely, touting eugenics as a civic religion or an international movement. Soon, however, it became clear that some ideological accommodation to the new regime was called for. Those from bourgeois family backgrounds and liberal politics were not especially well equipped for the task. Filipchenko held fast to his values. He continued his studies of the genealogy of the intelligentsia, refrained from political pronouncements, and retreated from the field altogether around 1925–1926 when his position became politically untenable. For his part, Kol'tsov tried as best he could to accommodate. Even so, some of his early articles in this vein seem a bit strained: one effort sought to demonstrate that Russian revolutionaries actually contained aristocratic genes; another, that historically renowned peasants were descended from illegitimate children of the nobility; yet another, that Communist party members were biologically superior and would not be fulfilling their social duty unless they had more children (e.g., Kol'tsov 1922a, 1923a, 1923c, 1924a, 1926). By the mid-1920s, the bourgeois founders of Soviet eugenics had retreated from articulations of broader social agendas.

In the early 1920s, however, a new group of young Marxist activists assumed increasing prominence in articulating the relevance of Russian science to socialist construction. Born around 1890, they had spent their teens during a period of increasing political activity and disaffection from the tsarist regime. They were in their twenties at the time of the Revolution and were entering their early thirties following the civil war. Most shared the faith of the new regime in the special status of science and believed that it offered the best hope of creating a better future.

During the 1920s this new generation helped to create a series of new Bolshevik institutions devoted to the development of Marxist science. Some were formed as circles and groups within existing institutions—for example, the Circle of Materialist Physicians at the Moscow University Medical School (renamed the Society of Materialist Physicians in November 1926), or the Society of Materialist

Biologists. Others were put forth as alternatives to existing Russian institutions. For example, in 1918 the Socialist Academy was created as a parallel to the Russian Academy of Sciences; in 1926 it was renamed the Communist Academy and opened a Section of Natural and Exact Sciences and a biological laboratory. The Communist Academy was abolished in 1936 after the Academy of Sciences had been "Bolshevized" (Graham 1967). Another example is the Timiriazev Biological Institute, which was founded after the Revolution; renamed the Timiriazev Scientific Research Institute in May 1924, it was later absorbed by the Communist Academy. Still other institutions sought to retool sympathetic specialists or train party members in academic subjects—for example, the Institute of Red Professors.

As the government and party consolidated power and began to establish policies and priorities in the mid-1920s, discussions of the sciences in these new Marxist institutions became more animated. Among the hotly debated topics were Freudian psychoanalysis and relativistic physics, but embryology, evolution, and genetics also attracted increasing attention from such figures as G. G. Bosse (1887–1965), S. S. Perov (1889–1967), Izrail I. Agol (1891–1937), Boris M. Zavadovsky (1895–1951), and M. S. Navashin (1896–1976) (Joravsky 1961; Gaissinovitch 1980). These discussions were often published in such new journals as *Pod Znamenem Marksizma* (Under the banner of Marxism) and *Estestvoznanie i Marksizm* (Science and Marxism). Generally, what 1920s Soviet Marxists liked about Darwinism, genetics, and eugenics was their experimentalist, materialist, scientific, nonreligious approach to the human condition; what they did not like were those aspects that appeared idealistic, suggested therapeutic impotence, and provided no basis for action.

From its earliest years Soviet eugenics attracted a number of young Marxist scientific activists centered in Moscow and specifically in the eugenics division of the Kol'tsov institute. One was M. V. Volotskoi (1893–1944); not much is known about his life (see Gaissinovitch 1980). In 1923 he became secretary of the Russian Eugenics Society, and in that capacity played an active role in its meetings and reported on them in its journal (Volotskoi 1924b). He published a number of research articles on the prehistory of Russian eugenics, including pieces on "the anthropotechnical projects" of Peter the Great and a study of V. M. Florinsky (Volotskoi 1923b, 1924a). He also became active in Marxist organizations, and spearheaded discussions of eugenics in the Society of Materialist Biologists and the Communist Academy (Volotskoi 1927). Another Moscow Bolshevik eugenicist was Aleksandr Serebrovsky (1892–1948), one of the five members of the governing bureau of the Russian Eugenics Society from its very first meeting. His earliest eugenic publications include pieces on human genetics and an account of the genealogy of the Aksakovs, a prominent family in eighteenth- and early-nineteenth-century Russian culture (Serebrovskii 1921b, 1922, 1923).

Serebrovsky's career illustrates the character of Bolshevik eugenics at its best—the transformation of brilliant scientific research and activist social commitment into a "socially responsible" science suitable for a revolutionary Russia. Born into the family of a leftist architect who was acquainted with the intellectual political revolutionaries A. A. Bogdanov and A. V. Lunacharsky, Serebrovsky entered

Moscow University in 1909 and became one of Kol'tsov's protégés in experimental biology. Kol'tsov assigned him to specialize in genetics. He served in the Great War and, in his home town of Tula, became the assistant president (1918–1921) of the local Soviet of Workers' and Soldiers' Deputies. Serebrovsky rapidly rose to prominence, first as head of poultry-breeding stations (at Tula 1918–1921, Anikovo 1921–1925, and Nazar'evo 1925–1928), chair of the poultry-breeding department of the Moscow Zootechnical Institute (1923–1930), and director of a laboratory of drosophila genetics at Kol'tsov's institute (1921–1927). He soon was one of the most prominent and internationally known Russian geneticists. But he was also an energetic enthusiast stimulated by the social and political changes occurring in his homeland. He was much taken with dialectical materialist philosophy in the 1920s, and used to chastise colleagues (including Dobzhansky) for not thinking dialectically. Beginning in 1925, after he was already well established in Soviet genetics, eugenics, and poultry breeding, he began to take an increasingly active role in various Marxist organizations in Moscow. In 1927–1929, possibly as a result of political disagreements with Kol'tsov, Serebrovsky moved his agricultural work from Kol'tsov's breeding station at Nazar'evo to the Moscow Zootechnical Institute, and transferred his research base from Kol'tsov's institute to the Timiriazev Institute, affiliated with the Communist Academy. By 1929 he was a candidate member of the Communist party (Adams 1989f).

Like Muller and Haldane, these Russian Bolshevik eugenicists were born around 1890; they, too, were smitten with the enormous practical potential of the new experimental biology and inspired by its visionary prospects. Unlike their Western counterparts, however, Volotskoi and Serebrovsky did not have the luxury of disinterested theorizing. For them, Bolshevik eugenics meant active engagement in both science and practice—doing research, arguing ideology, and ultimately formulating policy options. It meant translating eugenics into positive, progressive social action of immediate use to socialist construction.

For those who sought eugenic action, of course, the general policy options were apparent. As was the case with any animal population, the hereditary quality of future human populations could be improved in three ways. First, human heredity might be improved by negative selection or "negative eugenics," eliminating or limiting the breeding of the "unfit." Second, it might be improved by the direct induction of desirable heritable changes, whether by the control of mutation or by some Lamarckian mechanism. Finally, it might be improved by positive selection or "positive eugenics," increasing the number of offspring of people with desirable traits. In charting the development of Soviet Bolshevik eugenics in the 1920s, it is important to keep in mind that these three options defined the limits of the possible.

Option one: Sterilization

Soviet eugenicists were well informed about the public policies supported by foreign eugenics movements, and in such policies they sought models for social action. Some foreign prescriptions seemed irrelevant: there was hardly a flood of immigrants into Russia, so immigration restriction on the American model was pointless. But one measure seemed more promising: eugenic sterilization. Indeed, in the United States, vasectomies (and to a lesser extent salpingectomies) were being widely used, and eugenic sterilization laws had been enacted first in Indiana and subsequently in California and many other states. Even before the Revolution, psychiatrist Liublinsky had reported on the American program of eugenic sterilization, which he called "the Indiana Idea" (Liublinskii 1912).

The young Marxist activist who found the Indiana Idea most appealing was M. V. Volotskoi, who worked in the Eugenics Section of the Kol'tsov institute and, as noted, was secretary of the Russian Eugenics Society. In a series of talks, articles, and a 1923 book, Volotskoi urged that a sterilization program be undertaken in Russia (Volotskoi 1923a, 1923c). Pointing out that vasectomy was not castration, he argued that it was working well in America, that it could be put to immediate use, and that the success of the U.S. program would undoubtedly improve the biological quality of the American population in the very near future. Volotskoi's championing of eugenic sterilization met with considerable opposition: the reviews of his book by Filipchenko and others were hardly enthusiastic, and the discussions of sterilization in eugenic society meetings from Leningrad to Saratov were almost uniformly hostile.

In a 1924 book Filipchenko opposed the Indiana Idea on moral, programmatic, and scientific grounds. Morally, he declared "the compulsory sterilization of hundreds of thousands of citizens by some big government" to be a purely dystopian notion, a "crude assault on the human person." Programmatically, he noted that although "nothing indicates that such measures will have a significant result," they would undoubtedly be "harmful to the diffusion of eugenic ideas." Finally, he argued that, scientifically, the most efficient way of creating a desirable breed is positive, not negative, selection. For Filipchenko the creation of "especially favorable combinations of traits" was "the chief task of both human reproduction and all eugenics" (Filipchenko 1924a, pp. 156, 162, 186).

Filipchenko's position on sterilization fit his program and its patron. His Bureau of Eugenics was funded by the Academy of Sciences and, not surprisingly, he regarded the primary task of eugenics to be not social action, but rather education and research. Indeed, the research in which his bureau was engaged was an extensive Galtonian genealogical study of the Leningrad intelligentsia. Also, according to the testimony of contemporaries, Filipchenko felt some disdain for Kol'tsov's transparent attempts to cultivate political authorities. Volotskoi was, of course, one of Kol'tsov's protégés.

It might well seem curious that a Marxist like Volotskoi would embrace sterilization. In hindsight, we may tend to associate sterilization of the "defective" or "feebleminded" with extreme right-wing politics and might therefore expect to find strong political, moral, and scientific opposition to the idea in any moderately progressive setting. Given a choice between positive and negative eugenics, it might seem that encouraging "gifted" people to breed would be less problematic than compelling "defective" people not to breed; it certainly was so to Filipchenko. In Britain and America, however, it appears that the reverse was the case: as Diane Paul has demonstrated, until the 1940s there was a broad consensus among geneticists from the political Right, Left, and center that negative eugenics was desirable. If they could not agree on what were the most desirable human characteristics for which to breed, they could agree on what constituted "unfitness" and shared the conviction that humankind can and should act to eliminate it (Paul 1984). In light of that abiding consensus, then, it is natural that Volotskoi would consider sterilization a plausible form of Bolshevik action.

However, had Volotskoi not been so preoccupied with the need for action, he would have realized that there was an obvious reason for the widespread Russian opposition to sterilization that had nothing whatever to do with moral, programmatic, or scientific qualms about the procedure. As Gorbunov reported in 1922, Russia was experiencing a population implosion. In the years between 1917 and 1920, Moscow had lost 49.6 percent of its population and Petrograd, a staggering 71 percent. Nor was this simply the result of migration. In 1910 births had exceeded deaths in Moscow by 101 per 10,000; in 1920 deaths exceeded births by 243. In Petrograd the comparable figures were even more chilling: in 1910 births had exceeded deaths by 37 per 10,000; in 1920 deaths exceeded births by an awesome 484. Data from the provinces gave comparable figures (Gorbunova 1922). It is no wonder, then, that at meetings of the Saratov Eugenic Society held during the years 1922–1926, most of the discussion concerned the collapse of the local population. Participants urged the elimination of abortion on both health and demographic grounds (Kutanin 1927, 1928); sterilization was simply out of the question. Given the social realities, the common perception was that Russia needed not fewer births but many, many more.

During the remainder of the decade, the journal of the Russian Eugenics Society continued to carry detailed accounts of the eugenic programs and the eugenic legislation concerning sterilization in America, Scandinavia, and elsewhere (e.g., Filipchenko 1925c, Liublinskii 1927b, Prell 1927). But such accounts were often accompanied by critical comments noting the inappropriateness of such measures generally, and especially in Russia. By around 1925, then, the interest of eugenic activists in the prospects for sterilization had waned, and their attention shifted to another possibility for eugenic action.

Option two: Lamarckism

What, then, was to be done? By around 1925 it appeared to many that a eugenics based on Mendelian genetics had little to say about improving public health in the immediate future. At roughly the same time, biological issues became a subject of controversy in Marxist philosophical circles, where the impression was widespread that genetics itself was incompatible with Marxist revolutionary philosophy and dialectical materialism. One of the chief criticisms of genetics in Marxist circles was that its concept of the immortal germ plasm, subject only to rare, random, and generally harmful mutations, was contrary to a materialist view and rendered humankind impotent in directing mutational change. On philosophical grounds some Marxists preferred Lamarckism: if the Lamarckians were right, then hereditarily desirable traits might be induced deliberately by appropriate environmental or social conditions. The academic geneticists and eugenicists who founded the movement held the facts of science indisputable on the question; for them, Lamarckism had no scientific standing (e.g., Filipchenko and

Morgan 1925; Kol'tsov 1924b). But it became very popular among Marxists and social activists who were, after all, equally part of this biosocial science.

Interestingly, although this position was hostile to genetics, it was not necessarily hostile to eugenics. True, some Marxist Lamarckians saw genetics and eugenics as essentially the same, and rejected both (e.g., Slepkov 1925, 1927). But others followed the lead of Paul Kammerer. The Viennese biologist enjoyed great popularity in the Soviet Union in the mid-1920s, and half a dozen of his books were translated and published in large editions (e.g., Kammerer 1927). Lunacharsky had written a scenario for a film entitled *Salamandr,* which lionized Kammerer as a great scientist vilified in the West because of his Communist sympathies. The Communist Academy invited him to Moscow to head a laboratory (Gaissinovitch 1980). In one of his most influential books, Kammerer had claimed that Mendelian genetics makes us "slaves of the past" while Lamarckism makes us "captains of the future," and had devoted the last third of his text to a stirring call for a Lamarckian socialist eugenics (Kammerer 1924). Several Soviet Marxist Lamarckians responded to that call. Most notable among them was, once again, Volotskoi. Two years earlier he had published a book urging the creation of a sterilization program; now, in 1925, he published a second book—*Class Interests and Modern Eugenics*—echoing Kammerer's call for the creation of a socialist eugenics based on Lamarckism (Volotskoi 1925).

Once again Filipchenko took on the Moscow Marxist, this time by standing his central claim on its head. In a 1925 pamphlet attacking Lamarckism, he pointed out that "if acquired characters are inherited, then, obviously, all representatives of the proletariat bear in themselves the traces of all the unfavorable influences which their fathers, grandfathers, and a long series of distant ancestors have suffered over many, many years" (Filipchenko and Morgan 1925). Thus, it was not genetics but rather Lamarckism that would judge the proletariat inferior and render social action pointless. This argument reportedly caused quite a stir in Marxist circles. The consternation expressed by B. M. Zavadovsky may well have been generally shared: "Rarely have I felt so hurt and wounded," he wrote, "as when I read how our most intelligent bourgeois geneticist, Professor Filipchenko, armed with the facts of science, condescendingly lectures our ideological ally Volotskoi on how the proletariat should be taught to use the results of eugenics to their advantage" (Zavadovskii 1926, p. 76; Gaissinovitch 1980). Perhaps because of this reaction, Filipchenko, in the words of his biographers, "lost interest" in eugenics shortly after this controversy (Lepin 1930; Zavarzin 1930). Of course, however contrary in their science and their politics, both Volotskoi and Filipchenko shared the biosocial rhetoric so characteristic of the 1920s—and so essential to their common passion, eugenics.

Volotskoi was not alone in preferring Lamarckism to genetics because of his desire for action. He was joined by, among others, the physician Solomon G. Levit (1884–1938?), a central figure in Marxist biomedical circles. In a 1925 pamphlet responding to Filipchenko, Levit made the reason for his support of Lamarckism clear:

Does it make any sense to talk seriously about such undertakings [prophylactic medicine] if we accept the invariability of the genotype? . . . The arguments concerning this issue smack of desperate pessimism and impotence. If, indeed, pathology is determined by the genotype, while the latter develops solely under the influence of 'internal forces,' independent of the environment, what will become of human efforts to change pathological forms? (Levit 1925, 1927, pp. 21, 32; Gaissinovitch 1980)

Needless to say, such sentiments heightened the need for non-Lamarckian eugenics to demonstrate that it could be useful, and posed a serious problem for those attempting to Bolshevize it.

A way out of the problem was provided by Aleksandr Serebrovsky, another of Kol'tsov's protégés (Adams 1989f). In a paper delivered to the Society of Marxist Biologists on 26 January 1926, Serebrovsky asserted that it was wrong to argue that genetics was "unrevolutionary" or "counterrevolutionary" because it "attempts somehow to assign a value to different groups of the population, to various classes, and that its evaluation of the most revolutionary classes can be unfavorable." Western bourgeois eugenicists had been guilty of this, he agreed, but that did not damn genetics because "the genetic foundations are an entirely objective field." Only when we "move to the question of what is worse and what is better" do we "inevitably leave the precise ground of science and enter the field of opinions, of sympathies, and as with all sympathies, they inevitably reflect the class position of the author who expresses them. . . . Each class must create its own eugenics" (Serebrovskii 1926, p. 113). Thus, there could indeed be bourgeois and proletarian eugenics, but neither could be based on Lamarckism: both had to be based on the objective science of genetics. According to Gaissinovitch, who has inspected the records of the meeting, Serebrovsky was heavily criticized for his claim that genetics is "an entirely objective field, and as such independent of class . . . it cannot be revolutionary or unrevolutionary"—a phrase omitted from the published version. As a result, Serebrovsky's paper was published not in the journal of the Communist Academy, as was customary, but rather as an article for discussion in *Under the Banner of Marxism* (Adams 1979; Gaissinovitch 1980).

The matter came up at the Communist Academy again almost a year later, on 7 December 1926, when Volotskoi presented a paper entitled "Issues of Eugenics." Arguments broke out in the discussion period, leading Serebrovsky to restate his distinction in the strongest terms:

> Two entirely different things are constantly getting mixed up. On the one side is anthropogenetics, which is an exact science; on the other side is something constructed on it, something that has only a certain relation to science. Eugenics is not a science. It is an attempt to apply scientific data on human heredity in the discussion of ways of solving the problems the eugenicist chooses to address, and those problems are not biological, scientific problems. (Volotskoi 1927, pp. 240–41)

Thus, what a few months before had been the "genetic foundations" of eugenics had become a new scientific field, *antropogenetika*. Serebrovsky had first proposed this as a new discipline three years earlier, but he had drawn no clear demarcation between anthropogenetics and eugenics (Serebrovskii 1923); by

1926, however, the usefulness of the term lay precisely in its distinctness from eugenics.

In 1927 a scientific event occurred that helped win over a number of young Marxist Lamarckians to Serebrovsky's position. In a piece of research that would subsequently win him the Nobel Prize, the American geneticist H. J. Muller demonstrated that X rays cause genetic mutations (Muller 1927). For some Soviet Marxists, Muller's discovery redeemed Mendelian genetics by demonstrating that, far from being eternally fixed, genetic traits could be changed by environmental influences and might eventually be deliberately manipulated and controlled. Serebrovsky emphasized the scientific and ideological significance of Muller's discovery in an article in the 11 September 1927 issue of *Pravda* entitled "Four Pages That Shook the Scientific World" (Serebrovskii 1927). That autumn, in his laboratory at the Moscow Zootechnical Institute, he replicated Muller's findings.

As a result of these developments, and under the tutelage of fellow Marxist Serebrovsky, Levit abandoned Lamarckism and soon showed for genetics the enthusiasm of a recent convert. In 1927 Serebrovsky left the Kol'tsov institute and established a genetics laboratory at the Communist Academy's Timiriazev institute, where he pursued innovative studies of gene structure with a distinguished group of young Marxist biologists, including Levit (Shapiro 1966). However, much more research would be needed before techniques could be developed to induce desirable gene mutations in humans: most of the X-ray-induced mutations Muller had found in fruit flies were harmful, and some were lethal. Hence, for Bolshevik geneticists, a second option for therapeutic eugenic action had been ruled out.

The final option: Positive eugenics

In context, Serebrovsky's distinction between anthropogenetics (the science) and eugenics (the social construct) was itself revolutionary. It conflicted with the image of eugenics as a scientific discipline that Iudin and other advocates were laboring to establish. It also conflicted with the popular and widely held views set forth by contemporary Marxist luminaries that science itself was class-based and came in two different varieties—bourgeois and proletarian (Lecourt 1976). Nonetheless, Serebrovsky's distinction had critically important implications. First, if it denied "eugenics" the status of a scientific discipline, it nonetheless afforded that status to eugenic research, suitably relabeled "anthropogenetics." Second, it opened the opportunity for creating a Soviet style of eugenics different from foreign eugenics movements and untainted by them.

Yet, however useful, Serebrovsky's distinction left him in a dilemma of his own making. On the one hand, he had defended genetics from Lamarckism by declaring it to be objective, scientific, and utterly neutral on social questions—and thereby free of the unsavory taint of foreign bourgeois eugenic programs. On the other hand, he had defended genetics from the charge that it was counterrevolutionary by declaring that human genetics could be of great practical use as the basis for a distinctly proletarian eugenics—but he had not specified any distinctly proletarian eugenics that could be built on genetics, or any distinctly Bolshevik prescriptions for action that it would make possible. Consider his explicit references to the patron of eugenics, Narkomzdrav, which were intended to show that his ideas were "of more than academic interest." Noting that "almost every act of every commissariat affects in one degree or another the interest of the gene fund," he emphasized the role of Narkomzdrav, "which cures and supports the existence of the sick, which saves defective children, etc., which creates such an improvement of living conditions that infant mortality changes and those who formerly would have died have the possibility of living," thereby "serving as an agent which affects the fate of the gene fund" (Serebrovskii 1926). So far, so good, but precisely what did he propose that Narkomzdrav do, and how could "anthropogenetics" help do it?

This dilemma may well have prompted Serebrovsky to devise a truly Bolshevik "eugenics," based on his "anthropogenetics," that could aid the building of communism. In 1928 Serebrovsky set about creating a new institutional base for anthropogenetics together with Solomon Levit. Previously, Levit had rejected genetics because it left the physician therapeutically impotent; now, convinced of its scientific validity, its philosophical plausibility, and its potential for practical action, he threw himself into the study of anthropogenetics. In late 1928 he left the Moscow University Clinic and, together with Serebrovsky, established the Office of Human Heredity and Constitution at Narkomzdrav's Biomedical Institute. In 1929 Levit's office published its first volume of papers and, in a brief note, he laid out its research program: using case histories from anthropogenetic stations, twin studies, and genealogical analysis, it would seek to elucidate the topography of human chromosomes, human population genetics, and the genetic basis of human pathological forms (Levit 1929f).

What made the volume memorable, however, was its lead article—a startling piece by Serebrovsky entitled "Anthropogenetics and Eugenics in a Socialist Society" (Serebrovskii 1929). In 1928 the Party had called for discussions and suggestions for revisions in the new First Five-Year Plan. Long a Bolshevik enthusiast, Serebrovsky had recently become a candidate member of the Party, and his article was his enthusiastic response to the Party's call. Complaining that the plan's architects had taken into account gas, oil, and mineral resources, but had "completely left out the tabulation of the biological quality of the population of the Soviet Union," Serebrovsky commented that it was apparent "what a heavy burden is placed on man and his works by the accumulation of harmful genes in his gene fund." "If we calculate how much effort, time, and money would be freed if we succeeded in cleansing our country's population of various forms of hereditary ailments," he commented, "then probably it would be possible to fulfill the Five-Year Plan in two-and-a-half years" (Serebrovskii 1929, pp. 7, 12).

Serebrovsky then proposed a concrete Bolshevik form of eugenics that would fit with centralized planning: "the widespread induction of conception by means of artificial insemination using recommended sperm, and not at all necessarily from a beloved spouse." He then detailed the immediate technical possibilities. Given "the tremendous sperm-making capacity of men," and with the current state of artificial insemination technology (now widely used in horse and cattle breeding), one talented and valuable producer could have up to 1,000 children. . . . In these conditions, human selection would make gigantic leaps forward. And various women and whole communes would then be proud . . . of their successes and achievements in this undoubtedly most astonishing field—the production of new forms of human beings. (Serebrovskii 1929, p. 18)

He predicted that this time was "not far off."

Like Filipchenko and Kol'tsov before him, of course, Serebrovsky also used his vision to make a case for his own research enterprise and the new discipline it embodied. Having explained its scientific character, detailed its immediate usefulness, and justified its social importance, Serebrovsky emphasized that "anthropogenetics" must be developed energetically: "We must very rapidly broaden and deepen our work, make it maximally concrete, study our gene fund, study and analyze pedigrees, and proceed to the organization of experiments" so that when anthropogeneticists are asked for advice, they will be able to "go beyond general answers and give really scientific" information (1929, p. 19). Because it formed the scientific basis of practical eugenics, he concluded, anthropogenetics is a science geared to "helping us maximize the productive forces of our country."

How had Serebrovsky come upon such a daring idea for social action? We must remember that he was not only a drosophila geneticist, a eugenicist, and a Marxist: his central occupation was poultry breeding. Since 1918 he had headed experimental poultry stations and, beginning in 1926, he conducted expeditions to remote areas to investigate the "gene fund" of domesticated poultry in various tribal, mountainous regions in order to produce chickens with agriculturally desirable characteristics (Adams 1979).

As chairman of the department of poultry breeding at the Moscow Zootechnical Institute since 1923, Serebrovsky was well acquainted with the latest developments in *zootekhnika*. In particular, he had been impressed by the world-renowned pioneering work of Il'ia Ivanovich Ivanov (1870–1932) on artificial insemination. Ivanov, who had been Filipchenko's teacher and research collaborator from 1913 to 1920, had been teaching at the Moscow Zootechnical Institute since the early 1920s. In the period 1927–1930, his techniques for the artificial insemination of sheep and cows came into wide Soviet use. He had also published on artificial insemination in poultry, and Serebrovsky was seeking to use his techniques (Ivanov 1970).

In 1926–1927 Ivanov headed an expedition to West Africa with the purpose of attempting to hybridize different species of anthropoid apes by artificially inseminating chimpanzees (Ivanov 1970, p. 16). Although no results were published, Serebrovsky may have heard of the work when Ivanov returned to the Moscow Zootechnical Institute in 1928, the same year Serebrovsky moved his researches to his new laboratory at that institute. Ivanov's *zootekhnika* was almost certainly the inspiration for Serebrovsky's *antropotekhnika*—a word he used explicitly in connection with his plan.

But if Serebrovsky's proposal synthesized planning, Bolshevism, and *zootekhnika,* it also grew out of the logic of the Soviet eugenics movement. Over the preceding decade, Marxist eugenicists had been seeking a way to be useful. The first widely touted variant, Volotskoi's sterilization proposal, was the quintessential form of "negative eugenics," and had met with great resistance; by the late 1920s negative eugenics in the form of sterilization, marriage restrictions, and immigration control were the hallmarks of "bourgeois" eugenic programs. The second widely touted Marxist variant, Lamarckian eugenics, had offered little in the way of policy that was eugenic or new, and in any case had been partially discredited by the Kammerer scandal (Gaissinovitch 1980): in an exemplary toad whose coloration, Kammerer had claimed, proved the inheritance of acquired characteristics, the herpetologist G. K. Noble discovered India ink injected under the skin (Koestler 1971). Following Muller's discovery, it seemed possible that favorable genetic traits might eventually be induced, but that time was a long way off. With no support for negative eugenics, and no prospects for inducing desirable hereditary traits, there remained only one option: positive eugenics. Previously, all such proposals had seemed bourgeois, voluntaristic, and trivial in effect. Now, at last, zootechnology had come up with a technique that made positive eugenics a Bolshevik possibility (Adams 1979).

"Eugenics" and the Cultural Revolution, 1929–1932

Quite suddenly, in 1930, the Soviet eugenics movement ended. How suddenly the end came may be gauged from the last issue of the *Russian Journal of Eugenics*. In 1929 two issues of Volume 7 had appeared (No. 1 and No. 2/3). When issue No. 4 was published in early 1930, under the editorship of Kol'tsov, Liublinsky, and Filipchenko, it had a normal format and contained no mention whatever of any impending difficulty with the society or its journal. Indeed, its cover included an advertisement from the government publisher inviting subscriptions to the journal for 1930. The ad, which indicated that Levit and Serebrovsky had been added to the editorial board, informed readers that four issues would be published in 1930. However, that issue was to be the last. Shortly after it appeared, without any official announcement, the Russian Eugenics Society was disbanded and the Eugenics Section of Kol'tsov's Institute of Experimental Biology was abolished.

If there was a single public event that signaled the end, it was the publication in the 4 June 1930 *Izvestiia* of a stinging poem entitled "Evgenika" by the notorious popular satirist Demian Bednyi (a pseudonym). Bednyi began by noting the similarity between the Russian nobility's love of pedigree horse breeding and their preoccupation with their own genealogy. Then, quoting passages from Serebrovsky's 1929 article, he followed them with his own reductions to absurdity—for example, a future Moscow clogged by ten thousand carbon copies of the director of Gosizdat, the state publishing house that had issued the journal in which the article had appeared. Near the end of the poem, his tone becomes indignant:

> Our ancestors were all illiterate. They forgot to leave us a note on their pedigrees. . . . They were all mutilated by the old regime, injured by unbearable work, sent to the front—in short, it ruthlessly spoiled our gene fund. Thus contaminated, we finally started the struggle. . . . But our Eugenics is class eugen-

ics—proletarian—and it comes from the masses, not from an armchair in a stuffy room. (Bednyi 1930)

Serebrovsky must have read the poem with frustration. After all, Bednyi got Serebrovsky's position wrong. For one thing, Bednyi's image of eugenics appears to be Lamarckian: he suggests that the oppression of a group "spoils" its gene fund—a notion that Serebrovsky, Filipchenko, and others had been arguing against for almost a decade. For another, Bednyi seems to be advocating "proletarian" eugenics—the very thing that Serebrovsky thought he was proposing. Shortly after the poem appeared, Serebrovsky published a letter in the office's second research volume. In it he lamented the fact that some of his comrades had failed to understand his argument. He apologized for his criticism of the Five-Year Plan, agreeing with critics that the development of oil, gas, minerals, and the other natural resources was much more important for the immediate future of the Soviet state than the tabulation of the population's genes (Serebrovskii 1930). Thereafter, Serebrovsky published no more work in the proceedings of Levit's group.

The Great Break

Of course, Soviet eugenics was not done in by a poem: there were larger forces at work. The period of the First Five-Year Plan (1929–1932) is associated with a kind of sea change in Russian history that may well have had deeper and more long-lasting effects than the Revolution itself. The profound ideological, political, institutional, and cultural changes were termed by Stalin the *velikii perelom* and are often variously referred to by Western historians as the "Great Break" or the "cultural revolution."

The period 1929–1932 was the beginning of what has come to be called Stalinism. It saw the first show trials, the move to heavy industrialization, the collectivization of agriculture, the liquidation of the kulaks, a massive famine, the imposition of a narrow ideological "party line" in many areas, almost universal institutional harassment and reorganization, restrictions on foreign travel and communication, the Bolshevization of the Academy of Sciences, and widespread attacks on bourgeois experts. Two such experts were Kol'tsov and Filipchenko. In 1930 Kol'tsov was relieved of his teaching responsibilities at Moscow University, his Department of Experimental Biology was split apart, and the remarkable collection of young animal geneticists who worked at his institute was dispersed. Filipchenko was relieved of his teaching responsibilities at Leningrad University as of January 1930, and his Department of Experimental Zoology was disbanded (Filipchenko 1927–1930). He died suddenly from meningitis in May 1930.

These attacks on bourgeois experts also had a strongly antitechnocratic thrust. A growing literature on the history of technocracy early in this century in America and Europe has documented its characteristics: technocracy was a kind of "social engineering" that regarded social problems as technical problems with technical solutions, best understood by appropriate technical experts. The rise of the expert was accelerated and enforced by the world war, which played a key role in the development of economics, sociology, and psychology as "scientific" disciplines. During this period the apolitical "city manager" system became a popular form of municipal government. Fordism, Taylorism, scientific management, and other contemporary watchwords were part of a general movement that helped make Thomas Alva Edison a popular hero, H. G. Wells a renowned visionary, and Herbert Hoover the U.S. president.

As we might expect, in the 1920s "the Soviet experiment" seemed to many an ideal setting for technocracy. During the Great Break, as Kendall Bailes has convincingly demonstrated, the "Shakhty" and "Industrial Party Affair" trials singled out men who had been active in Russia's technocracy movement (Bailes 1978). The attack on the technocrats reflected growing intolerance for bourgeois experts and suspicions that their social agendas might well be subversive; henceforth, they were to be "on tap" but not "on top," while work was under way to create new cadres of "Red specialists" to replace them. The varied utopian futures of the 1920s, to be realized by experts, had given way to a more conservative vision of the Soviet future, to be formulated and enforced by political officials.

As is less often noted, however, technocracy was not limited to engineering or industry: there was also agricultural *tekhnika,* and *zootekhnika*—a term then in vogue for scientific animal breeding. During the collectivization of agriculture and the liquidation of the kulaks, of course, technical agricultural specialists were especially vulnerable: Ivanov was arrested in December 1930, was released in June 1931, and died in Alma-Ata in 1932. And, of course, there was *antropotekhnika.* By studying the hereditary basis of human physical and mental traits, some eugenicists argued that they had developed, or would soon develop, the necessary technical knowledge of human heredity to make possible a kind of social engineering that could be used to manage a society's human resources and mold its future citizens. Indeed, Serebrovsky's 1929 proposal was overtly technocratic in that it treated people as a resource like coal or gas, to be managed by those with technical expertise in human genetics, animal breeding, and the "zootechnology" of artificial insemination.

The Great Break also involved the ideological prescription of any attempt at theoretical links from the biological to the social. During the period a new pejorative word entered the Russian language—*biologizirovat'* (literally, "to biologize"), which was understood as one of the several sins collectively referred to during the period as "Menshevizing idealism." For example, at a March 1931 meeting of the Society of Materialist Biologists of the Communist Academy, B. Tokin alluded to "the perfectly clear attempts of Comrade Serebrovsky to biologize social phenomena" and noted with chagrin that "the exposure of the biologization of social phenomena" had been the work not of party members, but of the poet Demian Bednyi (Bondarenko et al. 1931, p. 20). As a result of the Great Break, the new biosocial fields that had grown up with such vigor in the previous decade were broken apart, dissolved, or renamed—for example, "plant sociology" became "phytocoenology." So far as I can determine, no field that linked the biological and the social survived the Great Break intact.

The proscription on links from the biological to the social seems to have been quite clear-cut, and its dangers perfectly apparent to Russian workers. We get a sense of this from a letter sent by the brilliant young Russian ecologist G. F. Gause to Raymond Pearl, a distinguished American biologist, a pioneer in demography, and a faculty member in the School of Hygiene and Public Health of The Johns Hopkins University in Baltimore. Pearl was arranging the American publication of Gause's book *The Struggle for Existence,* which would soon become a classic of modern ecology. In a letter to Pearl dated 21 September 1933, Gause wrote:

> I wish I could ask you a very great favor in regard to one particular point. I was trying to avoid in this book any mention of human competition and human populations. . . . I may assure you that there are particular reasons for me to ask you to avoid any mention of human beings in your introduction as well, and I hope that you will kindly fulfill this wish. This favor is for me of the highest importance. (Raymond Pearl Papers, Library of the American Philosophical Society)

Pearl responded in a letter of 5 October, "I shall take pains in writing the introduction to your book to make no mention of human affairs/beings." When the book appeared in 1934, he had been true to his word—although he could not resist allusions to Karl Pearson, whose eugenic interests were probably well known to most English-speaking readers.

As a field run by bourgeois experts, embued with technocratic values, and premised on a biosocial link, eugenics was doomed. Any lingering doubts about the status of the field were set to rest in 1931 with the appearance of the article on *evgenika* in the *Great Soviet Encyclopedia.* It was written by G. K. Batkis, since the mid-1920s a member of the same Marxist groups as Serebrovsky, Levit, and Volotskoi where eugenics in all its varieties had been discussed. The article defined "eugenics" as the "bourgeois doctrine of the biological improvement of the human race," and declared categorically that Filipchenko's eugenic ideas were "bourgeois," Kol'tsov's were "fascist," and Serebrovsky's constituted "Menshevizing idealism" (Batkis 1931, p. 812).

Mending the break

The Great Break had profound consequences for the institutions where Soviet genetics and eugenics had flourished. A historian comparing the situation in 1929 with that in 1934 can note several trends. In general, those from peasant, worker, or orphan backgrounds fared better than those from bourgeois, *intelligent,* or entrepreneurial families; Communists fared better than liberals. Nonetheless, by about 1934 much of Soviet genetics had been restored. We can trace this remarkable resurrection in the careers of the associates of Filipchenko and Kol'tsov.

In Leningrad Filipchenko was relieved of his teaching responsibilities as of January 1930, and his Leningrad University Department of Experimental Zoology was disbanded. In 1930 the Academy of Sciences's Bureau of Genetics became its new Laboratory of Genetics. Following Filipchenko's sudden death from meningitis in May 1930, the laboratory was taken over by the plant geneticist N. I. Vavilov, who headed a cadre of Filipchenko's students that included Iu. Ia. Kerkis, N. N. Medvedev, A. A. Prokof'eva[-Bel'govskaia], and M. E.

Lobashev. In 1932 two new substitute departments were created at Leningrad University: the department of plant genetics, chaired by Vavilov's protégé G. D. Karpechenko; and the department of animal genetics, chaired by the Marxist Lamarckian A. P. Vladimirsky. In 1933 the academy's Laboratory of Genetics was elevated to the status of an institute (IGEN), headed by Vavilov. At its core was a laboratory of genetics whose staff included most of the animal geneticists trained by Filipchenko. When the Academy of Sciences was relocated to Moscow in 1934–1935, most of this group moved with IGEN, leaving Lobashov and most of the plant specialists in Leningrad at the university. In general, then, the Leningrad genetics group suffered the disruptions of the Great Break only to emerge relatively intact thanks to Vavilov's patronage (Adams 1978, 1989d).

Comparable developments occurred in Moscow. During the 1920s genetic research in the Kol'tsov institute had centered in two groups headed by A. S. Serebrovsky and S. S. Chetverikov. Serebrovsky left in 1927. In 1929 Chetverikov was arrested and sent into exile for six years. In 1929–1930 the remarkable collection of young animal geneticists who had trained under Kol'tsov was dispersed. Three—N. K. Beliaev, B. L. Astaurov, and V. P. Efroimson—took up silkworm breeding in central Asia. In 1930 Kol'tsov was removed from teaching and his Department of Experimental Biology at Moscow University was disbanded, to be replaced by a new Department of Genetics chaired by Serebrovsky. By 1933, however, Kol'tsov was able to reestablish the genetics division of his institute by bringing in Nikolai Dubinin. Dubinin was a *bezprizornik* (an orphan who had been trained in special schools) and a *vydvizhenets* (someone who was promoted beyond his qualifications because of his political and class background). In the late 1920s Dubinin had taken Chetverikov's course at Moscow University and then had trained under Serebrovsky at the Zootechnical Institute in 1928–1932. In 1932 Kol'tsov invited Dubinin to head a new genetics division of his institute, and in the period 1933–1937 he highlighted Dubinin's background in public settings as a way of showing that his institute was in step with the times (Adams 1980a). Although officially headed by Dubinin, the genetics section also included D. D. Romashov, E. I. Balkashina, and V. V. Sakharov—three key figures in the division before the Great Break. By the mid-1930s Kol'tsov was able to reappoint Boris L. Astaurov, who returned from central Asia to join the institute's embryology division.

A similar pattern of recovery can be seen in other enterprises associated with eugenics. In 1928 the entrepreneurial Kol'tsov spearheaded the creation of a new Society for the Study of Racial Pathology and the Geographical Distribution of Diseases. In the spring of 1928, the idea for such a society was conceived by a group of ten important figures—A. I. Abrikosov, M. I. Averbakh, anthropologist V. V. Bunak, neurologist S. N. Davidenkov, N. K. Kol'tsov, A. V. Mol'kov, D. D. Pletnev, Commissar of Health N. A. Semashko, population geneticist S. S. Chetverikov, and G. I. Rossolimo. Its statutes were approved, and at its first organizational meeting, Kol'tsov became president. Its other eighteen administrators included S. N. Davidenkov (assistant), V. V. Bunak (treasurer), G. A. Batkis and T. I. Iudin (secretaries), S. G. Levit and V. V. Sakharov (candidates), and A. S. Serebrovsky (on its commission). In the spring of 1929 the society had four meetings, discussing papers on concepts of race, racial pathology, and the geographical distribution of endemic goiter. By the end of May 1929, it had fifty members ("Kratkii otchet" 1929).

No information is available on the fate of this society during the Great Break. However, in a pathbreaking article, Paul Weindling has reported evidence from German archives that a joint German-Soviet laboratory was created to study racial pathology. A conference held in Tiflis on 13 October 1930 discussed the establishment of a "racial institute" in Transcaucasia, and Kol'tsov spoke of its importance for research. Nothing came of these plans, but in March 1931 accommodations were found in Moscow for a Laboratory of Racial Research. Its status was precarious, however: the same month, a German worker there was ordered home. But two years later, after the Great Break was pretty much over, the German Foreign Office decided in March 1933 to continue the joint laboratory *aus kulturpolitischen Grunden;* in April the Soviet Commissariat of Public Health authorized it to continue. After the Russian Eugenics Society was terminated in 1930, its former secretary V. V. Sakharov—a member of the board of the racial pathology society—launched a series of studies of the geographical distribution and genetic dimensions of endemic goiter, and its relation to blood group frequencies, in Uzbekistan, under the auspices of the commissariat. This work may have been related to the work of both the society and the laboratory. Remarkably, the laboratory was transferred back to Germany only in 1938 (Weindling 1986).

Of course, the Great Break was not without costs: Leningrad and Moscow each lost its most outstanding young animal geneticist. In Leningrad Filipchenko's chief assistant Theodosius Dobzhansky left Russia in 1927 to work in the American laboratory of T. H. Morgan; as a consequence of the Great Break, he never returned to Russia. Instead, he became one of America's leading evolutionary theorists and population geneticists, and wrote a book that many regard as the founding work of the synthetic theory of evolution. Moscow suffered a comparable loss. Two members of Chetverikov's group were Nikolai and Elena Timofeeff-Ressovsky, who had left for Berlin in 1925 to work in the Kaiser Wilhelm Institute for Brain Research. As a result of the Great Break, they chose not to return to Moscow. In the 1930s N. Timofeeff-Ressovsky became the Continent's leading geneticist and evolutionary theorist.

Nonetheless, despite personnel losses and institutional disruption, by 1934 Soviet genetics had largely recovered. Furthermore, those who had led the Soviet eugenics movement, while ideologically chastened, were institutionally well situated; and the personal, institutional, and professional networks that had formed the tissue of the Soviet eugenics movement in the 1920s were still largely intact. For public and ideological purposes, however, the torch had been passed to a new generation of younger Marxists whom the founders had nurtured.

On the "demise" of Soviet eugenics

In retrospect, the demise of Soviet eugenics during the Cultural Revolution seems overdetermined. Eugenics was a utopian biosocial science, founded by bourgeois liberals, funded by the public health minister, a kind of human *antropotekhnika* that would have given eugenic experts control over social policy. As such, it could not survive a period that was characterized by the proscription of "biologization," attacks on bourgeois experts, the dismissal of its chief patron, and widespread antitechnocracy campaigns. Indeed, perhaps because of his belief that eugenics may be essentially incompatible with Marxist ideology, Loren Graham has declared, "By 1930 the eugenics movement in the Soviet Union was finished," adding: "The end of eugenics meant the end of discussions of human heredity" (Graham 1978, p. 1156). If this were so, an account of the history of Soviet eugenics might appropriately end here.

However, our analysis of the character of Soviet eugenics, set in a comparative context, should make us a bit skeptical. Without the luxury of knowing that it *should* have ended, we must inquire how it *could* have ended. True, political authorities had brought the Russian Eugenics Society into being in 1920 by giving it official approval and support, and they brought it to an official end by withdrawing that approval and support. But if Soviet eugenics had been officially sanctioned from the top down, it had nonetheless been created from the bottom up. As we have seen, its character was shaped by scientific entrepreneurship that was able to draw on interdisciplinary networks and to institutionalize a loose federation of varying institutional, professional, thematic, and ideological clusters. In principle, such a structure should have made eugenics adaptable. Certainly, in other countries this was the case: not only did eugenics manage to thrive in a variety of quite different political, cultural, social, and economic circumstances, but it was also able to adapt to rapid changes in those circumstances. The point is well illustrated by the example of German eugenics, which arose under the Kaiser, flowered during Weimar, and adapted to the Nazis.

In light of this, why did some part of Soviet eugenics not survive the *velikii perelom?* The simple answer is that some part of Soviet eugenics *did* survive— but not as "eugenics."

"Medical Genetics," Eugenics, and Lysenkoism, 1932–1940

We can trace the apex, the demise, and the tentative (and anonymous) resurrection of Soviet eugenics in the textbooks of P. F. Rokitsky, a student of both Kol'tsov and Serebrovsky. In 1928 he published a book entitled *Can the Human Race Be Improved?* which enthusiastically presented the standard eugenic views of his teachers (Rokitskii 1928). In 1930 the final issue of the *Russian Journal of Eugenics* contained two reports by Rokitsky, one on hereditary malformations of the extremities. In 1932, however, toward the end of the Great Break, he published the first edition of his general textbook *Genetika;* its pages contained not a single reference to human genetics, much less to eugenics.

But its second edition, sent to the printer in January 1934 and published in a edition of fifteen thousand copies, included a remarkable addition—a new chapter entitled "Genetics as Applied to Man and Its Bourgeois Perversion." To be sure, its text was critical of Western, and especially German, eugenics; it echoed Batkis in characterizing Filipchenko's position as bourgeois and in castigating Serebrovsky's plan for human artificial insemination. However, in the chapter's final section, Rokitsky once again argued for the importance of human genetics in a socialist society. He noted in conclusion that, although "eugenics in the USSR in large measure reflected the ideas of bourgeois eugenics," nonetheless, "only under the conditions of socialist construction is it possible to carry out those measures that will truly improve the health and quality of human heredity" (Rokitskii 1934, p. 246). Even more remarkably, the five suggested readings included Kol'tsov's programmatic eugenic statement (Kol'tsov 1922a); an ideological analysis by Levit (Levit 1932); and Serebrovsky's "Anthropogenetics and Eugenics in a Socialist Society" (Serebrovskii 1929). About the latter Rokitsky commented: "Providing an essentially correct but wholly inadequate critique of bourgeois eugenics, the author also develops the idea of human artificial insemination for eugenic purposes" (1934, p. 246).

Clearly, by 1934 the Great Break was over and, although "eugenics" remained illegitimate, something of its perspective and agenda had survived. But if this work could no longer be called "eugenics," what was it to be called, and where could it be carried out? The answers would be provided by the entrepreneurial efforts of a "Red specialist."

Levit and the birth of "medical genetics"

The effects of the Great Break on Serebrovsky and Levit were not entirely negative. One aspect of the Great Break was the replacement of bourgeois experts with Red specialists—and it would have been difficult to find better-qualified Red specialists than Serebrovsky and Levit. Serebrovsky had been active in Marxist circles since the early twenties and by 1930 had become a candidate member of the Party. Thus, when Kol'tsov's Department of Experimental Biology at Moscow University was abolished in 1930, a new Department of Genetics was established and Serebrovsky became its chair. True, Serebrovsky's eugenic views had been condemned as "Menshevizing idealism." This had not yet become a capital crime, but it did perforce remove Serebrovsky from a prominent public role in things associated with eugenics. As a result, he devoted increasing attention to poultry and animal breeding, and the central role was assumed by his colleague.

Solomon G. Levit (1884–1938?) was a Baltic Jew who had joined the Communist party in 1919 and was active in Marxist organizations. He earned his medical degree at Moscow University in 1921 and served on the administrative board of the university from 1922 to 1925. In October 1924, together with a group of young research assistants, he set up a circle of materialist physicians at the Moscow University medical school and became its permanent chairman. In November 1926 the group was renamed the Society of Materialist Physicians, and later came under the auspices of the Natural and Exact Sciences Section of the Communist Academy. There Levit began as a research assistant, was later appointed a vice-chairman, and from 1 February 1926 until 16 April 1930 was scientific secretary of the section.

During the period 1924–1927, Levit was a strong advocate of Lamarckism. In 1924 he called for a synthesis of Darwinism and Lamarckism, claiming that the influence of the external environment and the inheritance of acquired character-istics were the only factors capable of explaining the causes of variation. We have already noted the effect of H. J. Muller's 1927 discovery on changing Levit's views. Under the tutelage of fellow Marxist Serebrovsky, Levit abandoned Lamarckism and began to work in Serebrovsky's laboratory at the Timiriazev institute on *scute* mutations in drosophila, where he joined a younger group of Marxist biologists that included I. I. Agol, V. N. Slepkov, N. P. Dubinin, B. N. Sidorov, A. E. Gaisinovich, and N. I. Shapiro (Adams 1989f).

In late 1928 Levit left the Moscow University Clinic and joined the staff of the Biomedical Institute, which had been formed in 1924 under the direction of V. F. Zelenin (1881–1968), a specialist in internal medicine. On 21 December, together with Serebrovsky and N. N. Malkova, Levit established at the institute the Office of Human Heredity and Constitution and became its director (Levit 1929f). Four months later party officials decided that it was time to establish a new line in the philosophical disputes that had been raging in the Communist Academy over biology and other scientific issues. In the discussions of April 1929, Levit now strongly allied himself with such "militant" Morganists as Serebrovsky and Agol, who had called for Lamarckism to be expelled from the academy and its research institutions (Adams 1989e; Gaissinovitch 1980).

In March 1930, Levit replaced Zelenin as director of the Biomedical Institute. At the same time, his office was expanded into the institute's new Genetics Divi-sion, and its second volume of researches began with a new programmatic state-ment by Levit (Levit 1930). There he drew a strong distinction between eugenics (class-bound ideology and policy) and anthropogenetics (the science of human heredity). Admitting that the former was controversial, he emphasized that the latter was thoroughly objective, purely scientific in character, and beyond both class interests and controversy. Of course, in making this distinction, he was echo-ing the position Serebrovsky had repeatedly expressed since 1926. Serebrovsky had sharpened the distinction in order to protect genetics from Marxist Lamarck-ians; now, ironically, the same distinction served to protect anthropogenetics from Serebrovsky's Marxist eugenics.

In May 1930 the Soviet government nominated Levit, along with Agol, for Rockefeller Foundation grants to study genetics in the United States. Serebrovsky managed to place both Levit and Agol in the laboratory of H. J. Muller at the University of Texas in Austin, where they worked during 1931. Meanwhile, back at home, Agol, Levit, and Serebrovsky were under ideological attack as "Men-shevizing idealists." In February 1932 Levit returned to Moscow only to learn that he had lost his post as director and that his Genetics Division had been sus-pended. We do not know what Party politics transpired during the next six months. In August, however, Levit was once again director with a mandate to create a world-class research center for human genetics (Levit 1934a, p. 3).

Levit used his position to transform the institute. He assigned it a new mission: to study problems of human biology, pathology, and psychology from a genetic viewpoint, drawing upon cytology, embryology, and evolutionary theory. The reorganized institute included new divisions of cytology, internal secretions, and neurology, and offices of roentgenology, anthropometrics, and psychology. The research plans for 1933 indicate that Levit had managed to organize wide-scale twin studies, involving the cooperation of hospitals, orphanages, and clinics (Levit 1933). In addition, he headed a course in genetics for physicians, first offered 1933–1934, which included lectures on Mendelism, Morganism, sex determination, mutations, population genetics, evolutionary genetics, twin studies, human genetics, and medical cytology, and four lectures on "bourgeois eugenics and its class character" (Levit 1934a).

Levit's research program had emerged from the ideological turmoil and crackdowns of the Great Break intact and invigorated. Levit had managed to regain his ideological credentials following the Serebrovsky fiasco, and published a widely cited attack on Nazi biology entitled "Darwinism, Racial Chauvinism, and Social Fascism" (Levit 1932). However, the rise of Naziism in Germany created ideological problems in Stalinist Russia for an enterprise historically tied to eugenics. Although Serebrovsky had always distinguished between eugenics and the science of "anthropogenetics" or "human genetics," the latter was not well established as an international discipline in its own right. The closest parallel was "human heredity," as used in the name of the Kaiser Wilhelm Institute for Anthropology, Human Heredity, and Eugenics, founded in 1927 (Weindling 1985). In the German context, of course, the term *Eugenik* was associated with more liberal politics than its alternative, *Rassenhygiene.* But in the Soviet context of 1934, there was a need for a legitimate disciplinary name utterly distinct from eugenics, not one linked to it in the name of an institute in Nazi Germany.

An institute conference held on 15 March 1934 seemed to settle on the right language and ideological approach. A series of programmatic papers by Levit, H. J. Muller, N. K. Kol'tsov, neurologist S. N. Davidenkov, anthropologist V. V. Bunak, and cytologist A. G. Andres called for the establishment and expansion of "medical genetics" in the USSR as a way of improving the health of the working class and combating fascist pseudoscientific racism. The final resolution called for the wide expansion of the field through the opening of new clinics and medical school departments, increased research personnel and graduate students, the preparation of texts and teaching materials, new genetics courses for updating physicians, and the creation of regional and metropolitan offices (*Konferentsiia* 1934; Levit 1934d).

In March 1935 the institute was renamed the Maxim Gorky Research Institute of Medical Genetics. Its impressive fourth volume of researches, completed in 1935 and published in 1936, included twenty-five papers by over thirty authors, most on its staff (Levit and Ardashnikov 1936). The new psychology division was now headed by A. R. Luria, considered by many to be one of this century's greatest psychologists (Luria and Mirenova 1936; Luria 1979). An impressive series of studies by Levit analyzed the genetic component of a host of diseases, surveying and critiquing international findings on the genetics of human traits and human illnesses. In particular, Levit developed data indicating that a number of genetic diseases previously thought to have been caused by recessives appeared to be dominants (e.g., Levit 1936b; Levit and Pesikova 1936). A contemporary scientific polemic against German fascism made frequent positive reference to the institute's work, citing it to disprove Nazi claims (Finkel'shtein 1935). Much of this work was developed in consultation with Levit's mentor H. J. Muller, who worked in the USSR from 1933 to 1937 and was closely affiliated with the Levit institute.

As we have seen, the eugenics movement in the 1920s involved most of Russia's animal geneticists and forged links between them and psychologists, psychiatrists, and anthropologists. With the delegitimation of eugenics in the period 1929–1932, the movement dissolved, but the links it had created and the interest in common problems it had generated proved remarkably resilient. Beginning in 1933 part of the movement had begun to regroup in the Levit institute. Note that at the conference calling for the creation of medical genetics as a discipline, four of the five Russian papers were presented by men who had been active in eugenics in the 1920s: Kol'tsov, Bunak, Davidenkov, and Levit. A careful study of institute documents reveals the remarkable fact that, by 1936, Levit had managed to absorb a number of younger Moscow researchers from various fields who formerly had been eugenicists. For example, the collections of institute work published in 1934 and 1936 include three articles by the anthropologist Bunak on the sex ratio (1934a), population statistics (1936a), and twin studies (1936b). Volotskoi now worked at the institute and published a long study of the fingerprints of 234 pairs of twins (Volotskoi 1936); through his connections with Bunak, he also published a related article in the Soviet *Journal of Anthropology* (Volotskoi 1937). In addition, the neurologist Davidenkov was an institute consultant. We may recall that Davidenkov had been active in the movement, and both Bunak and Volotskoi had formerly worked in the Eugenics Division of the Kol'tsov institute before it was disbanded in 1930.

Soviet medical genetics at the Levit institute, then, constituted a modified version of the eugenics of the 1920s. It conserved essentially the same research agenda, using the same methods and carried out by many of the same people. But it had been methodologically refined, politically neutralized or co-opted, and ideologically sanitized. Safely ensconced within a biomedical institute in the health ministry, and run by men identified with the Communist party, Levit's enterprise grew and prospered. But its success depended on its antifascist ideology, its medical and research orientation, and its absolute dissociation from both "bourgeois eugenics" and the visionary utopian human breeding schemes of Serebrovsky's "Bolshevik eugenics."

Muller's eugenics in Russia

Western historians of biology have devoted considerable attention to H. J. Muller over the last few decades, and that interest is increasing. Most of articles in the fall 1987 issue of the *Journal of the History of Biology* deal with Muller. At least a dozen articles have been devoted to explicating Muller's eugenic views, and half a dozen books have discussed them. Most also allude briefly to Muller's struggles against Lysenko when he was in the Soviet Union in the 1930s. However, these treatments rely heavily on Muller's own accounts and interpretations of his ideas and experiences. The picture that often emerges involves three disjointed components: Muller formulated his eugenic ideas in 1910 and never essentially changed them; he went to Russia because of his leftist sympathies; and he left when the growing tide of Lysenkoism, against which he fought, made his work impossible. However, we get a rather different perspective if we carefully examine contemporary Russian-language materials in Muller's papers and from Soviet archives and publications.

Muller was interested in the Soviet experiment from the time of the Revolution, and first visited Russia in 1922. While there he visited the Kol'tsov institute and its Anikovo poultry-breeding station, headed by Serebrovsky, where he presented a survey of the chromosomal theory of heredity that was subsequently published in Russian (Muller 1923a). He also brought with him stocks of drosophila that became the basis of the remarkable genetics work by Chetverikov, Serebrovsky, Dobzhansky, and the entire Russian school (Adams 1980b). As he indicates in his account of his trip, he also met Ivanov and was impressed with his artificial insemination work (Muller 1923b). There is no indication of any substantial correspondence between Muller and Soviet geneticists in the 1920s in either Soviet or American archives, although he may have met with them at international scientific meetings. There was a large Soviet delegation at the Fifth International Congress of Genetics in Berlin in 1927, but Muller's paper on X-ray mutagenesis caused considerable excitement generally, and there is no indication of much exchange with Soviets.

Muller's thinking seems to have undergone a qualitative change in 1931, the year Levit and Agol worked with him in Texas. Both were members of the Soviet Communist party, and they no doubt painted a convincing picture of the socialist future being created in their homeland. In addition, Levit almost certainly acquainted Muller with Serebrovsky's eugenic views. It is likely that Levit brought with him a set of publications that included the 1929 volume, edited by Levit and Serebrovsky, in which Serebrovsky's controversial proposal for human artificial insemination had appeared.

Levit returned to the Soviet Union in early January 1932. On 10 January Muller apparently attempted suicide. In March he got into political difficulty for serving as the faculty sponsor of a Communist campus publication, *The Spark* (named after a revolutionary journal run by Lenin). In June Muller made a well-publicized break with the American eugenics movement in the paper "The Dominance of Economics over Eugenics," which reiterated some of Serebrovsky's arguments and even some of his language (Muller 1933a). That fall, Muller left for Europe and, after a short time in Berlin, moved to Leningrad.

Muller accepted an offer by Vavilov to head a laboratory in the new Institute of Genetics of the USSR Academy of Sciences. He had received offers to work at the Kol'tsov institute and the Levit institute, but his letters indicate that he accepted Vavilov's offer because he regarded Vavilov as more powerful and influential. He was right: at the time Vavilov was a member of the Central Executive Committee of the government (Adams 1978). Upon Filipchenko's death in 1930, his Bureau of Genetics became the academy's Laboratory of Genetics, headed by Vavilov; its elevation into the Institute of Genetics (IGEN) under Vavilov's direction coincided with Muller's arrival. Muller was elected corresponding member of the USSR Academy of Sciences in 1933 (Carlson 1981). The academy, together with IGEN, was moved to Moscow in 1933–1934.

There is strong evidence that Muller went to the Soviet Union in order to realize his eugenic program by adding a biological dimension to the "Soviet experiment." His letters indicate that he was very favorably impressed with the forced collectivization of agriculture during the Great Break, which apparently convinced him that Stalin was a man who knew how to get things done. While in the institute, much of his time was devoted to his laboratory drosophila studies, where he was assisted by Filipchenko's former students Kerkis, Medvedev, and Prokof'eva (-Bel'govskaia). But his presence in Moscow allowed him to be active in Levit's institute, and he also served as a consultant to the Kol'tsov institute.

In his many book reviews and articles for Soviet popular books and journals, Muller rarely passed up an opportunity to argue for the implementation of a socialist eugenic program. For example, in a volume devoted to the memory of Lenin, he contributed an article entitled "Lenin's Doctrines in Relation to Genetics" (Muller 1934b). The article is sometimes seen as proof that genetics was just as compatible with dialectical materialism as Lysenkoism, if not more so (Graham 1972, pp. 451–53). Actually, however, the article culminates in a thinly veiled argument that Marxism-Leninism requires using genetics to breed better people, transforming and perfecting human biology in a deliberate, planned way. Since "our genetic constitution" is "simply one among the material things of life," Muller wrote, "it is therefore up to us to change it, and to continue to change it, in all such ways as will best further the harmonious and effective development of the worker's society." Only "in this way," the article concludes, can every individual "reap the benefits of the biological fruits of socialism" and "enjoy increasingly that world conquest, on the path of which Lenin helped so much to set us" (Muller 1934b, p. 592).

In 1935, while still based in Moscow, Muller published in America and Britain his book *Out of the Night: A Biologist's View of the Future*. It was well received, especially in Britain. The most commented upon section of the book was its final chapter, "Birth and Rebirth," where Muller resurrected—without attribution— Serebrovsky's discredited human artificial insemination scheme.[2] Muller argued that with artificial insemination technology, "in the course of a paltry century or two . . . it would be possible for the majority of the population to become of the innate quality of such men as Lenin, Newton, Leonardo, Pasteur, Beethoven, Omar Khayyám, Pushkin, Sun Yat Sen, Marx . . . or even to possess their varied

---

2. In the book's preface, Muller claims to have written the final chapter in 1925; later he claimed to have expressed all the ideas in it around 1910. However, the evidence available in Muller's archives does not support either of these dates: the drafts dating from around 1910 mention only negative, not positive, eugenics, and, although there are 1925 drafts of several other chapters from the book, there are no drafts of the final chapter. Furthermore, a close comparison of the text of Muller's chapter and Serebrovsky's article shows great similarity of argument and even language. Muller's 1935 chapter is structured around distinctions he perceived between capitalism and socialism, expressed in a way that Muller first published in 1931, and almost certainly did not conceive or hold in 1925—a year before the Scopes trial, four years before the Depression.

faculties combined." However, he warned that in American capitalist society, tomorrow's population would be composed "of a maximum number of Billy Sundays, Valentinos, Jack Dempseys, Babe Ruths, even Al Capones" (Muller 1935, pp. 113–14). By contrast, in a socialist society, with its control of the means of production for the common good, Muller argued, we could also have confidence in its control of the most important form of production, "*re*production" (p. 117).

In May 1936, perhaps out of his growing impatience, Muller took the fateful step of sending Stalin a copy of the book, together with a long letter arguing for his plan. In his letter Muller explained that "as a scientist with confidence in the ultimate Bolshevik triumph throughout all possible spheres of human endeavor," he had decided to refer the matter "to you yourself, primarily" because of "your farsighted view and your strength in the realistic use of dialectic thought." After summarizing some basic genetics, Muller explained that

> it is quite possible, by means of the technique of artificial insemination which has been developed in this country, to use for such purposes the reproductive material of the most transcendently superior individuals, of the one in 50,000, or one in 100,000, since this technique makes possible a multiplication of more than 50,000 times.

Then he told Stalin of the power of eugenics to ensure the triumph of socialism:

> A very considerable step can be made even within a single generation. And the character of this step would in fact begin to be evident after only a few years, for by that time many children have already developed enough to be distinctly recognizable as backward or advanced. After 20 years, there should already be very noteworthy results accruing to the benefit of the nation. And if at that time capitalism still exists beyond our borders, this vital wealth in our youthful cadres, already strong through social and environmental means, but then supplemented even by the means of genetics, could not fail to be of very considerable advantage for our side. . . . We hope that you will wish to take this view under favorable consideration and will eventually find it feasible to have it put, in some measure at least, to a preliminary test of practice. (Muller 1936b)

Muller was also active in plans for the International Genetics Congress, scheduled for Moscow in 1937. Muller pressed to have Levit made secretary of the congress. By 1936, however, high officials of the Central Committee, especially Bauman, in charge of science, were anxious about the forthcoming congress.

Finally, Serebrovsky has left a considerable "paper trail" that shows the evolution of his ideas and their sources in his breeding work, his contacts with Ivanov, his involvement in the eugenics movement, and his regular contributions to the evolving Marxist discussions, whereas Muller's paper trail, which is also considerable, would seem to show no evidence of the idea before 1929 and no clear sources for it. Furthermore, there were good reasons for Muller not to cite Serebrovsky: since the Great Break, the idea had been a considerable embarrassment to him and, at the height of the purges, the attribution certainly would have undermined him and might even have physically endangered him. Serebrovsky died only in 1948. By that time eugenics was largely discredited in America, Lysenkoism had taken hold, Muller had broken with Stalinism, the cold war was under way, and Muller would soon be experiencing difficulties with McCarthyism. In this context it is hardly surprising that Muller did not dwell on the links discussed here.

Levit, Muller, Kol'tsov, and Serebrovsky were all members of the organizing committee, and all had at one time advocated eugenic ideas. Muller and his colleagues were planning to give special attention to sections on human genetics, where they wished not only to show off the impressive Russian work, but also to confront the Germans over Nazi racial views. Molotov and others interceded to try to prevent the inclusion of human genetics, at a time when they did not wish biologists from fascist countries to have a forum in Moscow. The timing of Muller's letter to Stalin, then, was most unfortunate.

On 13 November 1936 Ernst Kol'man, the party official in charge of science in Moscow, staged a public meeting to denounce Levit as an abetter of Nazi doctrines. The testimony of various principals, including Luria and Muller, suggests that Levit was in serious trouble in the fall of 1936, was despondent, and was considering leaving genetics altogether. Matters came to a head in December 1936 at the fourth session of the Lenin All-Union Academy of Agricultural Sciences. On the first day of the meeting, it was announced in the press that Agol—Muller's student in Texas—had been arrested as an "enemy of the people."

The four main speakers were Vavilov, Lysenko, Serebrovsky, and Muller. All participants had been instructed to avoid mention of human heredity. However, Muller defied the political instructions and, against the urgings of both Vavilov and Serebrovsky, attacked the Lysenkoists by triumphantly declaring that a belief in the inheritance of acquired characteristics was not only unscientific, but also played into the hands of fascism since it would mean that peoples oppressed for eons would thereby have become hereditarily inferior. As many of the listeners must have known, this was the exact argument that had gotten Filipchenko into trouble in 1925.

Once Muller had broached the issue, Lysenko's supporters warmed to it. For the first time in the controversy, Lysenkoists drove home the argument that genetics, eugenics, and fascism were all of a piece. Michurinists L. K. Greben', I. G. Eikhfel'd, I. I. Prezent, and others found Serebrovsky an especially apt target and castigated him for his eugenic views; one remarked that Soviet women would never forgive him for his human breeding scheme. Academician S. S. Perov suggested ominously that "Levit has already been unmasked." On 27 December, the final day of the conference, Serebrovsky rose on a point of personal privilege and admitted that his 1929 article "presents a whole series of the crudest political, anti-scientific, and anti-Marxist mistakes, which I now find painful to remember." These events and remarks were published in the daily bulletins of the meeting, but were omitted from the official version published the following year (*Biulleten'* 1936; Targul'ian 1937).

Only after the meeting did Muller learn that Stalin had been reading a translation of Muller's communications, apparently in the summer or fall of 1936. Evidently, Stalin did not like what he read. Events followed precipitately. Word reached Muller that he might be arrested, and Vavilov managed to arrange his passage out of Moscow through the quickest possible route—as a member of the International Brigade headed for the Spanish civil war. The day Muller left, Agol was shot. Shortly thereafter N. K. Beliaev, Kol'tsov's student, was arrested in Central Asia and shot as an "enemy of the people." Even the secretary who translated Muller's book into Russian was reportedly arrested and shot. Levit was removed as institute director in July 1937. He was arrested 11 January 1938 and was probably shot in May. After a series of investigations, the Institute of Medical Genetics was disbanded, and several of its less controversial divisions were incorporated into the new expanded Maxim Gorky Institute (Fedorov 1939).

In the months and years following his departure from Russia in 1937, Muller remained silent about Stalinism for fear of alienating Western leftists from his eugenics. We know in some detail about Muller's experiences because, on the train out of Russia, he wrote two letters, dated 9 and 11 March, to his friend Julian Huxley. Detailing recent events, Muller noted unhappily that the Soviet Union "is hardly the place, at present and probably for some years to come, where one can hope to develop genetics effectively—let alone the application of genetics to man which I had hoped might gradually be introduced." Nonetheless, Muller urged Huxley to keep the information in his letter confidential: "Haldane, especially, must not be informed—not now, anyway—for I judge from the tone and content of his letters to me that he is at present having his political opinions impressed upon him with a rubber stamp (greatly as I admire his intellect and person) and would be influenced in the reverse direction from that which I intended" (Muller archives, Lilly Library).

Genetics, eugenics, and Lysenkoism

It is difficult to avoid the conclusion that, however inadvertently, Muller had played directly into the hands of the Lysenkoists. By resurrecting Filipchenko's and Serebrovsky's forgotten arguments—so fitting in the 1920s, so inappropriate in 1936—he helped reestablish the ideological links between eugenics and genetics. In the process he compromised the Soviet Union's leading geneticists: his senior colleague Kol'tsov, his patron Vavilov, his friend Serebrovsky, his student Levit. In addition, he also inadvertently undermined the field he had helped to create—medical genetics.

The security of the Institute of Medical Genetics depended on its antifascist ideology, its medical and research orientation, and its absolute dissociation from both "eugenics" and visionary utopian human breeding schemes. As one of the world's great geneticists who had gone over to the Bolshevik cause, as one of the pioneers of human genetics, and as Levit's former teacher, Muller naturally played a visible role in the institute's work. Indeed, at the 1934 meeting at which the field had been christened, Muller had been one of the honored speakers (Muller 1934a). Unfortunately, then, when he insisted on resurrecting Serebrovsky's seminal artificial insemination scheme in his book, directly approaching Stalin with it, and deliberately interjecting his discussions of human breeding into contexts where the party chiefs had specifically instructed him not to do so, the institute's "new face" became inextricably reminiscent of its old one and its legitimacy in the new Soviet context became fatally compromised. If in 1935 "medical genetics" seemed independent and prosperous by virtue of having its own institute run by a party member, in late 1936 and 1937 its institutional centralization made it an easy target.

The years 1936–1937 also proved a turning point ideologically. Before that time, of course, Lysenkoists had claimed that genetics was ineffective in rendering help to agricultural production, but their rhetoric had been largely devoid of references to human genetics or allegations that genetics had fascist links; after December 1936, however, the charge was a dominant Lysenkoist motif and became virtually a cliché in subsequent propaganda.

The charge was used to good effect in 1938 in helping to secure Lysenko's election to the USSR Academy of Sciences. In 1938 Kol'tsov and Lysenko were the two candidates for the same academy slot in genetics. Kol'tsov had been a corresponding member since 1915 and was the clear favorite in the academy. In the months before the election, a press campaign was launched against Kol'tsov that attacked him for his earlier eugenic views and intimated that he had imported fascist ideas into the socialist motherland. A meeting was organized at the Kol'tsov institute in 1939 to "discuss" his eugenic mistakes. According to Dubinin, who presented the case against him, Kol'tsov declared that he repudiated neither the idea of eugenics nor a single word he had ever written on the subject (Dubinin 1973, p. 71). In 1939 Kol'tsov was relieved as director of the institute he had created, and his institute was absorbed into the Academy of Sciences and renamed the Institute of Cytology, Histology, and Embryology. Kol'tsov died of a heart attack in December 1940; the next day his wife committed suicide.

During the quarter century from 1940 until 1965, the Soviet state gave Lysenko strong support, and his version of "Michurinism" dominated official policy and rhetoric. In 1937 Lysenko became assistant to the president of the council of the Supreme Soviet. In 1938 he became president of the Lenin All-Union Academy of Agricultural Sciences (VASKhNIL) and used that position to move against his arch-rival N. I. Vavilov and his associates. With the help of the secret police, his supporters organized young party workers to harass the Vavilovites at both the Institute of Applied Botany and the Institute of Genetics. In 1940, apparently with Lysenko's complicity, Vavilov was arrested; as the case was being prepared, several of his closest colleagues, notably G. D. Karpechenko and G. A. Levitsky, were arrested as well. All three died in prison in 1942–43. As a result of these arrests, the Institute of Applied Botany was gutted and effectively ceased to be a center of genetics research. In 1939, following the press attack on the eugenic past of his rival, Kol'tsov, Lysenko was elected to full membership in the Academy of Sciences and made a member of its presidium. Upon Vavilov's arrest in 1940, Lysenko became director of the Institute of Genetics. Some geneticists lost their jobs immediately; with the coming of the war, others managed to hang on until 1945. Remarkably, most professional geneticists refused to go along with Lysenko's theories. By war's end, Medvedev, Kerkis, and Prokof'eva-Bel'govskaia had all lost their jobs.

However, despite these purges and institutional takeovers, Lysenko did not manage to uproot the discipline of genetics entirely. Within the Academy of Sciences, the Kol'tsov institute proved a center of resistance (Adams 1980a). Following Kol'tsov's death G. K. Khrushchov was eventually appointed director. Khrushchov was a histologist and a compromise candidate; so far as one can tell, his rhetoric supported Lysenko, but his administrative actions supported genetics. As a result, in the postwar years, the genetics staff at the Kol'tsov institute was the best in the USSR, and included Dubinin, Romashov, Sakharov, and Astaurov. The institute even managed to provide a haven for Muller's colleague Prokof'eva-Bel'govskaia when she was fired by Lysenko from IGEN; she worked at the Kol'tsov institute from 1945 to 1948. The universities also proved to be centers of resistance. At Moscow University Serebrovsky continued to head the Department of Genetics and trained a number of young workers, including the party member S. I. Alikhanian, who took over when Serebrovsky died of a stroke in June 1948 (Adams 1989f). At Leningrad University Lobashov ran genetics. A former student of Filipchenko, he had joined the party in 1941, emerged in 1945 a war hero, and by 1948 was chairman of the department of genetics and dean of the biological faculty. Thus, despite his triumphs over Vavilov, Lysenko had only limited success in undermining the geneticists who had been active in eugenics during the 1920s and in medical genetics during the 1930s.

Nor did Lysenko succeed in uprooting human genetics from the affiliated disciplines where it had flourished. When the Levit institute was abolished in 1937, Bunak moved his activities to the academy's Anthropological Institute and to Moscow University's Department of Anthropology, where he pursued his work on craniometry, heredity, and race unmolested, and even managed to train a school of students in these subjects (Roginskii 1940; Bunak 1938, 1940, 1941). Davidenkov continued as a professor at his Leningrad medical institute and, in the late 1930s, headed a genetic neurological clinic associated with the vast medical research complex, the Maxim Gorky All-Union Scientific Research Institute for Experimental Medicine (Fedorov 1939). He wrote the article on human heredity for the *Great Soviet Encyclopedia* (Davidenkov 1939a) and continued to publish on the genetics of neurological conditions (Davidenkov 1936b, 1939b). Elected a founding member of the Academy of Medical Sciences in 1945, he rose to become one of the elite corps of physicians treating Kremlin officials. In 1947 he even published a book on medical genetics (Davidenkov 1947). As director of the Pavlov institute and academic secretary of the Biological Sciences Division of the Academy of Sciences, Orbeli continued to support genetics, and in 1946 sought to create a new academy institute of genetics to be headed by Dubinin (Dubinin 1973).

In 1948 Lysenko's position was approved by Stalin and the party. During the most intense period of Lysenkoism (1948–1952), Michurinists had a field day at the geneticists' expense; one of their principal themes was that genetics, eugenics, and fascism were essentially the same. Two examples from the period may suffice to give the flavor of such rhetoric. In a pamphlet published in an edition of 110,000 copies, Minister of Education S. V. Kaftanov wrote that "the propositions of Morganism-Mendelism led in our country, just as they did abroad, to eugenic ideas and the ideology of Fascism" (Kaftanov 1948, p. 10). The next year, Lysenko's outspoken supporter A. N. Studitsky published an article in *Ogonek* that dwelt on both the uselessness of geneticists (because they worked on drosophila) and their fascist orientation, appropriately calling them and his article

"Fly-Lovers and Man-Haters." Yet, even then, genetics was not destroyed: it moved underground. The links geneticists had forged with psychologists, physiologists, and physicians continued to pay dividends. For example, the geneticists in Leningrad fared somewhat better thanks to the help of Orbeli, who gave Lobashov haven to do research in behavioral genetics in the Pavlov institute from 1949 to 1957.

Conclusion

In the mid-1950s, with Stalin's death and Khrushchev's subsequent de-Stalinization campaign, Lysenkoism began to wane and genetics began to be reestablished (Adams 1977–1978). The necessities of the atomic age stimulated the interest of biological and physical scientists alike in the effects of radiation on humans, and by 1963 medical genetics began to be reborn. Beginning in 1965, Lysenkoism was repudiated and the senior surviving geneticists, trained in the period 1910–1935, set about rebuilding their discipline. Because Lysenko's work focused on plants, most of these geneticists were animal geneticists and therefore had studied with Kol'tsov, Filipchenko, Serebrovsky, or Muller.

Naturally, their conception of the new Soviet genetics was shaped by their historical experience. The leading Soviet medical geneticist, Efroimson (a student of Kol'tsov), has written an article on hereditary altruism, suggesting that respect for the elderly and other moral and social virtues are inherited. Kerkis (a student of Filipchenko and Muller) has suggested that chronic criminality may be inherited. A new Soviet genetics textbook, written by M. E. Lobashov (a student of Filipchenko), has called for the rebirth of eugenics and has characterized it as one of the most important areas of genetics. And anthropologists, psychologists, and psychiatrists (often historically associated with Bunak, Iudin, Pavlov, and Davidenkov) have joined in the discussion (Adams 1990).

In the West the eugenics movement lost support among many geneticists around the time of the Great Break, but it was never repressed. Western geneticists gradually dissociated themselves from eugenic ideas, and as the discipline of genetics grew and subspecialties arose, positions could be openly debated, and new generations of students took up new lines of research, maintained interdisciplinary links, and were able to recast and refocus discussion in their own ways. The "repudiation" of eugenics in America, Britain, Germany, and elsewhere generally occurred only after World War II and the revelations of Nazi death-camp atrocities. "Medical genetics" and "human genetics" arose as disciplinary alternatives shortly thereafter. Indeed, today many medical geneticists are unaware of, and sometimes deny outright, any historical link between their field and eugenics.

By contrast, no such evolution could occur in Stalinist times. In the USSR, genetics, eugenics, and medical genetics were repressed at roughly the same time, by the same people, and for the same reasons. The discipline of genetics, with all of its eugenic and human aspects, went into a kind of "deep freeze" as a result of the rise of Lysenkoism and the repression of genetics and geneticists. When it "thawed" in the 1960s, what emerged were concepts of the discipline, its agenda, and its mission closely resembling those that had existed on the eve of Lysenkoism.

Thus, the networks that underlay the development of eugenics in the Soviet Union have proved remarkably enduring. They survived the Revolution and sought to create eugenics in the 1920s. They survived the demise of "eugenics" in the Great Break only to create "medical genetics" in 1934. They even managed to survive the purges and the rise of Lysenkoism. Repressed in the Soviet Union at the same time, by the same people, and for the same reasons, eugenics, medical genetics, and genetics were reborn together in the 1960s. Since that time the debates have continued (Adams 1990).

These facts should not surprise us. Eugenics movements, like scientific institutions and disciplines, are intellectual, professional, and personal networks shaped in a societal cauldron. Such networks have proved remarkably resilient and enduring, and they have their own momentum. Nor should the differences in the political and ideological dimensions of eugenics and human genetics surprise us: these, too, are not written in stone, but are historically contingent linkages shaped in very particular intellectual and social settings. In the age of Gorbachev, of *glasnost'* and *perestroika,* new information is appearing almost daily about the history of Soviet genetics and Lysenkoism. It will be interesting to follow both the scientific and historical discussions of human heredity in the USSR, to see whether private, professional, and scientific networks reassert themselves, and to what extent the ideas they once fostered reemerge into public discourse.

Bibliography

This work is based in part on materials in the Archives of the USSR Academy of Sciences in Moscow (Fund 450, Kol'tsov papers), the Manuscript Division of the Saltykov-Shchedrin Public Library in Leningrad (Fund 813, Filipchenko papers), the Lilly Library in Bloomington (Muller papers), and the Library of the American Philosophical Society in Philadelphia (Davenport and Dobzhansky papers). For help in gaining access to materials, special thanks are due to A. E. Gaissinovitch, Gunnar Broberg, A. I. Kolchinsky, Tilia S. Levit, B. V. Levshin, Thea Muller, and William Provine. Thanks are also due for support provided by grants from the History and Philosophy of Science Program of the Social Sciences Division of the National Science Foundation and from the Program on Science, Technology, and the Humanities of the National Endowment for the Humanities.

Adams, Mark B. 1965. Measurements of the Growth of Russian Science. Manuscript.

————.1977–1978. Biology after Stalin: A Case Study. *Survey: A Journal of East/West Studies* 23: 53–80.

————.1978. Nikolai Ivanovich Vavilov. In *Dictionary of Scientific Biography,* vol. 15, suppl. 1. New York: Charles Scribners Sons. 505–13.

————. 1979. From 'Gene Fund' to 'Gene Pool': On the Evolution of Evolutionary Language. In *Studies in the History of Biology,* vol. 3, ed. William Coleman and Camille Limoges, 241–85. Baltimore and London: Johns Hopkins University Press.

————. 1980a. Science, Ideology, and Structure: The Kol'tsov Institute 1900–1970. In *The Social Context of Soviet Science,* ed. Linda Lubrano and Susan Gross Solomon, 173–204. Boulder, Colo.: Westview Press.

————. 1980b. Sergei Chetverikov, the Kol'tsov Institute, and the Evolutionary Synthesis. In *The*

Evolutionary Synthesis, ed. Ernst Mayr and William B. Provine, 242–78, Cambridge, Mass.: Harvard University Press.

————. 1989a. Eugenics as Social Medicine: Prophets, Patrons, and the Dialectics of Discipline-Building. In Solomon and Hutchinson 1989, in press.

————. 1989b. Astaurov, Boris L'vovich. In *Dictionary of Scientific Biography,* suppl. 2, s.v. New York: Charles Scribners Sons, in press.

————. 1989c. Chetverikov, Sergei Sergeevich. In *Dictionary of Scientific Biography,* suppl. 2, s.v. New York: Charles Scribners Sons, in press.

————. 1989d. Filipchenko, Iurii Aleksandrovich. In *Dictionary of Scientific Biography,* suppl. 2, s.v. New York: Charles Scribners Sons, in press.

————. 1989e. Levit, Solomon Grigorevich. In *Dictionary of Scientific Biography,* suppl. 2, s.v. New York: Charles Scribners Sons, in press.

————. 1989f. Serebrovsky, Aleksandr Sergeevich. In *Dictionary of Scientific Biography,* suppl. 2, s.v. New York: Charles Scribners Sons, in press.

————. 1989g. The Politics of Human Heredity in the USSR. *Genome* 31, no. 2, in press.

————. 1990. Nature and Nurture in the USSR: Historical Roots of Current Controversies. In *Science and the Soviet Social Order,* ed. Loren Graham. Cambridge, Mass.: Harvard University Press, forthcoming.

Alekseev, A. 1914. Alkogolizm i nasledstvennost' [Alcoholism and heredity]. *Priroda,* no. 3: 362.

Astaurov, B. L. and P. F. Rokitskii. 1975. *Nikolai Konstantinovich Kol'tsov.* Moscow: Izdatel'stvo "Nauka."

Babkov [Babkoff], V. 1975. Sakharov, Vladimir Vladimirovich. In *Dictionary of Scientific Biography,* vol. 12: 76–77.

Bailes, Kendall E. 1978. *Technology and Society under Lenin and Stalin.* Princeton, N.J.: Princeton University Press.

Batkis, G. 1931. Evgenika [Eugenics]. *Bol'shaia sovetskaia entsiklopediia* [Great Soviet encyclopedia], vol. 23: 812–19.

Bednyi, Dem'ian. 1930. Evgenika [Eugenics]. *Izvestiia,* 4 June, 4.

Bekhterev, V. M. 1908. Voprosy vyrozhdeniia i bor'ba s nim [Questions of degeneration and the struggle against it]. *Obozrenie Psikhiatrii i Nevropatologii,* no. 9.

Biulleten' IV sessii VASKhNILa [Bulletin of the fourth session of the Lenin All-Union Academy of Agricultural Sciences]. 1936. Moscow: Izd. VASKhNILa.

Blium, A. 1909. *Etika i evgenika* [Ethics and eugenics]. Petrograd: Izd. Suvorina.

Bogdanov, E. 1914. *Mendelizm.* Moscow: Izdatel'stvo "Priroda."

Bondarenko, P. P., et al., eds. 1931. *Protiv mekhanisticheskogo materializma i men'shevistvuiushchego idealizma v biologii* [Against mechanistic materialism and Menshevizing idealism in biology]. Kommunisticheskaia Akademiia. Moscow and Leningrad: Gosudarstvennoe meditsinskoe izdatel'stvo.

Bowler, Peter. 1984. E. W. MacBride's Lamarckian Eugenics and Its Implications for the Social Construction of Scientific Knowledge. *Annals of Science* 41: 245–60.

Bumke, Oswald. 1926. *Kul'tura i vyrozhdenie* [Culture and degeneration]. Translation edited by P. B. Gannushkin. Preface by V. Volgin and P. Gannushkin. Moscow: Izdatel'stvo Sabashnikovykh.

Bunak, V. V. 1917. Rasovy tip donskikh kazakov [The racial type of the Don cassacks]. *Priroda,* no. 4: 520–21.

————. 1922a. Evgenicheskie opytnye stantsii, ikh zadachi i plan rabot [Eugenic experimental stations, their tasks and work plans]. *Russkii Evgenicheskii Zhurnal* 1: 82–97.

————. 1922b. O deiatel'nosti Russkogo Evgenicheskogo Obshchestva [On the activities of the Russian Eugenics Society]. *Russkii Evgenicheskii Zhurnal* 1: 99–101.

————. 1923a. Metody izucheniia nasledstvennosti u cheloveka: Kriticheskoe issledovanie s prakticheskimi ukazaniiami dlia spetsialistov [Methods of studying human heredity: A critical investigation with practical illustrations for specialists]. *Russkii Evgenicheskii Zhurnal* 1: 137–200.

————. 1923b. Novye dannye k voprosu o voine, kak biologicheskom faktore [New data on war as a biological factor]. *Russkii Evgenicheskii Zhurnal* 1: 223–32.

_____. 1923c. Ob ispol'zovanii biosanitarnykh dannykh arkhivov pedagogicheskikh lechebnykh i t. p. uchrezhdenii [On the use of biosanitary data from the archives of teaching clinics and similar institutions]. *Russkii Evgenicheskii Zhurnal* 1: 238–39.

_____. 1923d. K antropometricheskoi kharakteristike potomstva sifilitikov [Anthropometric profile of the offspring of syphilitics]. *Russkii Evgenicheskii Zhurnal* 1: 348–57.

_____. 1923e. Neskol'ko dannykh po bioantropologii mari (cheremisov) [Some data on the biological anthropology of residents of Mariiskii (Cheremiiskii) province]. *Russkii Evgenicheskii Zhurnal* 1: 358–62.

_____. 1924a. Iz otcheta o deiatel'nosti Russkogo Evgenicheskogo Obshchestva za 1923 g [From the 1923 report of the activities of the Russian Eugenics Society]. *Russkii Evgenicheskii Zhurnal* 2: 66–67.

_____. 1924b. Materialy dlia sravnitel'noi kharakteristiki sanitarnoi konstitutsii evreev [Material for the comparative profile of the sanitary constitution of Jews]. *Russkii Evgenicheskii Zhurnal* 2: 142–52.

_____. 1925. O smeshenii chelovecheskikh ras [On the mixing of human races]. *Russkii Evgenicheskii Zhurnal* 3: 121–38.

_____. 1925–1926. Einige Daten über die Isohaemagglutination bei verschiedene asiatischen Stämmen. *Archiv für Rassen- und Gesellschafts-Biologie* 17: 316–18.

_____. 1926. O morfologicheskikh osobennostiakh odno- i dvuiatsevykh bliznetsov [On the morphological characteristics of identical and fraternal twins]. *Russkii Evgenicheskii Zhurnal* 4: 23–52.

_____. 1930. Termin 'rasa' v zoologii i antropologii [The term 'race' in zoology and anthropology]. *Russkii Evgenicheskii Zhurnal* 7: 117–32.

_____. 1934a. Materialy dlia opredeleniia istinnogo sootnosheniia polov [Materials for determining the true sex-ratio]. In *Trudy MBI*, 195–212. Summary in English. *See* Levit 1934a.

_____. 1934b. Konstitutsional'nye tipy v geneticheskom osveshchenii [Constitutional types in the light of genetics]. In *Konferentsiia po meditsinskoi genetike*, 49–52.

_____. 1936a. O nekotorykh sluchaiakh izmeneniia srednei velichiny priznakov v smeshivaiushchikhsia populiatsiiakh [On some changes in the mean values of characters in mixed populations]. In *Trudy MGI*, 237–53. Summary in English. *See* Levit and Ardashnikov 1936.

_____. 1936b. Rol' nasledstvennosti i sredy v izmenchivosti struktury kozhnykh kapiliarov (Issledovanie 91 pary bliznetsov) [The role of heredity and environment in the variability of cutaneous capillaries (Investigation of 91 pairs of twins)]. In *Trudy MGI*, 383–403. Summary in English. *See* Levit and Ardashnikov 1936.

_____. 1938. Rasa kak istoricheskoe poniatie [Race as a historical concept]. In *Nauka o rasakh i rasizm* [Racism and the science of races], Nauchno-issledovatel'skii institut MGU, Trudy, no. 4, 5–40. Moscow and Leningrad: Izd. Akademii nauk SSSR.

_____. 1940. Morfologiia cheloveka v Moskovskom universitete [Human morphology at Moscow University]. *Uchenye zapiski Moskovskogo gosudarstvennogo universiteta, Biologiia, Iubileinaia seriia*, no. 53: 208–15.

_____. 1941. *Antropometriia* [Anthropometry]. Moscow: Uchpedgiz.

Bunak, V. V., and G. V. Soboleva. 1925. Issledovanie elementov okraski raduzhiny u cheloveka [Investigation of the color elements of the human iris]. *Zhurnal Eksperimental'noi Biologii*, Series A, 1: 145–76. Summary in German.

Carlson, Elof Axel. 1981. *Genes, Radiation, and Society: The Life and Work of H. J. Muller*. Ithaca, N.Y., and London: Cornell University Press.

Chulkov, N. P. 1923. Rod grafov Tolstykh [The family of the counts Tolstoi]. *Russkii Evgenicheskii Zhurnal* 1: 308–20.

_____. 1927. Genealogiia dekabristov Murav'evykh [Geneaology of the Murav'ev Decembrists]. *Russkii Evgenicheskii Zhurnal* 5: 3–20.

Davenport, Charles B. 1913. *Evgenika kak nauka ob uluchshenii prirody cheloveka* [Eugenics as the science of improving the human breed]. Translation from *Heredity in Relation to Eugenics*. Moscow: G. Vuttke.

Davidenkov, S. N. 1925a. *Nasledstvennye bolezni nervnoi sistemy* [Hereditary diseases of the nervous system]. Kharkov: Gosudarstvennoe izdatel'stvo Ukrainy.

———. 1925b. K klassifikatsii i genetike semeinykh nevrodistrofii [On the classification and genetics of familial neurodystrophy]. *Russkii Evgenicheskii Zhurnal* 3: 45–60.

———. 1928. Geneticheskoe biuro pri M. O. N. i P. [The genetic bureau of the Moscow Society of Neuropathologists and Psychiatrists]. *Russkii Evgenicheskii Zhurnal* 6: 55–56.

———. 1931. Genetika v nevropatologii [Genetics in neuropathology]. *Zhurnal Nevrologii i Psikhiatrii imeni S. S. Korsakova,* no. 5: 60–66.

———. 1932. *Nasledstvennye bolezni nervnoi sistemy* [Hereditary diseases of the nervous system]. 2d ed. Moscow: Gosudarstvennoe meditsinskoe izdatel'stvo.

———. 1934a. Genetika i klinika [Genetics and the clinic]. In *Konferentsiia po meditsinskoi genetike,* 34–42.

———. 1934b. *Problema polimorfizma nasledstvennykh boleznei nervnoi sistemy: Kliniko-geneticheskoe issledovanie* [The problem of polymorphism in hereditary diseases of the nervous system: A genetic clinical investigation]. Leningrad: Vsesoiuznyi institut eksperimental'noi meditsiny.

———. 1936a. Genetika i meditsina [Genetics and medicine]. In *Sovremennye problemy teoreticheskoi meditsiny* [Current problems of theoretical medicine], vol. 1: 171–84. Moscow and Leningrad.

———, ed. 1936b. *Nevrologiia i genetika* [Neurology and genetics]. Vol. 2. Leningrad: Vsesoiuznyi institut eksperimental'noi meditsiny.

———. 1939a. Nasledstvennost' cheloveka [Human heredity]. *Bol'shaia sovetskaia entsiklopediia* [Great Soviet encyclopedia], Vol. 41, 273–76.

———. 1939b. Nervnaia klinika imeni akademika I. P. Pavlova (Leningrad) [The Pavlov nerve clinic (Leningrad)]. In *Otchet VIEM,* 199–208. *See* Fedorov 1939.

———. 1947. *Evoliutsionno-geneticheskie problemy v nevropatologii* [Problems of evolutionary genetics in neuropathology]. Leningrad: Izdatel'stvo GIDUV.

D'iakonov, D. M., and Ia. Ia. Lus. 1922. Raspredelenie i nasledovanie spetsial'nykh sposobnostei [The distribution and inheritance of special abilities]. *Izvestiia Biuro po Evgenike,* no. 1: 72–112. Summary in English.

Dubinin, N. P. 1973. *Vechnoe dvizhenie* [Perpetual motion]. Moscow: Izdatel'stvo politicheskoi literatury.

Efroimson, V. P. 1967. K istorii izucheniia genetiki cheloveka v SSSR [On the history of the study of human genetics in the USSR]. *Genetika,* no. 10: 114–27.

Fedorov, L. N., ed. 1939. *Otchet o nauchno-issledovatel'skoi rabote Vsesoiuznogo instituta eksperimental'noi meditsiny imeni A. M. Gorkogo za 1933–1937 gg.* [Survey of research of the A. M. Gorkii All-Union Institute of Experimental Medicine for 1933–37]. Moscow and Leningrad: Gosudarstvennoe izdatel'stvo meditsinskoi literatury.

Filipchenko, Iurii Aleksandrovich [Philiptschenko, Jur.]. 1913. O vidovykh gibridakh [On species hybrids]. *Novye Idei v Biologii,* no. 4: 122–49.

———. 1915. *Izmenchivost' i evoliutsiia* [Variation and evolution]. Petrograd: Biblioteka naturalista.

———. 1916. *Proiskhozhdenie domashnykh zhivotnykh* [The origin of domesticated animals]. Petrograd: Izdatel'stvo Bieka. 2d ed., 1924. Leningrad: Izdatel'stvo "Seiatel'" E. V. Vysotskogo.

———. 1916–1917. Izmenchivost' i nasledstvennost' cherepa u mlekopitaiushchikh [Variation and heredity of mammalian skulls]. *Russkie Arkhivy Anatomii, Gistologii i Embriologii* 1: 311–404, 747–818.

———. 1917. *Nasledstvennost'* [Heredity]. Moscow: Izdatel'stvo "Priroda."

———. 1918. Evgenika [Eugenics]. *Russkaia Mysl',* nos. 3–4 (March–June): 69–95.

———. 1919. Khromozomy i nasledstvennost' [Chromosomes and heredity]. *Priroda,* nos. 7–9: 327–50.

———. 1921a. *Chto takoe Evgenika* [What is eugenics]. Petrograd: Izdatel'stvo KEPSa, Biuro po Evgenike.

_____. 1921b. *Kak nasleduiutsia razlichnye osobennosti cheloveka* [How various human traits are inherited]. Petrograd: Izdatel'stvo KEPSa, Biuro po Evgenike.

_____. 1921c. Anketa po nasledstvennosti sredi uchenykh Peterburga [Questionnaire on inheritance among Petersburg scientists]. *Nauka i ee Rabotniki*, no. 2: 33–35.

_____. 1921d. Rezul'taty ankety po nasledstvennosti sredi uchenykh Peterburga [Results of the questionnaire on inheritance among Petersburg scientists]. *Nauka i ee Rabotniki*, no. 6: 3–9.

_____. 1922a. Biuro po Evgenike [The Bureau of Eugenics]. *Izvestiia Biuro po Evgenike*, no. 1: 1–4. Summary in English.

_____. 1922b. Statisticheskie rezul'taty ankety po nasledstvennosti sredi uchenykh Peterburga [Statistical results of a questionnaire on inheritance among Petersburg scientists]. *Izvestiia Biuro po Evgenike*, no. 1: 5–21. Summary in English.

_____. 1922c. Nashi vydaiushchiesia uchenye [Our eminent scientists]. *Izvestiia Biuro po Evgenike*, no. 1: 22–38. Summary in English.

_____. 1922d. Zakon Mendelia i zakon Morgana [Mendel's law and Morgan's law]. *Priroda*, nos. 10–12: 51–66.

_____. 1923a. *Izmenchivost' i metody ee izucheniia* [Variation and methods for its study]. Petrograd: Gosudarstvennoe izdatel'stvo.

_____. 1923b. *Obshchedostupnaia biologiia* [Biology for the general reader]. Petrograd: Knigoizdatel'stvo "Seiatel'" E. V. Vysotskogo.

_____. 1923c. *Evoliutsionnaia ideia v biologii* [The evolutionary idea in biology]. Moscow: Izd. Sabashnikovykh.

_____. 1924a. *Puti uluchsheniia chelovecheskogo roda (evgenika)* [Ways of improving the human race (eugenics)]. Leningrad: Gosudarstvennoe izdatel'stvo.

_____. 1924b. Rezul'taty obsledovaniia leningradskikh predstavitelei iskusstva [Results of a survey of representatives of the arts from Leningrad]. *Izvestiia Biuro po Evgenike*, no. 2: 5–28. Summary in English.

_____. 1924c. Nekotorye rezul'taty ankety po nasledstvennosti sredi leningradskikh studentov [Some results of a questionnaire on inheritance among Leningrad students]. *Izvestiia Biuro po Evgenike*, no. 2: 29–48. Summary in English.

_____. 1924d. *Nasledstvennost'* [Heredity]. 2d ed. Moscow: Gosudarstvennoe izdatel'stvo.

_____. 1925a. *Frensis Gal'ton i Gregor Mendel'* [Francis Galton and Gregor Mendel]. Biograficheskaia biblioteka. Moscow: Gosudarstvennoe izdatel'stvo.

_____. 1925b. Evgenika v shkole [Eugenics in school]. *Russkii Evgenicheskii Zhurnal* 3: 31–35.

_____. 1925c. Obsuzhdenie norvezhskoi evgenicheskoi programmy na zasedaniiakh Leningradskogo otdeleniia R. E. O. [The discussion of the Norwegian eugenics program at a meeting of the Leningrad division of the Russian Eugenic Society]. *Russkii Evgenicheskii Zhurnal* 3: 139–43.

_____. 1925d. Novye zhurnali i novye knigi po evgenike [New journals and new books on eugenics]. *Russkii Evgenicheskii Zhurnal* 3: 145–52.

_____. 1925e. Intelligentsiia i talanty [Talent and the intelligentsia]. *Izvestiia Biuro po Evgenike*, no. 3: 83–96.

_____. 1925–1926. Die russische rassenhygienische Literatur 1921–1925. *Archiv für Rassen- und Gesellschafts-Biologie* 17: 346–48.

_____. 1926. Biuro po Genetike i Evgenike [Bureau of Genetics and Eugenics]. *Izvestiia Biuro po Genetike i Evgenike*, no. 4: 3–4.

_____. 1927. *Variabilität und Variation*. Berlin: Borntraeger.

_____. 1927–1930. Letters to Theodosius Dobzhansky. Dobzhansky archives, Library, American Philosophical Society.

_____. 1928. Nasledovanie odarennosti [The inheritance of talent]. *Chelovek*, no. 1: 10–22.

_____. 1929. *Genetika* [Genetics]. Leningrad: Gosudarstvennoe izdatel'stvo.

_____. 1932. *Eksperimental'naia zoologiia* [Experimental zoology]. Edited and with an introduction by I. Prezent. Leningrad and Moscow: Gosudarstvennoe meditsinskoe izdatel'stvo.

Filipchenko, Iu, A., and I. I. Ivanov [Philiptschenko and Ivanow]. 1916. Beschriebung von Hybriden zwischen Bison, Wisent und Hausrind. *Zeitschrift für Abstammungs- und Vererbungslehre* 16: 1–48.

Filipchenko, Iu. A., and T. K. Lepin. 1922. K voprosu o nasledovanii tsveta glaz i volos [On the inheritance of eye and hair color]. *Izvestiia Biuro po Evgenike,* no. 1: 39–63. Summary in German.

Filipchenko, Iu. A., and T. H. Morgan. 1925. *Nasledstvenny li priobretennye priznaki?* [Are acquired characteristics inherited?]. Leningrad: Knigoizdatel'stvo "Seiatel'" E. V. Vysotskogo.

Finkel'shtein, E. A. 1935. Evgenika i fashizm [Eugenics and fascism]. In *Rasovaia teoriia na sluzhbe fashizma* [Race theory in service of fascism], 58–88. Kiev: Gosudarstvennoe meditsinskoe izdatel'stvo.

Florinskii, [F., V. M., V. N.] 1866. *Usovershenstvovanie i vyrozhdenie chelovecheskago roda* [The improvement and degeneration of the human race]. St. Petersburg.

———. 1926. *Usovershenstvovanie i vyrozhdenie chelovecheskago roda* [The improvement and degeneration of the human race]. 2d ed. Vologda: Izd. "Severnyi pechatnik."

Frank-Kamenetskii, Z. G. 1927. O svoeobraznoi nasledstvennoi forme glaukomy v Irkutskoi gubernii [On a unique hereditary form of glaucoma in the province of Irkutsk]. *Russkii Evgenicheskii Zhurnal* 5: 25–36.

Gaissinovitch, A. E. 1980. The Origins of Soviet Genetics and the Struggle with Lamarckism, 1922–1929. Trans. Mark B. Adams. *Journal of the History of Biology* 13: 1–51.

Galach'ian, A. G., and T. I. Iudin. 1923. Opyt nasledstvenno-biologicheskogo analiza odnoi man'iakal'no-depressivnoi sem'i [Example of the hereditary biological analysis of a single manic depressive family]. *Russkii Evgenicheskii Zhurnal* 1: 321–42.

Galton, F. 1875. *Nasledstvennost' talanta, ee zakony i posledstviia* [Translation of *Hereditary Talent*]. St. Petersburg: Izd. zhurnala "Znanie."

Gamaleia, N. F. 1912. Ob usloviiakh, blagopriatstvuiushchikh uluchsheniiu prirodnykh svoistv liudei [On conditions favoring the improvement of the natural characteristics of humans]. *Gigiena i Sanitariia,* 6: 340–61.

Gates, R. Ruggles [R. A.] 1926. *Nasledstvennost' i evgenika* [Heredity and eugenics]. Trans. A. A. Filipchenko. Ed. Iu. A. Filipchenko. Leningrad: Knigoizdatel'stvo "Seiatel'" E. V. Volotskogo.

Gorbunov, A. V. 1922. Vliianie mirovoi voiny na dvizhenie naseleniia Evropy [The influence of the world war on population movement in Europe]. *Russkii Evgenicheskii Zhurnal* 1: 39–63.

———. 1928. Razmnozhaemost' moskovskoi intelligentsii po dannym ankety russkogo evgenicheskogo obshchestva [The fertility of the Moscow intelligentsia according to questionnaires of the Russian Eugenics Society]. *Russkii Evgenicheskii Zhurnal* 6: 3–53.

Graham, Loren R. 1967. *The Soviet Academy of Sciences and the Communist Party 1927–1932.* Princeton, N.J.: Princeton University Press.

———. 1972. *Science and Philosophy in the Soviet Union.* New York: Knopf.

———. 1977. Science and Values: The Eugenics Movement in Germany and Russia in the 1920s. *American Historical Review* 82: 1135–64.

Gurvich, K. 1928. Ukazatel' literatury po voprosam evgeniki, nasledstvennosti i selektsii i sopredel'nykh oblastei, opublikovannoi na russkom iazyke do 1/I 1928 godu [List of literature on questions of eugenics, heredity, and selection and related subjects, published in the Russian language before 1 January 1928]. *Russkii Evgenicheskii Zhurnal* 6: 121–43.

Haecker [Gekker], V. 1923. *Genetika: Etapy mendelizma* [Genetics: Principles of Mendelism]. Edited and introduced by A. A. Sapegin. Odessa: Izd. Narkozema Ukrainy.

———. 1924. O nasledovanii muzykal'nykh sposobnostei [On the inheritance of musical talent]. *Russkii Evgenicheskii Zhurnal* 2: 103–16.

Iudin, T. I. 1907. Psikhozy u bliznetsov [Psychoses in twins]. *Zhurnal Nevropatologii i Psikhiatrii imeni Korsakova* 7: 68–83.

———. 1922. Nasledstvennost' dushevnykh boleznei (Istoriia i sovremennoe sostoianie voprosa)

[The inheritance of mental diseases (history and current status of the question)]. *Russkii Evgenicheskii Zhurnal* 1: 28–38.

———. 1923. Uchenie o konstitutsiiakh v patologii i ego znachenie dlia evgeniki [The study of constitutions in pathology and its significance for eugenics]. *Russkii Evgenicheskii Zhurnal* 1: 117–36.

———. 1924. Skhodstvo bliznetsov i ego znachenie v izuchenii nasledstvennosti [The similarity of twins and its significance for the study of heredity]. *Russkii Evgenicheskii Zhurnal* 2: 28–49.

———. 1925. *Evgenika: Uchenie ob uluchshenii prirodnykh svoistv cheloveka* [Eugenics: The doctrine of the improvement of natural human traits]. Moscow: Izdanie Sabashnikovykh.

———. 1926. *Psikhopaticheskie konstitutsii: Vydelenie tipov kharaktera (tipov slozhnykh psikhicheskikh reaktsii) na osnovanii nasledstvenno-biologicheskogo i klinicheskogo analiza psikhozov* [Psychopathological constitutions: The delineation of character types (types of complex mental reactions) on the basis of the hereditary, biological, and clinical analysis of psychoses]. Moscow: Izdanie Sabashnikovykh.

———. 1934. Konstitutsiia i genetika [Constitution and genetics]. In *Konferentsiia po meditsinskoi genetike*, 43–48.

Iudin, T. I., and F. F. Detengof, 1924. Opyt geneticheskogo analiza skhizoidnogo kompleksa [An attempt to analyze the schizoid complex genetically]. *Russkii evgenicheskii zhurnal* 2: 126–41.

Iudin, T. I., and M. N. Ksenokratov. 1929. Zavisimost' klinicheskogo techeniia maniakal'no-depressivnogo psikhoza ot osobennostei nasledstvennoi struktury lichnosti (kliniko-genealogicheskoe issledovanie) [The dependence of the clinical course of manic depressive psychosis on the particulars of hereditary personality structure]. *Russkii Evgenicheskii Zhurnal* 7: 39–62.

Ivanov, I. I. 1970. *Izbrannye trudy* [Selected works]. Moscow: Izdatel'stvo "Kolos".

Joravsky, David. 1961. *Soviet Marxism and Natural Science 1917–1932*. New York: Columbia University Press.

———. 1970. *The Lysenko Affair*. Cambridge, Mass.: Harvard University Press.

Kaftanov, S. V. 1948. *Za bezrazdel'noe gospodstvo michurinskoi biologicheskoi nauki* [For the undivided rule of Michurinist biology]. Vsesoiuznoe obshchestvo po rasprostraneniiu politicheskikh i nauchnykh znanii. Moscow: Izdatel'stvo "Pravda."

Kammerer, Paul. 1912. K voprosu o nasledovanii priobretennykh priznakov [On the question of the inheritance of acquired characteristics]. *Priroda*, no. 2: 239–76.

———. 1924. *The Inheritance of Acquired Characteristics*. New York: Boni and Liveright.

———. 1927. *Zagadka nasledstvennosti: Osnovy ucheniia o nasledstvennosti* [The riddle of heredity: Foundations of hereditary science]. Translated from German and edited by B. S. Kuzin. Preface by I. Agol.

Karaffa-Korbut. 1910. Ocherki po evgenike [Essays on eugenics]. *Gigiena i Sanitariia*, no. 1: 41; no. 2: 138; no. 4: 276.

———. 1922. Evgenicheskoe znachenie voiny [The eugenic significance of war]. Moscow: Izd. "Prakticheskaia meditsina."

Klodnitzky, I. 1925. Letter. *Eugenical News* 10: 59–60.

Koblitz, Ann Hibner. 1988. Science, Women, and the Russian Intelligentsia: The Generation of the 1860s. *Isis* 79: 208–26.

Koestler, Arthur. 1971. *The Case of the Midwife Toad*. London: Hutchinson.

Kol'tsov [Koltzoff], N. K. 1916a. Alkogolizm i nasledstvennost' [Alcoholism and inheritance]. *Priroda*, no. 4: 502–5.

———. 1916b. K voprosu o nasledovanii posledstvii alkogolizma [On the question of the heritability of alcoholism]. *Priroda*, no. 10: 1189.

———. 1917. Genealogiia doma Romanovykh [Genealogy of the house of the Romanovs]. *Priroda*, no. 4: 509.

———. 1921–1922. O nasledstvennykh khimicheskikh svoistvakh krovi [On hereditary chemical characteristics of the blood]. Parts 1, 2. *Priroda*, nos. 4–6: 2–7; *Uspekhi Eksperimental'noi Biologii* 1: 333–61.

————. 1922a. Uluchshenie chelovecheskoi porody [Improving the human breed]. *Russkii Evgenicheskii Zhurnal* 1: 3–27.

————. 1922b. Genealogiia Ch. Darvina i F. Gal'tona [The genealogy of Ch. Darwin and F. Galton]. *Russkii Evgenicheskii Zhurnal* 1: 64–73.

————. 1923a. *Uluchshenie chelovecheskoi porody* [Improving the human breed]. Petrograd: Izd. "Vremia."

————. 1923b. Poteri v sostave naseleniia Evropy v gody Mirovoi voiny 1914–1917 gg. [Losses in the composition of the population of Europe in the years of the world war 1914–1917]. *Russkii Evgenicheskii Zhurnal* 1: 233–34.

————. 1923c. Geneticheskii analiz psikhicheskikh osobennostei cheloveka [Genetic analysis of human mental characteristics]. *Russkii Evgenicheskii Zhurnal* 1: 253–307.

————. 1924a. Vliianie kul'tury na otbor v chelovechestve [The influence of culture on selection in the human race]. *Russkii Evgenicheskii Zhurnal* 2: 3–19.

————. 1924b. Noveishie popytki dokazat' nasledstvennost' blagopriobretennykh priznakov [The latest attempts to prove the inheritance of favorable acquired characteristics]. *Russkii Evgenicheskii Zhurnal* 2: 159–67.

————. 1924c. Razmnozhaemost' vo Frantsii [Fertility in France]. *Russkii Evgenicheskii Zhurnal* 2: 171–74.

————. 1924d. Experimental Biology and the Work of the Moscow Institute. *Science* 59: 497–502.

————. 1925a. Evgenicheskie s"ezdy v Milane v sentiabre 1924 goda [Eugenic meetings in Milan in September 1924]. *Russkii Evgenicheskii Zhurnal* 3: 73–78.

————. 1925b. Evgenicheskie s"ezdy v Evrope v 1924 g. [Eugenic conferences in Europe in 1924]. *Nauchnyi Rabotnik*, no. 2: 115–19.

————. 1925c. Razmnozhaemost' semei Amerikanskikh studentov [Fertility of the families of American students]. *Russkii Evgenicheskii Zhurnal* 3: 82–84.

————. 1925–1926. Die rassenhygienische Bewegung in Russland. *Archiv für Rassen- und Gesellschafts-Biologie* 17: 96–99.

————. 1926. Rodoslovnye nashikh vydvizhentsev [Pedigrees of our vydvizhentsy]. *Russkii Evgenicheskii Zhurnal* 4: 103–43.

————. 1927. Experimental'naia biologiia v SSSR [Experimental biology in the USSR]. In *Nauka i tekhnika SSSR 1917–1927* [Science and technology in the USSR 1917–1927], vol. 2, 37–64. Moscow: "Rabotnik prosveshcheniia."

————. 1928a. O potomstve velikikh liudei [On the descendants of great people]. *Russkii Evgenicheskii Zhurnal* 6: 164–77.

————. 1928b. Dva sluchaia nasledstvennoi anomalii pal'tsev [Two cases of hereditary anomalies of the digits]. *Russkii Evgenicheskii Zhurnal* 6: 196–202.

————. 1928c. Kak izuchaiutsia zhiznennye iavleniia [How living phenomena are studied]. Moscow: Izdatel'stvo Narkomzdrava RSFSR.

————. 1929a. Evfenika [Euthenics]. *Bol'shaia meditsinskaia entsiklopediia*, vol. 9, 689–92.

————. 1929b. Zadachi i metody izucheniia rasovoi patologii [Tasks and methods in studying racial pathology]. *Russkii Evgenicheskii Zhurnal* 7: 69–87.

————. 1929c. O rabotakh Instituta eksperimental'noi biologii v Moskve [On the works of the Institute of Experimental Biology in Moscow]. *Uspekhi Eksperimental'noi Biologii* 8: 15–28.

————. 1934. Rol' genetiki v izuchenii biologii cheloveka [The role of genetics in the study of human biology]. In *Konferentsiia po meditsinskoi genetike*, 29–33.

Kondorskii, I. K. 1922. *Vyrozhdenie, ego prichiny i evgenicheskoe dvizhenie* [Degeneration, its causes and the eugenics movement]. Simferopol: Izdatel'stvo Krymnarkomzdrava.

Konferentsiia po meditsinskoi genetike: Doklady i preniia [Conference on medical genetics: Proceedings]. 1934. Suppl. to *Sovetskaia klinika* 20, nos. 7–8. Also published separately, Moscow: Izdanie Polikliniki Komissii Sodeistviia Uchenym pri SNK SSSR.

Kostiamin, N. 1925. Letter. *Eugenical News* 10: 57.

Kovalevskii, P. I. 1899. Vyrozhdenie i vozrazhdenie [Degeneration and regeneration]. St. Petersburg: Izd. "Russkii Meditsinskii Vestnik."

Kravets, L. I. 1914a. Nasledstvennost' u cheloveka [Human heredity]. *Priroda*, no. 6: 721.

———. 1914b. Evgenetika [Eugenetics]. *Priroda*, no. 10: 1229.

Krontovskii [Krontovsky], A. A. 1925a. *Nasledstvennost' i konstitutsiia: Prakticheskoe posobie k issledovaniiu patologicheskoi nasledstvennosti i konstitutsii cheloveka* [Heredity and constitution: Practical handbook for the investigation of pathological heredity and constitution in humans]. Kiev: Gosudarstvennoe izdatel'stvo Ukrainy.

———. 1925. *Über das Sammeln des Materials über pathologische Heredität beim Menschen*. Kiev.

Kutanin, M. P. 1927. Otchet o rabote Saratovskogo otdeleniia Russkogo Evgenicheskogo Obshchestva [Survey of the work of the Saratov branch of the Russian Eugenics Society]. *Russkii Evgenicheskii Zhurnal* 5: 93–96.

———. 1928. Otchet o deiatel'nosti Saratovskogo otdeleniia Russkogo Evgenicheskogo Obshchestva za 1927 god [Survey of activities of the Saratov branch of the Russian Eugenics Society for 1927]. *Russkii Evgenicheskii Zhurnal* 6: 54–55.

Lass, D. I. 1927. Brak i detorozhdenie (Po anketnym materialam odesskikh vuzov) [Marriage and childbirth (On questionnaire material from higher educational institutions of Odessa)]. *Russkii Evgenicheskii Zhurnal* 5: 90–92.

Lebedev, N. N. 1911. Bor'ba s pristupnost'iu v Amerike [The struggle against crime in America]. *Vestnik Obshchego Gigieny* (January).

Lecourt, Dominique. 1976. *Lyssenko: Histoire réelle d'une "science prolétarienne."* Paris: François Maspero.

Lepin, T. K. 1924. K voprosu o nasledovanii blizorukosti [On the inheritance of myopia]. *Izvestiia Biuro po Evgenike*, no. 2: 60–66. Summary in German.

———. 1930. Iurii Aleksandrovich Filipchenko. *Priroda*, no. 7–8: 683–98.

Lepin, T. K., Ia. Ia. Lus, and Iu. A. Filipchenko. 1925. Deistvitel'nye chleny Akademii Nauk za poslednie 80 let (1846–1924) [Full members of the Academy of Sciences for the last eighty years (1946–1924)]. *Izvestiia Biuro po Evgenike*, no. 3: 3–82. Summary in English.

Levit, S. G. 1925. Evoliutsionnye teorii v biologii i marksizm [Marxism and theories of biological evolution]. *Vestnik Sovremennoi Meditsiny*, no. 9.

———. 1927. *Problema konstitutsii v meditsine i dialekticheskii materializm* [Dialectical materialism and the problem of constitution in medicine]. Trudy kruzhka Vrachei-materialistov Pervogo MGU za 1925–1926 gg., no. 2. Moscow.

———. 1929a. O poniatii bolezni [On the concept of disease]. *Estestvoznanie i Marksizm*, no. 1: 93–105.

———. 1929b. *Gemorragicheskie diatezi* [Hemorrhagic diathesis]. Moscow.

———. 1929c. Genetika i patologiia (v sviazi s sovremennym krizisom meditsiny) [Genetics and pathology (in relation to the current crisis in medicine)]. In *Trudy kabineta*, 20–39. *See* Levit and Serebrovskii 1929.

———. 1929d. Materialy k voprosu o stseplenii genov u cheloveka [Material on gene linkage in humans]. In *Trudy kabineta*, 40–50. Summary in German. *See* Levit and Serebrovskii 1929.

———. 1929e. K voprosu o nasledovanii ateromy [Concerning the inheritance of atheroma]. In *Trudy kabineta*, 90–93. Summary in German. *See* Levit and Serebrovskii 1929.

———. 1929f. Otchet o rabote Kabineta nasledstvennosti i konstitutsii cheloveka pri Mediko-biologicheskom institute za 1928–29 akad. g. [Survey of the work of the Office of Human Heredity and Constitution associated with the Biomedical Institute for academic year 1928–29]. In *Trudy kabineta*, 115–16. *See* Levit and Serebrovskii 1929.

———. 1930. Chelovek kak geneticheskii ob"ekt i izuchenie bliznetsov kak metod antropogenetiki [The human as a genetic object and the study of twins as a method of anthropogenetics]. In *Trudy Geneticheskogo Otdeleniia*, 273–87. *See* Levit and Serebrovskii 1930a.

———. 1932. Darvinizm, rasovyi shovinizm, sotsial-fashizm [Darwinism, racial chauvinism, social fascism]. In *Uchenie Darvina*, 107–25. *See* Valeskaln and Tokin 1932.

———. 1933. Work plans for the Biomedical Institute, Muller Papers, Lilly Library.

———, ed. 1934a. *Trudy Mediko-biologicheskogo nauchno-issledovatel'skogo Instituta imeni*

Maksima Gor'kogo (MBI) [Works of the Maxim Gorky Biomedical Scientific Research Institute]. Vol. 3. Moscow and Leningrad: Gosudarstvennoe izdatel'stvo biologicheskoi and meditsinskoi literatury.

————. 1934b. Nekotorye itogi i perspektivy bliznetsovykh issledovanii [Some results and prospects of twin studies]. In *Trudy MBI*, 5–17. *See* Levit 1934a.

————. 1934c. Kriticheskie zamechaniia po povodu raboty Goldena 'O khromosomnykh aberratsiiakh u cheloveka' [Critical comments on Haldane's 'On cytological abnormalities in man']. In *Trudy MBI*, 235–38. *See* Levit 1934a.

————. 1934d. Ot Mediko-Biologicheskogo Instituta [From the Biomedical Institute]. In *Konferentsiia po meditsinskoi genetike*, 1.

————. 1934e. Antropogenetika i meditsina [Anthropogenetics and medicine]. In *Konferentsiia po meditsinskoi genetike*, 3–16.

————. 1936a. Predislovie [Preface]. In *Trudy MGI*, 5–16. *See* Levit and Ardashnikov 1936.

————. 1936b. Problema dominantnosti u cheloveka [The problem of dominants in humans]. In *Trudy MGI*, 17–40. Summary in English. *See* Levit and Ardashnikov 1936.

Levit, S. G., and S. N. Ardashnikov, eds. 1936. *Trudy Mediko-geneticheskogo nauchno-issledovatel'skogo instituta imeni Maksima Gorkogo* (MGI) [Works of the Maxim Gorky Scientific Research Institute of Medical Genetics]. Vol. 4. Moscow and Leningrad: Gosudarstvennoe izdatel'stvo biologicheskoi and meditsinskoi literatury.

Levit, S. G., and N. N. Malkova. 1929. Novaia mutatsiia u cheloveka (geterogemofiliia) [New human mutation (heterohemophilia)]. *Russkii Evgenicheskii Zhurnal* 7: 106–12.

Levit, S. G., and L. N. Pesikova. 1934. Genetika sakharnogo diabeta [The genetics of sugar diabetes]. In *Trudy MBI*, 132–47. *See* Levit 1934a.

————. 1936. Obuslovlen li Diabetes insipidus 'khoroshim' dominantnym genom [Is diabetes insipidus caused by a good dominant gene?]. In *Trudy MGI*, 149–58. Summary in English. *See* Levit and Ardashnikov 1936.

Levit, S. G., and A. S. Serebrovskii, eds. 1929. *Trudy kabineta nasledstvennosti i konstitutsii cheloveka pri Mediko-biologicheskom institute* [Works of the office of human heredity and constitution of the Biomedical Institute]. Vol. 1. Moscow: Gosudarstvennoe izdatel'stvo Glavnauka. Suppl. to *Mediko-biologicheskii zhurnal*, 1929, no. 5.

————. 1930a. *Trudy Geneticheskogo otdeleniia (b. kabineta nasledstvennosti i konstitutsii cheloveka) pri Mediko-biologicheskom institute* [Works of the genetic division (formerly the office of human heredity and constitution) of the Biomedical Institute]. Vol. 2. Issued as *Mediko-biologicheskii Zhurnal*, 1930, nos. 4–5.

————. 1930b. K voprosu o geneticheskom analize cheloveka [On the question of human genetic analysis]. In *Trudy Geneticheskogo Otdeleniia*, 321–28. *See* Levit and Serebrovskii 1930a.

Levit, S. G., and G. V. Soboleva. 1935. Comparative Intrapair Correlations of Fraternal Twins and Siblings. *Journal of Genetics* 30: 389–96.

Lifshits, B. M. 1924. *Uchenie o konstitutsiiakh cheloveka* [Study of human constitution]. Kharkov: Gosudarstvennoe izdatel'stvo Ukrainy.

Liubimov, S. V. 1928. Predki grafa S. Iu. Vitte [The ancestry of Count S. Witte]. *Russkii Evgenicheskii Zhurnal* 6: 203–13.

Liublinskii [Liublinsky], P. I. 1912. Novaia mera bor'by s vyrozhdeniem i prestupnost' iu [New measure for fighting degeneration and crime]. *Russkaia Mysl'*, no. 3 (March): 31–56.

————. 1925. Evgenicheskie tendentsii i noveishee zakonodatel'stvo o detiakh [Eugenic tendencies and the latest legislation on children]. *Russkii Evgenicheskii Zhurnal* 3: 3–29.

————. 1926a. Sovremennoe sostoianie evgenicheskogo dvizheniia [The current state of the eugenics movement]. *Russkii Evgenicheskii Zhurnal* 4: 63–75.

————. 1926b. Rozhdaemost' i problema naseleniia v sovremennom obshchestve [Fertility and the population problem in modern society]. *Russkii Evgenicheskii Zhurnal* 4: 144–77.

————. 1927a. Brak i evgenika [Marriage and eugenics]. *Russkii Evgenicheskii Zhurnal* 5: 49–89.

————. 1927b. Novoe v voprose o sterilizatsii defektivnykh [New information on the question of the sterilization of defectives]. *Russkii Evgenicheskii Zhurnal* 5: 155–62.

_____. 1929. Okhrana materinstva i razvod [Divorce and the protection of maternity]. *Russkii Evgenicheskii Zhurnal* 7: 3–38.

Lobashev, M. E. 1969. *Genetika* [Genetics]. 2d ed. Leningrad: Izdatel'stvo Leningradskogo Universiteta.

Lundborg, G. 1924. Shvedskii institut rasovoi biologii [The Swedish Institute of Race Biology]. *Russkii Evgenicheskii Zhurnal* 2: 61–62.

Luriia [Luria], A. R. 1936–1937. The development of mental functions in twins. *Character and Personality,* 5: 35–47.

_____. 1979. *The Making of Mind: A Personal Account of Soviet Psychology.* Ed. Michael Cole and Sheila Cole. Introduction and epilogue by Michael Cole. Cambridge, Mass.: Harvard University Press.

Luriia [Luria], A. R., and A. N. Mirenova. 1936a. Issledovanie eksperimental'nogo razvitiia vospriiatiia metodom diferentsial'nogo obucheniia odnoiaitsevykh bliznetsov [Investigation of the experimental development of perception following the method of the differential training of monozygotic twins]. *Nevrologiia i Genetika* 1: 407–43.

_____. 1936b. Eksperimental'noe razvitie konstruktivnoi deiatel'nosti: Diferentsial'noe obuchenie odnoiaitsevykh bliznetsov, Soobshchenie III [Experimental development of constructive activity: Differential training of monozygotic twins, 3]. In *Trudy MGI,* 487–506. Summary in English. *See* Levit and Ardashnikov 1936.

Lus, Ia. Ia. 1924. K voprosu o nasledovanii rosta i slozheniia [On the inheritance of stature and constitution]. *Izvestiia Biuro po Evgenike,* no. 2: 48–59. Summary in English.

Lysenko, T. D., ed. 1949. *The Situation in Biological Science.* Moscow: Foreign Languages Publishing House.

Malkova, N. N. 1929. Kratkie predvaritel'nye dannye o rabote po statsionarnomu izucheniiu populiatsii cheloveka [A short preliminary communication of the work of stations in studying human populations]. In *Trudy kabineta,* 72–78. Summary in German. *See* Levit and Serebrovskii 1929.

Mandrillon, Marie-Hélène. 1987. Eugénisme et hygiènisme en URSS: La 'Revue russe d'eugénique' 1920–1929. Paper presented at the Colloque international d'Histoire de la Génétique, 19–22 May 1987, Paris.

Medvedev, N. N. 1978. *Iurii Aleksandrovich Filipchenko 1882–1930.* Moscow: Izdatel'stvo "Nauka."

Medvedev, Zhores A. 1969. *The Rise and Fall of T. D. Lysenko.* Trans. and ed. I. Michael Lerner. New York: Columbia University Press.

Membership and organization of the International Commission of Eugenics. 1924. *Eugenical News,* no. 2: 17–20.

Mints, Ia. V. 1925. Materialy k patografii Pushkina [Materials on the pathography of Pushkin]. *Klinicheskii Arkhiv Geneal'nosti i Odarennosti (Evropatologii)* 1: 29–46.

Morgan, T. H. 1909. *Eksperimental'naia zoologiia* [Experimental zoology]. Trans. N. Zograf. Moscow.

_____. 1924. *Strukturnye osnovy nasledstvennosti* [Material basis of heredity]. Translation edited by V. N. Lebedev. Moscow and Petrograd: Gosudarstvennoe izdatel'stvo.

_____. 1925. Nasledstvennost' u cheloveka [Heredity in humans]. Trans. E. K. Emme and M. E. Emme. *Russkii Evgenicheskii Zhurnal* 3: 99–114.

Morgan, T. H., and Iu. A. Filipchenko. 1925. *Nasleduiutsia li priobretennye priznaki.* Leningrad: Izdatel'stvo "Seiatel'" E. V. Vysotskogo.

Muller, H. J. [G. G. Meller]. 1923a. Rezul'taty desiatiletnikh geneticheskikh issledovanii s Drosophila [Results of a decade of genetic research on *Drosophila*]. *Uspekhi Eksperimental'noi Biologii* 1: 292–321.

_____. 1923b. Observations of Biological Science in Russia. *Scientific Monthly* 16: 539–52.

_____. 1927. Artificial Transmutation of the Gene. *Science* 66: 84–87.

_____. 1933a. The Dominance of Economics over Eugenics, *Scientific Monthly* 37: 40–47.

_____. 1933b. Evgenika v usloviiakh kapitalisticheskogo obshchestva [Eugenics under the conditions of capitalist society]. *Uspekhi Sovremennoi Biologii* 2: 3–11.

_____. 1922. Genealogiia roda Aksakovykh (po posmertnym bumagam S. I. Gal'perin) [The genealogy of the Aksakov family (from the posthumous papers of S. I. Galperin)]. *Russkii Evgenicheskii Zhurnal* 1: 74–81.

_____. 1923. O zadachakh i putiakh antropogenetiki [On the tasks and approaches of anthropogenetics]. *Russkii Evgenicheskii Zhurnal* 1: 107–16.

_____. 1926. Teoriia nasledstvennosti Morgana i Mendelia i marksisty [Marxists and the theory of heredity of Morgan and Mendel)]. *Pod Znamenem Marksizma*, no. 3: 98–117.

_____. 1927. Chetyre stranitsy, kotorye vzvolnovali uchenyi mir [Four pages that shook the scientific world]. *Pravda,* 11 September.

_____. 1929. Antropogenetika i evgenika v sotsialisticheskom obshchestve [Anthropogenetics and eugenics in a socialist society]. In *Trudy kabineta,* 3–19. *See* Levit and Serebrovskii 1929.

_____. 1930. Pis'mo v Redaktsiiu [Letter to the editor]. In *Trudy Geneticheskogo otdeleniia,* 447–48. *See* Levit and Serebrovskii 1930a.

_____. 1937. Genetika i zhivotnovodstvo [Genetics and animal husbandry]. In *Spornye voprosy,* 72–113, 443–51. *See* Targul'ian 1937.

_____. 1976. *Izbrannye trudy po genetike i selektsii kur* [Selected works on the genetics and selection of poultry]. Moscow: Izdatel'stvo Nauka.

Shapiro, N. I. 1966. Pamiati A. S. Serebrovskogo [In memory of A. S. Serebrovskii]. *Genetika,* no. 9: 3–17.

Sholomovich. 1913. *Nasledstvennost' i fizicheskie priznaki vyrozhdeniia u dushevno-bol'nykh i zdorovykh* [Inheritance and physical signs of degeneration in the mentally ill and the healthy]. Kazan.

Shtekher [Stecher], G. G. 1927. *Vyrozhdenie i evgenika* [Degeneration and eugenics]. Priroda i Kul'tura, no. 26. Moscow and Leningrad: Gosudarstvennoe izdatel'stvo.

Siemens, H. W. [Simens, G. V.] 1927. *Vvedenie v patologiiu nasledstvennosti cheloveka* [Introduction to hereditary pathology in humans]. Trans. V. A. Pavlov and L. A. Andreev. Moscow and Leningrad: Gosudarstvennoe izdatel'stvo. (Translation of 2d German edition of *Einfuhrung in die Vererbungspathologie des Menschen.*)

Slepkov, V. 1925. Nasledstvennost' i otbor u cheloveka: Po povodu teoreticheskikh predposylok evgeniki [Heredity and selection in humans: On the theoretical assumptions of eugenics]. *Pod Znamenem Marksizma*, no. 4: 102–22.

_____. 1927. *Evgenika* [Eugenics]. Moscow and Leningrad: Gosudarstvennoe izdatel'stvo.

Slovtsov, B. I. 1923. *Uluchshenie rasy* [Improving the race]. Leningrad: Akademicheskoe izdatel'stvo Leningrada.

Soboleva, G. V. 1923. Neskol'ko dannykh iz bio-sanitarnoi statistiki detskikh domov g. Moskvy [Some data from the biosanitary statistics from Moscow orphanages]. *Russkii Evgenicheskii Zhurnal* 1: 236–37.

_____. 1924. Iz posemeinykh obsledovanii v Zvenigorodskom uezda [From research on families in Zvenigorod]. *Russkii Evgenicheskii Zhurnal* 2: 168–70.

_____. 1926. Rezul'taty obsledovaniia 105 par bliznetsov g. Moskvy [Results of a study of one hundred five pairs of twins from the city of Moscow]. *Russkii Evgenicheskii Zhurnal* 4: 3–22.

_____. 1929. Zaikanie kak nasledstvenoe zabolevanie [Stuttering as a hereditary disease]. *Russkii Evgenicheskii Zhurnal* 7: 88–105.

_____. 1930. Vrozhdennoe otsutstvie zubov [Inborn absence of teeth]. *Russkii Evgenicheskii Zhurnal* 7: 144–46.

_____. 1934. Issledovanie konstitutsional'nogo tipa iazvennykh bol'nykh [Investigation of the constitutional type of patients with ulcers]. In *Trudy MBI,* 154–68. Summary in English. *See* Levit 1934a.

Solomon, Susan G., and John F. Hutchinson, eds. 1989. *Social Medicine in Revolutionary Russia.* Bloomington, Ind.: Indiana University Press.

Stein, V. M. 1922. *Die Professur von Odessa: Statistisch-eugenische Skizze* [Odessa professors: A statistical eugenic overview]. Odessa.

Strogaia, E. Z. 1926. K voprosu o nasledovanii muzykal'nykh sposobnostei [On the heritability of musical talent]. *Russkii Evgenicheskii Zhurnal* 4: 85–88.

Studitskii [Studitsky], A. N. 1949. *Mendelevsko-morganovskaia genetika na sluzhbe amerikanskoi reaktsii* [Mendelist-Morganist genetics in service of American imperialism]. Vsesoiuznoe obshchestvo po rasprostraneniiu politicheskikh i nauchnykh znanii. Moscow: Izdatel'stvo Pravda.

Targul'ian, O. M., ed. 1937. *Spornye voprosy genetiki i selektsii: Raboty IV sessii Akademii 19–27 dekabria 1936 goda* [Issues in genetics and selection: Proceedings of the fourth session of the academy 19–27 December 1936]. Moscow and Leningrad: Izd. Vsesoiuznoi Akademii Sel'sko-khoziaistvennykh Nauk imeni V. I. Lenina.

Tereshkovich, A. M. 1926. Nasledstvennost' u pristupnikov [Heredity in criminals]. *Russkii Evgenicheskii Zhurnal* 4: 76–84.

Todes, Daniel P. 1984. Biological Psychology and the Tsarist Censor: The Dilemma of Scientific Development. *Bulletin of the History of Medicine* 58: 529–44.

Ukshe, S. 1915. Vyrozhdenie, ego rol' v prestupnosti i mery bor'by s nim [Degeneration, its role in crime and measures for fighting it]. *Vestnik Obshchei Gigieny*, no. 6.

Utkina, N. F. 1975. *Positivizm, antropologicheskii materializm i nauka v Rossii* [Positivism, anthropological materialism, and science in Russia]. Moscow: "Nauka."

Valeskaln, P. I., and B. P. Tokin. 1932. *Uchenie Darvina i marksizm-leninizm* [Darwinism and Marxism-Leninism]. Moscow: Partiinoe izdatel'stvo.

Vermel', S. S. 1924. Prestupnost' evreev [Criminality among Jews]. *Russkii Evgenicheskii Zhurnal* 2: 153–58.

Vishnevskii, B. 1924. Vtoraia mezhdunarodnaia vystavka po evgenike [The Second International Eugenics Exhibition]. *Russkii Evgenicheskii Zhurnal* 2: 63–65.

———. 1926. Zabytyi russkii evgenik [A forgotten Russian eugenist]. *Priroda*, nos. 3–4, 100.

Volotskoi [Wolotzkoi], M. V. 1923a. O polovoi sterilizatsii nasledstvenno defektivnykh [On the sexual sterilization of the hereditarily defective]. *Russkii Evgenicheskii Zhurnal* 1: 201–22.

———. 1923b. Antropotekhnicheskie proekty Petra I-go: Istoricheskaia spravka [Anthropotechnological projects of Peter I: Historical note]. *Russkii Evgenicheskii Zhurnal* 1: 235–36.

———. 1923c. *Podniatie zhiznennykh sil rasy: Novyi put'* [Raising the vital forces of the race: A new approach]. Biologicheskaia biblioteka, vol. 1. Moscow: Kooperativnoe izdatel'stvo Zhizn' i Znanie.

———. 1924a. K istorii evgenicheskogo dvizheniia [On the history of the eugenics movement]. *Russkii Evgenicheskii Zhurnal* 2: 50–55.

———. 1924b. Evgenicheskie zametki [Eugenic notes]. *Russkii Evgenicheskii Zhurnal* 2: 58–60.

———. 1925. *Klassovye interesy i sovremennaia evgenika* [Class interests and modern eugenics]. Gosudarstvennyi nauchno-issledovatel'skii institut imeni K. A. Timiriazeva, Otdelenie biologicheskikh faktorov sotsial'nykh iavlenii, Sektsiia antropoekologii, no. 1. Biologicheskaia biblioteka, vol. 4. Moscow: Kooperativnoe izdatel'stvo Zhizn' i Znanie.

———. 1926. *Voprosy biologii i patologii evreev* [Questions of the biology and pathology of Jews]. Vol. 1. Leningrad: Izd. Prakticheskaia meditsina.

———. 1927. Spornye voprosy evgeniki [Issues of eugenics]. *Vestnik Kommunisticheskoi Akademii*, no. 20: 212–54.

———. 1936. K voprosu o genetike papilliarnykh uzorov pal'tsev: Issledovanie 234 par bliznetsov [On the genetics of fingerprints: Investigation of 234 pairs of twins]. In *Trudy MGI*, 404–39. Summary in English. *See* Levit and Ardashnikov 1936.

———. 1937. Bliznetsovyi metod i problema izmenchivosti genov [The twin method and the problem of variability of genes]. *Antropologicheskii Zhurnal*, no. 2: 3–26.

Weindling, Paul. 1985. Weimar Eugenics: The Kaiser Wilhelm Institute for Anthropology, Human Heredity and Eugenics in Social Context. *Annals of Science* 42: 303–18.

———. 1986. German-Soviet Cooperation in Science: The Case of the Laboratory for Racial Research, 1931–1938. *Nuncius: Annali di Storia della Scienza* 1: 103–9.

Weiss, Sheila. 1986. Wilhelm Schallmayer and the Logic of German Eugenics. *Isis* 77: 33–46.
Zavadovskii [Zavadovsky], B. M. 1926. *Darvinizm i marksizm.* Moscow and Leningrad: Gosizdat.
Zavarzin, A. 1930. Iurii Aleksandrovich Filipchenko. *Trudy Leningradskogo obshchestva estestvoispytatelei* 60: 3–16.
Zhbankov, D. N. 1928. Potomstvo vydaiushchikh liudei [The descendants of great people]. *Russkii Evgenicheskii Zhurnal* 6: 145–63.
Zolotarev, V. 1927. Rodoslovnye A. S. Pushkina, gr. L. H. Tol'stogo, P. Ia. Chaadaeva, Iu. F. Samarina, A. I. Gertsena, kn. P. A. Kropotkina, kn. S. N. Trubetskogo [Genealogy of A. S. Pushkin, Count L. N. Tolstoi, P. Ia. Chaadaev, Iu. F. Samarin, A. I. Gertsen, Prince P. A. Kropotkin, Prince S. N. Trubetskoi]. *Russkii Evgenicheskii Zhurnal* 5: 113–32.
———. 1928. Dekabristy [The Decembrists]. *Russkii Evgenicheskii Zhurnal* 6: 178–95.

CHAPTER 6

Toward a Comparative History of Eugenics

Mark B. Adams

At the International Congress of the History of Science held at Bucharest in August 1981, only one or two papers were presented on the history of eugenics. Four years later, at the next congress, held in Berkeley, California, in August 1985, there were some fifteen papers on the history of eugenics, covering ten national cases, presented in seven sessions by scholars from six countries. Since then five books have appeared dealing with Germany, and studies are under way on the eugenics movements in Denmark (Hansen), Finland (Hietala) Norway (Roll-Hansen), Sweden (Broberg), Russia (Adams), France (Schneider, Lemaine, Clark), Austria (Hubensdorf), Brazil, Mexico, Argentina, and Cuba (Stepan). The studies in this volume and elsewhere are giving us a much richer picture of eugenics throughout the world. Taken together, they are forcing us to rethink our earlier impressions of eugenics, and are beginning to reveal some intriguing patterns that are worth exploring.

Myths and Realities

Until recently, our perception of eugenics has been dominated by stereotypes that have persisted since World War II and are still remarkably pervasive in public and even scholarly settings. Now, by uncovering the diversity of historical eugenics, the newly burgeoning literature allows us to set to rest four interconnected myths.

The first myth is that eugenics was a single, coherent, principally Anglo-American movement with a specifiable set of common goals and beliefs. This essentialist view of eugenics helps to explain the fact that the vast majority of recently published studies on the history of eugenics have focused on Britain and America, and that, despite their geographical limitations, such studies have sometimes generalized the specific characteristics of their own particular national case into an all-embracing concept. Investigators have tended to see eugenics as "really" about either race or class, depending on whether they were studying the United States or Britain.

Recent studies have begun to dispel this myth. In his article on eugenics in Germany and Russia in the 1920s, Loren Graham has pointed out that eugenics in both countries encompassed a rather wider range of opinion than might be thought, and has urged us to see each eugenic community as developing certain orientations and values by virtue of its place in its own particular social and historical matrix (Graham 1978). Although he dealt only with Britain and America, Kevles in his recent book has pointed up the diversity of views put forth "in the name of eugenics" and has not only compared, but also contrasted, the British and American cases (Kevles 1985). The chapters in this volume about Germany, France, Brazil, and Russia, as well as recent studies on the United States, Germany, and Britain, point up the remarkable diversity of ideas that passed for eugenics in quite different national, professional, social, racial, economic, religious, and cultural settings. Such studies have shown that, however important, the movements in the United States and Britain are certainly not archetypal and may not even be especially typical of the thirty or so movements worldwide. We are beginning to understand eugenics as a complex population of ideas, professionals, and institutions that became rather different things as it evolved in many diverse settings.

The second myth is that eugenics was somehow intrinsically bound up with Mendelian genetics. According to this view, general hereditarian thought was reinforced and rigidified by the rediscovery of Mendel's laws and the growth of genetics, which replaced the more socially ameliorative versions of "soft" or "Lamarckian" inheritance with a scientific and experimentally verified "particulate" inheritance that supported "biological determinism." According to the myth, Mendelism was the scientific basis of the eugenicist's harsh and pessimistic view of the human genetic future.

This myth, too, has been largely refuted by recent historical scholarship. It is now clear that the absence of Mendelian genetics from a culture did not mean the absence of eugenics. In some countries where Lamarckism dominated, leading Lamarckian eugenicists based their movement on an entirely non-Mendelian view of the inheritance of acquired characteristics. Indeed, in cases where Mendelism eventually made inroads in these countries, its partisans sometimes found well-established and active eugenics movements dominated by their opponents. As the essays on France, Brazil, and Russia in this volume demonstrate, Lamarckian eugenics had energetic proponents elsewhere (e.g., Kammerer 1924). As a rule, the forecasts of the Lamarckians for the human future were no more hopeful, and their solutions no less draconian, than those of Mendelians. As the examples of Kehl, Volotskoi, and MacBride show, Lamarckians could be strident advocates of eugenic sterilization. Had we not been so mystified by myth, we might have seen the obvious sooner: if undesirable acquired traits (such as alcoholism or criminality) are assumed to be passed on hereditarily, then a case for sterilization

tion can be based on diagnosis alone, without having to establish through genealogical analysis that the trait follows the laws of Mendel or Morgan. What advocates of compulsory eugenic sterilization shared was not a theory of heredity, but a view of "socially responsible" scientistic activism—a willingness to force a medical procedure on certain other human beings, denying them further progeny, in the hope of improving humanity. Furthermore, as a comparison of Hitlerism and Stalinism sadly demonstrates, both "biological determinism" and "socioeconomic determinism" can be used to legitimate mass extermination.

The third myth is that eugenics was essentially a pseudoscience. Here the arguments take various forms: that eugenics was incompatible with up-to-date genetics, which undermined its central premises; that its interpretations were based on prejudice and bias, and on personal and nonobjective views; that on occasion eugenicists even faked their data (consciously or unconsciously) to conform to their own biases. In this view the eugenicists' pseudoscientific extensions of genetics into the social realm distinguish eugenics from other, truly scientific disciplines and put it in league with mesmerism, Lysenkoism, and other so-called pseudosciences, "fads and fallacies in the name of science." According to this myth, then, although eugenics grew out of Mendelism, legitimate and unbiased geneticists of the 1920s came to recognize its pseudoscientific nature. Such a view entails historiographic consequences, since it legitimates histories of genetics that are devoid of references to its misbegotten and mean-spirited cousin, eugenics. Such a view underlies the mocking tone that often creeps into descriptions of the eugenics movement and the highlighting of particulars that sound absurd to the modern ear or are offensive to postwar sensibilities.

This myth may once have had a certain utility as a way of acquitting the science of genetics and freeing it from any socially unsavory eugenic associations, but it can find relatively little support in the historical record. For example, much work on R. A. Fisher was published before note was taken of the fact that his classic work on population genetics, which helped found the evolutionary synthesis, devoted its last third to eugenics (Fisher 1930). Nor was this bias limited to studies of orthodox geneticists: Koestler, who sympathized with Lamarckism, wrote an admiring biography of Kammerer that ignored his great interest in eugenics (Kammerer 1924; Koestler 1971).

But studies of the early development of eugenics in both Britain and America, where Mendelism-Morganism flourished, show that eugenics was often inextricably intertwined with the development of genetics—in courses, textbooks, institutional names, monographs, and the concerns of investigators (e.g., Kimmelman 1983, Selden 1985). Diane Paul's work has demonstrated the degree to which American and British geneticists sustained a commitment to eugenics until the late 1940s and, in some cases, the early 1950s (Paul 1984). Furthermore, there is now a wealth of historical evidence that the thinking of legitimate scientists, doing legitimate science, has often been influenced by "nonobjective," "extrascientific" considerations—including religious beliefs, class values, political concerns, metaphysical commitments, and even popular culture. In this light there would seem to be no clear grounds to distinguish eugenics from any other science according to these criteria. Judgments of this sort are often post hoc and almost always involve some retroactive application of our own ideas about what is "scientific." Such an approach is not always helpful in understanding the historical development of science. Finally, the term "pseudoscience" itself must give us pause. To the best of my knowledge, no one has ever claimed to be one of its practitioners; the term is generally reserved for castigating one's opponents. As a polemical word widely deployed in past struggles for legitimacy and power—and in historical accounts of those struggles written by the victors—"pseudoscience" is less interesting as a mode of historical explanation than as an object of historical study; it is not part of the solution, but part of the problem.

This brings us to the fourth myth: that, politically, eugenics was essentially right-wing or "reactionary." According to this view, depending on whom you are talking to, eugenics variously grew out of and supported racism, sexism, anti-Semitism, or capitalist exploitation of an oppressed working class, and led naturally to fascism, Naziism, and ultimately (either directly or by natural extension) to the "Aryanism," barbarous human experimentation, genocide, and the death camps of the Third Reich. There can be no doubt that the postwar revelations of Nazi atrocities deprived eugenics of much public and scientific support, as Paul, Provine, and others have tellingly demonstrated.

But such postwar revelations have also undoubtedly colored and distorted our conception of eugenics as a historical phenomenon. Here and elsewhere, Weiss has shown that the "Aryanists" constituted only one strain of German race hygiene, that Schallmayer and other German founders saw race hygiene as a form of progressive "scientific management," and that the movement encompassed important leaders of diverse political orientations (Weiss 1986, 1987). Weindling's work has also demonstrated the diversity of German eugenics, and has highlighted the degree to which, even in Nazi times, German eugenics was a "divided" science (Weindling 1984b, 1985a, 1985b). Finally, as to eugenics being politically right-wing, we should not forget that a number of leading leftist scientists were strong adherents of eugenics. The Soviet A. S. Serebrovsky, the American H. J. Muller, and the Briton J. B. S. Haldane, three of the most distinguished geneticists of our century, advocated communism *and* exhibited a lifelong commitment to eugenic ideals. Those who continue to maintain the myth that eugenics was "essentially" reactionary pseudoscience will have to explain how it could have been supported by leading leftist scientists. The label "reactionary" also seems particularly inapt in describing a movement many of whose members envisioned a scientifically engineered, highly progressive future that had never existed before.

As to the other political charges, although unquestionably true of some members of some eugenics movements, they too appear largely mythical. Concerning sexism, it would appear that by contemporary standards eugenics was one of the least sexist fields of the day in a number of countries. Not only were many women active in eugenics, not only did Margaret Sanger strongly support eugenics, but in addition, eugenics could provide a biological rationale for feminism: since women contributed half of the heredity of the offspring (in some renderings, at least half!), they had to be given every educational opportunity to develop their mental and physical potentials before their eugenic worth could be properly evaluated (Bettes

1977). As for anti-Semitism, not only did leading eugenicists include Jews, but many eugenics texts singled out Jews as a race of special talent, ability, and achievement, often explaining this in terms of generations of strong selection. As for being anti–working class, eugenics texts frequently included workers' and craft skills in their lists of desirable hereditary talents, and some eugenics movements especially emphasized the eugenic value of the proletariat. Finally, as for racism, some movements were more concerned with race than others, and those in several countries—notably Mexico—emphasized the value of racial mixing for the production of an improved national stock. Here too, breeding science could cut both ways: those opposed to interracial marriage could point to the lowered viability observed in distant hybrid crosses; those favoring it could point to examples of "hybrid vigor." Of course, some scientists may well have used their science to legitimate their own social biases; no doubt others were driven to adopt social positions by what they took to be scientific necessity. My point is that there was nothing in eugenic science *per se* that compelled any particular policy position: the links between science and society were not intrinsic but contingent.

Of course, in debunking one set of myths, we should not replace it with another. In arguing that eugenics was not an essentialist monolith, I do not wish to suggest that it did not have a certain self-conscious coherence; to what extent that coherence extended beyond the rhetoric of human improvement remains to be seen. Likewise, in arguing that eugenics was not a pseudoscientific spawn of Mendelism, I do not wish to suggest that eugenics in Britain, America, and perhaps elsewhere had nothing to do with Mendelism genetics, nor, alternatively, that all eugenics programs were scientifically well founded. Finally, in recognizing that eugenics was not *essentially* on the right, that *not all* eugenicists were reactionary, racist, sexist, anti-Semitic, or anti–working class, we should not think (with the same kind of essentialism) that *no* eugenicist was. Rather, we should ponder the degree to which essentialist treatments of eugenics have been unhelpful and misleading.

These four myths share the tendency to see eugenics in essentialist terms, and to see the eugenics movements of the United States or Britain as central, exemplary, and archetypal. What these myths fail to grapple with is the social, disciplinary, and intellectual character of eugenics as it was understood and practiced by those who identified themselves with it, and, the enormous variety of ideas, researches, and viewpoints that fell under its rubric. To come to terms with the historical eugenics, we must take into account its full scientific, political, and geographical variability.

Issues and Agendas

In the light of recent studies, then, the old myths about eugenics have begun to lose their appeal. Is all coherence gone? Fortunately not, since the recent analyses that have helped to undercut these myths have also given us a new perspective— one that is amenable to the analytic styles that have been developing in other areas of the history of science over the last two decades. What the many varieties of eugenics seem to share is not essence but evolution, not a single nature obscured by variability, but rather a process of formation, divergence, adaptation, and change that this same variability can help us to understand. We can see these patterns most clearly if we look at the comparative anatomy of the various movements not country by country but feature by feature, and these patterns can serve, in turn, as a basis of broader comparative work.

First, *scientific* dimensions of eugenics would appear to warrant reexamination. Recent work by Jan Sapp, Jonathon Harwood, and others has broadened the history of genetics in important ways. Instead of maintaining the historical perspective derived from the history of genetics in the countries where Mendelism and Morganism developed rapidly and triumphed early on, they have illuminated other strands of hereditary science—extrachromosomal heredity, developmental genetics, and so forth—that have been largely omitted from standard histories but assumed distinct importance in France and Germany (Sapp 1983, 1987; Harwood 1985). The persistence of Lamarckian perspectives in the biological and medical communities in many countries is becoming evident. This new historiography suggests, contrary to the view of Roll-Hansen (1988), that what counted as science (and pseudoscience) varied in different disciplines and locales. This in turn may mean that science played a larger role in eugenics than we have supposed—not the "good" science we see so clearly in retrospect, but the real science of people representing competing schools, traditions, and approaches, seeing nature differently, and hoping that the future will prove them right.

Second, *disciplinary* approaches promise to be especially useful in comparing national case studies. In most countries it would appear that eugenics was a field trying to come into being, sharing some of the characteristics of civic religions, social movements, applied science, and would-be independent scientific disciplines. In at least some countries, the development of genetics, and especially animal genetics, was inextricably intertwined with eugenics as the study of human heredity. Like all new fields, eugenics was often structured as a loose federation of clusters of workers, institutions, and concerns; to bring it into being as a separate and legitimate field, eugenicists laid claims to distinctive methods, problems, capabilities, and missions that set it apart from existing fields. From the papers on Brazil, France, Germany, and Russia in this volume, it is clear that creative entrepreneurship was vital to the formation of journals, societies, and other organizations.

Disciplinary dimensions of eugenics assume special importance in regard to the relationship between eugenics, human genetics, and agricultural genetics. Russia institutionalized "medical genetics" first, and it was also the first to "ban" eugenics. Has a similar replacement occurred elsewhere? How generally were genetics and eugenics linked in the spread of genetic science? A recent book by Kohler has demonstrated the gradual process by which the discipline of biochemistry was formed—and the important role played by professionals, disciplinary entrepreneurs, and institutions (Kohler 1982). Of course, biochemistry succeeded in becoming a discipline and eugenics did not, but a perspective that explains why one candidate succeeded may be helpful in understanding why another failed.

Third, *professional* dimensions also should prove useful for comparative purposes. Recent accounts of the history of eugenics in the United States and Britain have focussed on the role of biologists, and particularly geneticists. Although Müller-Hill (1984, 1988) has examined the role of geneticists under the Nazis, other new works on German eugenics and race hygiene (by Baader, Proctor, Weindling, Weingart, and Weiss) have presented a fundamentally different picture, one dominated by physicians and shaped by strong statist traditions of social medicine and public health. The central role of physicians in France, Brazil, and Russia suggests parallels. By paying attention to the mix of physicians, civil servants, lawyers, and academics in different settings, we may begin to see the history of eugenics in light of the history of professions.

In a recent article dealing with American medical and agricultural reformers, Rosenberg has encouraged us to see professions and academic disciplines as occupying places along a continuum defined by the nature of their social-support system (Rosenberg 1979). The relationship of the intellectual, institutional, and social format of the medical profession and its various specialties in different countries may help us to understand the different ways in which eugenics was shaped. Recent works by Weindling point in this direction in the German case (Weindling 1987, 1989); Solomon's recent work (1989, 1990) on Soviet social hygiene and public health, which supported Russian eugenics, suggests parallels.

Fourth, *institutional* analysis may play an especially important and interesting role in comparative studies. Recent works covering several countries and subjects have shown the power of institutional analysis and the momentum that institutions, once established, can manifest (e.g., Adams 1980; Kohler 1982; Hughes 1983). The ways eugenics became institutionalized in scientific societies and research institutions and in biological education differed greatly. An encouraging recent trend has emphasized the institutional foundations of eugenics, focusing on funding, patrons, and institutional rhetoric and dynamics. For example, a recent study by Allen has detailed the history of the Eugenics Record Office at Cold Spring Harbor (Allen 1986); Weindling has studied the history of eugenics in the context of the Kaiser Wilhelm Institute for Anthropology, Human Heredity, and Eugenics (Weindling 1985a); and Weingart has also studied the institutionalization of German eugenics (Weingart 1987). The situations are very different, of course, as are sources of funding and relations to political authorities; yet these studies show the degree to which institutional considerations and institutional inertia influenced the fate of eugenics movements in various countries.

Institutional analysis invites us to examine the role of entrepreneurs in the establishment of the field, the structure of the institutions in which it flourished, its relationship to other fields represented there, and the role it played in the organization's overall mission. Social influences on scientific work often can be mediated and translated through institutions and their patrons, which suggests that it may be useful to examine the ways in which the pertinent institutions were funded and supported and the ways in which the field's public profile and policy commitments may have been influenced by the patronage on which those institutions relied. Of course, as a recent book on early-nineteenth-century French and

British pathology reminds us, traditions can be established among groups without formal institutionalization (Maulitz 1987), but where such institutions exist, their character obviously must play an important role.

Fifth, *popular and pedagogical* dimensions of eugenics should also provide a useful mode of comparison. Although they differed in the policies they endorsed and the biological theories they favored, almost all of the world's eugenics movements assigned public education a high priority. How was eugenics taught in biological courses and secondary schools? Then, too, as a movement eugenics was one among many. The heyday of eugenics was also a period in which Taylorism, Fordism, scientific management, technocracy, and the efficiency movement flourished. The relationship between eugenics and the progressive movement in the United States is well documented (e.g., Pickens 1968); Freeden sees something comparable in Britain (Freeden 1979); Weiss sees Schallmayer's race hygiene as a form of biological technocracy (Weiss 1986). The relation of eugenics to other similar but distinct contemporary movements and to popular culture more broadly needs analysis. Popular culture is extremely difficult to study in a rigorous and illuminating way, but serious comparative study of the public dimensions of eugenics may well be manageable.

Sixth, *ideological* and *political* dimensions of eugenics assume special importance when our perspective becomes broadly international. The fact that eugenics could flourish in both Weimar and Nazi Germany, in Coolidge's America and Lenin's Russia, and that it could count among its adherents renowned Communists, Socialists, liberals, conservatives, and Fascists, suggests that any simplistic political classification of the movement cannot sustain analysis. In comparing German and Russian eugenics in the 1920s, Graham has made the point that eugenics was neither intrinsically right-wing nor left-wing, but that it acquired particular political dimensions in particular national and historical contexts (Graham 1978). Later he retreated from that position (e.g., Graham 1981, 1983), holding that perhaps eugenics is more essentially "Right" and Lamarckism more essentially "Left"; but I think more recent studies demonstrate that he was right the first time.

The different ideological shape eugenics assumed over time in individual countries where the political conditions changed opens a number of interpretive possibilities. Did political changes tend to selectively favor certain trends and people within the eugenics movement, while marginalizing others? Such analysis might allow us to understand the ways in which individuals, institutions, and ideologies adapted to fit changing political realities.

Finally, properly conducted national case studies may well allow us to discern certain characteristic *regional, national, or cultural styles* of eugenics. Hughes has developed the concept of regional technological style to good effect, showing how common technical information could be mobilized into distinctly different styles of technology in different regions depending on local geography, economics, laws, and politics (Hughes 1983). Schneider (1985) has characterized a "French style" of eugenics; Harwood (1987) has distinguished between the American and German "styles" of genetics. But comparative approaches may take us even further. Given the similarities between eugenics in France and Brazil, do they belong to a common "Latin" style of eugenics shaped by common Catholic and other cultural roots? Would they share these characteristics with the movement in Italy, for example? Or would other classifications prove more natural—for example, as Stepan queries, a "New World" Latin American style of eugenics, with the movements in Cuba, Brazil, Mexico, and Argentina having more in common with one another than with movements in Latin Europe? Is there a Scandinavian style? The analysis of such questions would invite us to trace the international patterns of cultural influence and diffusion.

Such "regional" studies are under way. Since preparing her chapter for this volume, Nancy Stepan has undertaken a comparative study of the eugenics movements in Argentina, Brazil, Cuba, and Mexico. A group of researchers is exploring Scandinavian eugenics in Norway, Sweden, Finland, and Denmark (Roll-Hansen 1987; Broberg 1987; Hietala 1987; Hansen 1987). They have found that sterilization laws were adopted in Scandinavia in the 1930s in the context of the formation of the welfare states; they have already identified an intriguing regional pattern—the appropriation of eugenic sterilization in the 1930s to serve the purposes of the modern welfare state. Where regions encompass diverse languages, such cooperative work will be needed.

Prospects

In this light, other eugenics movements that await analysis take on special interest. Despite the vast literature on eugenics in Britain and America, a number of important movements throughout the English-speaking world and the former British Empire remain to be studied. Generally unpublished research has been done on eugenics in Canada and Australia. Considering the centrality of questions relating to class and race in India and South Africa, studies of their eugenics movements would be of special value. Are all these best seen as British exports, or do they more closely fit regional or colonial patterns? For example, given Canada's geography and history, what characteristics does its movement share with those in Britain (to whose empire it belonged), the United States (its powerful North American neighbor), France (through Quebec), and Australia (another British colony of vast size and sparse aboriginal population)? Similar questions may be asked of Cuba; preliminary work by Avalos (1973), Hoepfner (1984), and Stepan suggests that its movement mixed North American, Latin American, and Spanish motifs. We are beginning to know something of Russian eugenics, but what of the Austro-Hungarian Empire and Slavic eastern Europe—Czechoslovakia, Hungary, Bulgaria, and the Ukraine? As a Catholic slavic country, Poland should be an especially intriguing test case. Lemaine, Schneider, Clark, and others are clarifying the character of eugenics in France; what of other Latin cultures of Europe, what of eugenics in Italy, Spain, Portugal, Romania? And what of Latin America? Stepan's study of Brazil should make us even more curious about the important eugenics movements that developed in Argentina, Chile, Colombia, Cuba, Mexico, El Salvador, Uruguay, and Venezuela. Finally, for obvious reasons having to do with race, class, culture, religion, and the selective importation of science and technology from the West, the eugenics movements of Japan and China are worthy of substantial studies.

For reasons of language, archival access, and the sheer volume of effort required, an understanding of international eugenics will require much work by historians of many nations. But by using what we are learning about the formation and institutionalization of disciplines and research programs, we may begin to come to terms with what eugenics was and how it developed. Such an emerging picture should enrich our understanding of the history of science and of its historical interaction with society. And only such an understanding will make our historical knowledge truly useful in dealing with contemporary issues.

Often, in current work, the term *eugenics* is used pejoratively, sometimes to make a political point against an idea or program. Indeed, Daniel Kevles has described his earliest work on eugenics as "coming to terms with a dirty word" (Kevles 1983), and Crow's review of his book (1988) raises the same point. But, given the diversity of views within early eugenics, of course, many things can rightly be seen as resurrecting eugenics that are quite laudable—such as research on blood diseases and blood chemistry, or genetic counseling, or the understanding of certain genetic diseases. Indeed, given the importance of Lamarckism and puericulture in the eugenics movements in France, Brazil, and elsewhere, one might well see jogging or aerobics as a throwback to eugenics with some justice. If we are to develop a useful and sophisticated understanding of the complex social and ethical choices we face in matters relating to human biology—of dangers, benefits, and likely consequences—we will have to put aside the "mythical" history of eugenics, and learn more about its real history.

Such a comparative history of eugenics may also illuminate our understanding of the history of science in broader terms. For several decades, understanding the relationship between the "internalist" or intellectual history of science and its "externalist" or social history has been a central theme in the field of the history of science. In this context eugenics may serve as a strategic research site for several reasons—first, because it is a field at the interface of science and society; second, because its formation and international spread may allow us to trace in some detail the diffusion of knowledge, the formation of disciplines, and their adaptation to particular national settings; finally, because eugenics is an extraordinarily well-documented, temporally limited, but geographically pervasive phenomenon, it may serve as a kind of international tracer or marker for approaching broader historical issues. Because of this, the new work on eugenics may contribute to our understanding of the evolution of science and the ecology of knowledge in a broad sense.

Some two decades ago, when it was suggested that a survey of the comparative history of eugenics be included in a conference on genetics and society, the organizer responded: "We already know too much about the history of eugenics." Much literature has been published since then on that history, but from the point of view of a historian, the subject is becoming more interesting than ever, and some of the most important questions remain unanswered. We may hope that the international research now under way in a dozen countries may allow us to know much more in the not-too-distant future.

Bibliography

Adams, Mark B. 1979. From 'Gene Fund' to 'Gene Pool': On the Evolution of Evolutionary Language. In *Studies in History of Biology,* Vol. 3, ed. William Coleman and Camille Limoges, 241–85. Baltimore and London: Johns Hopkins University Press.

———. 1980. Science, Ideology, and Structure: The Kol'tsov Institute, 1900–1970. In *The Social Context of Soviet Science,* ed. Linda L. Lubrano and Susan Gross Solomon, 173–204. Boulder, Colo.: Westview Press.

———. 1989. Eugenics as Social Medicine: Prophets, Patrons, and the Dialectics of Discipline-Building. In *Social Medicine in Revolutionary Russia,* ed. Susan Gross Solomon and John Hutchinson, Bloomington, Ind.: University of Indiana Press, in press.

———. 1990. The Soviet Nature-Nurture Debate. In *Science and the Soviet Social Order,* ed. Loren Graham. Cambridge, Mass.: Harvard University Press, in press.

Allen, Garland E. 1983a. The Misuse of Biological Hierarchies: The American Eugenics Movement, 1900–1940. *History and Philosophy of the Life Sciences* 5: 105–28.

———. 1983b. Eugenics and American Social History 1880–1960. Paper presented at conference, History of Eugenics: Work in Progress, 16–19 May 1983, at the University of Pennsylvania, Philadelphia.

———. 1986. The Eugenics Record Office at Cold Spring Harbor, 1910–1940. *Osiris,* 2d ser., 2: 225–64.

Avalos, Dolores Maria. 1973. Eugenical Movement in Spain. Manuscript.

Baader, Gerhard. 1984. Die Medizin im Nationalsozialismus: Ihre Wurzeln und die erste Periode ihrer Realisierung 1933–1938. In *Nicht misshandeln,* Stätten der Geschichte Berlins, Vol. 5, 61–107. Berlin: Edition Hentrich.

———. 1985a. Sozialhygienische Theorie und Praxis in der Gewerbehygiene im Berlin der Weimarer Zeit. *Koroth* 8, no. 11–12: 24–37.

———. 1985b. Social Hygiene and Eugenics in Germany 1900–1932. Paper given at the XVIIth International Congress of the History of Science, 31 July–8 August 1985, University of California, Berkeley.

Bajema, Carl J., ed. 1976. *Eugenics: Then and Now.* Benchmark Papers in Genetics, No. 5. Stroudsburg, Pa.: Dowden, Hutchinson, and Ross.

Basalla, George. 1988. *The Evolution of Technology.* Cambridge: Cambridge University Press.

Bennett, J. H., ed. 1983. *Natural Selection, Heredity, and Eugenics.* Oxford: Clarendon Press.

Bettes, Martha. 1977. Women and the American Eugenics Movement. A. M. thesis, University of Pennsylvania.

Blacker, C. P. 1952. *Eugenics: Galton and After.* London: Gerald Duckworth and Co.

Blustein, Bonnie. 1973. Eugenics and Race Hygiene in Germany. Manuscript.

Bowler, Peter J. 1984. E. W. MacBride's Lamarckian Eugenics and Its Implications for the Social Construction of Scientific Knowledge. *Annals of Science* 41: 245–60.

Box, Joan Fisher. 1978. *R. A. Fisher: The Life of a Scientist.* New York: John Wiley and Sons.

Boyer, Samuel H., IV, ed. 1963. *Papers on Human Genetics.* Englewood Cliffs, N.J.: Prentice-Hall.

Broberg, Gunnar. 1987. The Swedish Debate About Sterilization. Manuscript.

Carlson, Elof Axel. 1981. *Genes, Radiation, and Society: The Life and Work of H. J. Muller.* Ithaca and London: Cornell University Press.

Carroll, P. Thomas. 1973. The Swedish Eugenics Movement: An Overview. Manuscript.

———. 1983. The Swedish Eugenic Movement 1910–1943: An Overview. Science and Technology Studies, Working Paper 83-4, Rensselaer Polytechnic Institute.

Clark, Linda. 1983. Eugenics in France. Paper presented at conference, The History of Eugenics: Work in Progress, 16–19 May 1983, University of Pennsylvania, Philadelphia.

Crow, James F. 1988. Eugenics: Must It Be a Dirty Word? *Contemporary Psychology* 33: 10–12.

Dunn, L. C., ed. 1951. *Genetics in the 20th Century: Essays on the Progress of Genetics during Its First 50 Years.* New York: Macmillan.

———. 1962. Cross Currents in the History of Human Genetics. *American Journal of Human Genetics* 14: 1–13.

Efroimson, V. P. 1967. K istorii izucheniia genetiki cheloveka v SSSR [On the history of the study of human genetics in the USSR]. *Genetika,* no. 10: 114–27.

Farrall, Lyndsay A. 1970. The Origins and Growth of the English Eugenics Movement, 1865–1925. Ph.D. dissertation, Indiana University.

———. 1979. The History of Eugenics: A Bibliographical Review. *Annals of Science* 36: 111–23.

———. 1983. Reflections on the history of eugenics. Paper presented at conference, The History of Eugenics: Work in Progress, 16–19 May 1983, at the University of Pennsylvania, Philadelphia.

———. [1983]. Wilfred Eade Agar: An Australian Eugenist. Manuscript.

Fisher, R. A. 1930. *The Genetical Theory of Natural Selection.* Oxford: Oxford University Press.

Freeden, Michael. 1979. Eugenics and Progressive Thought: A Study in Ideological Affinity. *Historical Journal* 22: 645–71.

Gaissinovitch, A. E. 1980. The Origins of Soviet Genetics and the Struggle with Lamarckism, 1922–1929. Trans. Mark B. Adams. *Journal of the History of Biology* 13: 1–51.

Glass, Bentley. 1981. A Hidden Chapter of German Eugenics between the Two World Wars. *Proceedings of the American Philosophical Society* 125: 357–67.

Graham, Loren R. 1978. Science and Values: The Eugenics Movement in Germany and Russia in the 1920s. *American Historical Review* 83: 1135–64.

———. 1981. *Between Science and Values.* New York: Columbia University Press.

———. 1983. Is Eugenics Conservative or Radical? Paper presented at conference, The History of Eugenics: Work in Progress, 16–19 May 1983, at the University of Pennsylvania, Philadelphia.

Haller, Mark H. 1963. *Eugenics: Hereditarian Attitudes in American Thought.* New Brunswick, N.J.: Rutgers University Press.

Hansen, Bent Sigurd. 1987. Eugenics in Denmark. Manuscript.

Harwood, Jonathon. 1985. Geneticists and the Evolutionary Synthesis in Interwar Germany. *Annals of Science* 42: 279–301.

———. 1987. National Styles in Science: Genetics in Germany and the United States between the World Wars. *Isis* 78: 390–414.

Hietala, Marjatta. 1987. The Eugenics Movement in Finland. Manuscript.

Hoepfner, Christine. 1984. The Pan-American Conferences of Eugenics and Homiculture: 1920–1935. Manuscript.

Houck, Donna M. 1982. Kammerer's Eugenics: The Fusion of Politics and Science. Manuscript.

Hughes, Thomas P. 1983. *Networks of Power: Electrification in Western Society, 1880–1930.* Baltimore and London: Johns Hopkins University Press.

Hull, David L. 1988. *Science as a Process: An Evolutionary Account of the Social and Conceptual Development of Science.* Chicago: Chicago University Press.

Jones, Greta. 1980. *Social Darwinism and English Thought: The Interaction between Biological and Social Theory.* Sussex: Harvester Press; Atlantic Highlands, N.J.: Humanities Press.

———. 1982. Eugenics and Social Policy between the Wars. *Historical Journal* 25: 717–28.

———. 1983. The Eugenics Movement in Britain 1907–1960. Paper presented at conference, The History of Eugenics: Work in Progress, 16–19 May 1983, at the University of Pennsylvania, Philadelphia.

Joravsky, David. 1970. *The Lysenko Affair.* Cambridge, Mass.: Harvard University Press.

Kammerer, Paul. 1924. *The Inheritance of Acquired Characteristics.* New York: Boni and Liveright.

Kevles, Daniel J. 1983. Eugenics in the United States and Great Britain: Confronting the History of a Dirty Word. Paper presented at conference, History of Eugenics: Work in Progress, 16–19 May 1983, at the University of Pennsylvania, Philadelphia.

———. 1985. *In the Name of Eugenics.* New York: Knopf.

Kimmelman, Barbara A. 1983. The American Breeders' Association: Genetics and Eugenics in an Agricultural Context, 1903–13. *Social Studies of Science* 13: 163–204.

Koestler, Arthur. 1971. *The Case of the Midwife Toad.* London: Hutchinson.

Kohler, Robert E. 1982. *From Medical Chemistry to Biochemistry: The Making of a Biomedical Discipline.* Cambridge: Cambridge University Press.

Kudlien, Fridolf, ed. 1985. *Ärzte im Nationalsozialismus.* Berlin: Kiepenheur and Witsch.

Lemaine, Gérard, and Benjamin Matalon. 1985. *Hommes supérieurs, hommes inférieurs?* Paris: Armand Colin.

Leonard, J. 1983. Eugénisme et darwinisme: Espoirs et perplexités chez des médecins français du XIXe siècle et du debut du XXe siècle. 187–207.

Lipman, Eric. 1973. Eugenics in Austria and Czechoslovakia. Manuscript.

Lorwin, John. 1973. Eugenics in East Asia: China and Japan. Manuscript.

Ludmerer, Kenneth M. 1972. *Genetics and American Society: A Historical Appraisal.* Baltimore and London: Johns Hopkins University Press.

Machiz, Marc. 1973. Eugenics in Argentina. Manuscript.

MacKenzie, Donald A. 1976. Eugenics in Britain. *Social Studies of Science* 6: 499–532.

_____. 1981. *Statistics in Britain, 1865–1930: The Social Construction of Scientific Knowledge.* Edinburgh: Edinburgh University Press.

MacPherson, George S. 1973. Aspects of the Italian Eugenics Movement. Manuscript.

Maline, Joseph M., and Daniel P. Todes. 1973. Canadian Public Health Organizations: The Problem of the Feeble-Minded. Manuscript.

Mandrillon, Marie-Hélène. 1987. Eugénisme et hygiénisme en URSS: La 'Revue russe d'eugénique' 1920–1929. Paper presented at the Colloque international d'Histoire de la Génétique, 19–22 May 1987, Paris.

Mann, Günter. 1978. Neue Wissenschaft im Rezeptionsbereich des Darwinismus: Eugenik—Rassenhygiene. *Berichte zur Wissenschaftsgeschichte* 101–11.

Maulitz, Russell C. 1987. *Morbid Appearances: The Anatomy of Pathology in the Early Nineteenth Century.* Cambridge: Cambridge University Press.

Mautner, Nancy. 1973. Eugenics in France. Manuscript.

Mazumdar, Pauline M. 1980. The Eugenists and the Residuum: The Problem of the Urban Poor. *Bulletin of the History of Medicine* 54: 204–15.

Membership and Organization of the International Commission of Eugenics. 1924. *Eugenical News,* 2: 17–20.

Muller, H. J. 1933. The Dominance of Economics over Eugenics. *Scientific Monthly* 37: 40–47. Published in Russian as Evgenika v usloviiakh kapitalisticheskogo obshchestva. *Uspekhi sovremennoi biologii* 2: 3–11.

Müller-Hill, Benno. 1984. *Tödliche Wissenschaft: Die Aussonderung von Juden, Ziegeunern und Geisteskranker 1933–45.* Hamburg: Rowolth.

_____. 1988. *Murderous Science: Elimination by Scientific Selection of Jews, Gypsies, and Others, Germany 1933–1945.* Trans. George R. Fraser. Oxford and New York: Oxford University Press.

Paul, Diane. 1983. Contribution to conference, History of Eugenics: Work in Progress, 16–19 May 1983, at the University of Pennsylvania, Philadelphia.

_____. 1984. Eugenics and the Left. *Journal of the History of Ideas* 567–90.

_____. 1985. The "Real Menace" of the Feebleminded: Scientists and Sterilization, 1917–1930. Paper presented at the XVIIth International Congress of the History of Science, 31 July–8 August 1985, University of California, Berkeley.

Paul, Diane, and Barbara Kimmelman. 1988. Mendel in America: Theory and Practice, 1900–1919. In *The American Development of Biology,* ed. Ronald Rainger, Keith R. Benson, and Jane Maienschein. Philadelphia: University of Pennsylvania Press.

Pickens, Donald K. 1968. *Eugenics and the Progressives.* Nashville, Tenn.: Vanderbilt University Press.

Proctor, Robert. 1988. *Racial Hygiene: Medicine Under the Nazis.* Cambridge, Mass.: Harvard University Press.

Richards, Robert J. 1987. *Darwin and the Emergence of Evolutionary Theories of Mind and Behavior.* Chicago: Chicago University Press.

Roll-Hansen, Nils. 1980a. Eugenics before World War II: The Case of Norway. *History and Philosophy of the Life Sciences* 2: 269–98.

_____. 1980b. Den norske debatten om rasehygiene. *Historisk Tidsskrift* 59: 259–83.

_____. 1987. Sterilization and Norwegian Eugenics. Manuscript.

———. 1988. The Progress of Eugenics: Growth of Knowledge and Change in Ideology. *History of Science* 26: 295–331.

Rosenbaum, Eli. 1973. Eugenics in Norway: A Historical Account. Manuscript.

Rosenberg, Charles. 1961. Charles Benedict Davenport and the Beginning of Human Genetics. *Bulletin of the History of Medicine* 35: 266–76.

———. 1979. Toward an Ecology of Knowledge: On Discipline, Context, and History. In *The Organization of Knowledge in Modern America,* ed. Alexandra Oleson and John Voss, 440–55. Baltimore and London: Johns Hopkins University Press.

Sapp, Jan. 1983. The Struggle for Authority in the Field of Heredity, 1900–1932: New Perspectives on the Rise of Genetics. *Journal of the History of Biology* 16: 311–42.

———. 1987. *Beyond the Gene: Cytoplasmic Inheritance and the Struggle for Authority in Genetics.* New York and Oxford: Oxford University Press.

Schneider, William. 1973. A History of Eugenics in France, 1886–1940: or, What's In a Name? Manuscript.

———. 1982. Toward the Improvement of the Human Race: The History of Eugenics in France. *Journal of Modern History* 54: 268–91.

———. 1985. Puericulture and the Style of Eugenics in France. Paper presented at the XVIIth International Congress of the History of Science, 31 July–8 August 1985, University of California, Berkeley.

Searle, G. R. 1976a. *Eugenics and Politics in Britain 1900–1914.* Leyden: Noordhoff International Publishing.

———. 1976b. Eugenics and Class. *Social Studies of Science* 6: 217–42.

———. 1979. Eugenics and Politics in Britain in the 1930s. *Annals of Science* 36: 159–69.

Selden, Steven. 1985. Education Policy and Biological Science: Genetics, Eugenics, and the College Textbook, c. 1908–1931. *Teachers College Record* 87: 35–51.

Snyder, Lawrence H. 1951. Old and New Pathways in Human Genetics. In *Genetics in the 20th Century: Essays on the Progress of Genetics during Its First 50 Years,* ed. L. C. Dunn, 369–92. New York: Macmillan.

Solomon, Susan G. 1989. David and Goliath in Soviet Public Health. *Soviet Studies* 41: 254–75.

———. 1990. Social Hygiene in Soviet Public Health. In *Social Medicine in Revolutionary Russia,* ed. Susan Gross Solomon and John F. Hutchinson. Bloomington: Indiana University Press, in press.

Stepan, Nancy. 1976. *The Beginnings of Brazilian Science.* New York: Science History Publications.

———. 1982. *The Idea of Race in Science.* London: Macmillan.

Weindling, Paul. 1984a. Soziale Hygiene: Eugenik und Medizinische Praxis—Der Fall Alfred Grotjahn. In *Kritische Medizin in Argument.* Berlin: Argument-Verlag.

———. 1984b. Die Preussische Medizinalverwaltung und die "Rassenhygiene": Anmerkunger zur Gesundheitspolitik der Jahre 1905–1933. *Zeitschrift für Sozialreform,* 675–87.

———. 1985a. Weimar Eugenics: The Kaiser Wilhelm Institute for Anthropology, Human Heredity and Eugenics in Social Context. *Annals of Science* 42: 303–18.

———. 1985b. Eugenics in Nazi Germany: A Divided Science. Paper presented at the XVIIth International Congress of the History of Science, 31 July–8 August 1985, University of California, Berkeley.

———. 1986. German-Soviet Co-operation in Science: The Case of the Laboratory for Racial Research, 1931–1938. *Nuncius: Annali di Storia della Scienza* 1, no. 2: 103–9.

———. 1987. Medical Practice in Imperial Berlin: The Casebook of Alfred Grotjahn. *Bulletin of the History of Medicine* 61: 391–410.

———. 1989. *Health, Race, and German Politics between National Unification and Nazism, 1870–1945.* Cambridge: Cambridge University Press.

Weingart, Peter. 1987. The Rationalization of Sexual Behavior: The Institutionalization of Eugenic Thought in Germany. *Journal of the History of Biology* 20: 159–93.

Weingart, Peter, and Harold Kranz. 1985. Eugenics under the Naziregime. Paper presented at the XVIIth International Congress of the History of Science, 31 July–8 August 1985, University of California, Berkeley.

Weingart, Peter, Jürgen Kroll, and Kurt Bayertz. 1988. *Rasse, Blut und Gene: Geschichte der Eugenik und Rassenhygiene in Deutschland.* Frankfurt am Main: Suhrkamp Verlag.

Weiss, Sheila Faith. 1983. Race Hygiene and the Rational Management of National Efficiency: William Schallmayer and the Origins of German Eugenics, 1890–1920. Ph.D. dissertation, The Johns Hopkins University.

———. 1985. Race and Class in Fritz Lenz's Eugenics. Paper presented at the XVIIth International Congress of the History of Science, 31 July–8 August 1985, University of California, Berkeley.

———. 1986. Wilhelm Schallmayer and the Logic of German Eugenics. *Isis* 77: 33–46.

———. 1987a. The Race Hygiene Movement in Germany. *Osiris,* 2d ser., 3: 193–236.

———. 1987b. *Race Hygiene and National Efficiency: The Eugenics of Wilhelm Schallmayer.* Berkeley, Ca.: University of California Press.

Name Index

Subject Index